W9-BKU-925

THE GOOD BOOK OF HUMAN NATURE

Also by Carel van Schaik

Among Orangutans: Red Apes and the Rise of Human Culture
(with Perry van Duijnhoven)

The Primate Origins of Human Nature

THE GOOD BOOK OF HUMAN NATURE

An Evolutionary Reading of the Bible

Carel van Schaik

and

Kai Michel

BASIC BOOKS

A Member of the Perseus Books Group

New York

Published by Basic Books, Member of the Perseus Books Group

Designed by Linda Mark

Library of Congress Cataloging-in-Publication Data
Names: Schaik, Carel van, author.
Title: The good book of human nature : an evolutionary reading of the Bible / Carel van Schaik and Kai Michel.
Description: New York City, NY : Basic Books, a member of Perseus Books Group, 2016. | Includes bibliographical references and index.
Identifiers: LCCN 2016009059 (print) | LCCN 2016010322 (ebook) | ISBN 9780465074709 (hardcover) | ISBN 9780465098675 (e-book)
Subjects: LCSH: Bible.--Criticism, interpretation, etc. | Bible and anthropology. | Bible--Psychology. | Bible--Philosophy.
Classification: LCC BS511.3 .S365 2016 (print) | LCC BS511.3 (ebook) | DDC 20.6--dc23
LC record available at http://lccn.loc.gov/2016009059

10 9 8 7 6 5 4 3 2 1

CONTENTS

INTRODUCTION

COULD IT BE THAT THE BIBLE HAS NEVER REALLY RECEIVED THE attention it truly deserves? To many readers, this must sound like a preposterous question. After all, we are dealing here with the Book of Books, the holy scripture of Judaism and Christianity, which millions have scrutinized for countless centuries. We are nonetheless convinced that the Bible has not been given its due. It has not received recognition for the only thing that can be said with absolute certainty: it is the most important book of all of humankind. No other work has so much to reveal about us. The Bible is the Good Book of Human Nature.

The numbers alone support this proposition. The Bible was written over the course of more than 1,000 years, and for nearly 2,000 years it determined the fate of a large part of the world's population. Today over 2 billion people still revere it as a holy text. And with a total print run of an estimated 5 billion copies, the Bible occupies an uncontested first place on the world's eternal best-seller list.[1]

We do not intend for these numbers to prove the value of the Good Book, either in absolute terms or in comparison to texts from other religions. We only want to stress that the Bible—which has fascinated so many people from so many different cultures over such a long period—has a great deal of essential information concerning human nature hidden in its pages. The Bible is much more than a religious testimonial: it also documents humanity's cultural evolution.

This realization led us to read the Bible again, keeping in mind the rich harvest of new findings in cognitive science and evolutionary biology to see what it has to say about human nature. We began our journey with Adam and Eve in the Garden of Eden, made our way through the Old Testament's vast expanses, met Moses and his chosen people along the way, encountered all the kings and prophets, immersed ourselves in the psalms, climbed up into heaven and down into hell, and finally ended our quest in the city of Jerusalem, where the Romans nailed Jesus of Nazareth to the cross—but not without taking one last look at Armageddon, where the final apocalyptic battle against the powers of evil is set to take place.

We soon discovered that a number of the Bible's stories appear in a whole new light when viewed from a biological-anthropological perspective. Onetime mysteries suddenly began to make sense. We even began to understand some of God's more unusual characteristics. What's more, biblical anthropology also offers insight into the process of cultural evolution. It allows us to understand how the belief in supernatural beings helped us to overcome tremendous challenges, how human nature channeled cultural change into the trajectories we recognize today, and, last but not least, how the Bible offers a key to a better understanding of ourselves in the here and now. In the Bible we find answers to humanity's greatest questions. We do not mean this in a religious sense. Rather, it teaches us why we fear death, how we deal with great misfortunes, and where our deep-seated desire for justice originated. The Bible shows us how we learned to survive in large, anonymous societies, why our modern lives sometimes seem so pointless, and why we are so often nagged by what we would describe as a longing for Paradise. When viewed without its halo, the Bible has something important to say to all of humanity.

WHY ARE WE THE FIRST TO SEE IT THIS WAY?

Remarkably, no other authors have ever taken this approach to the Bible—a fact that troubled us for quite some time. Why should we be the first to have discovered the unknown side of the Bible that has so much to say about human evolution?

For most of its history the Bible has been read as a holy text, as the word of God, but the last few decades have also witnessed an increasing number of attempts to uncover what else it holds for us. It has served as a historical source to reconstruct the birth of monotheism. Books such as Werner Keller's *The Bible as History* and Israel Finkelstein and Neil Silberman's *The Bible Unearthed* have reached millions of readers with their accounts of the archaeology behind the tales of the exodus from Egypt or about David and Solomon. The Bible was even placed on the couch: thirty-four psychologists interpreted its pages from Freudian or Jungian perspectives or examined it for traces of post-traumatic stress syndrome. But to date no one has dared to undertake the quest for the Bible's anthropological treasures and the insights they might yield into human nature and cultural evolution. Why?[2]

Of course the theologians—the true experts of the Bible—also occasionally engage in biblical anthropology. They can be forgiven for focusing their efforts on reconstructing the biblical perspective on humanity in light of philosophy, or at best history,[3] rather than evolution. Indeed, they remain focused on exegetical studies in which they try, with a detective-like flair, to uncover the Bible's complex origins and fathom the seemingly inexhaustible ways in which it can be interpreted. Many of them still endeavor to distill the divine spirit from these ancient stories. This is no easy task, for the will of the Almighty is not at all clearly formulated. In fact—and not every believer is aware of this—no generally recognized theological interpretations exist for most episodes in the Bible. Given that we're dealing with texts from the ancient Near East that were written millennia ago and subsequently reworked repeatedly, it's no wonder that they are anything but self-explanatory for today's readers. An increasingly wide gulf has therefore opened up between what the average churchgoer claims to know about the Bible and what theologians have uncovered through years of meticulous research.

The public at large also tends to know little about the work of the second major group to study the Bible, religious scholars. In their comparative studies of cultures, these researchers have discovered that many of the stories once believed to have been biblical originals are actually deeply rooted in the cultures of the ancient Near East. And archaeologists have uncovered an ancient Israel that looks quite different from the one we read about in the Bible. Even Yahweh himself was not always as lonely as a quick read of the Old Testament might imply—most likely there was once even a "Mrs. God."[4] Because this work aims to reconstruct the historical conditions in which the Bible and its religions arose, here, too, the Book of Books is not read as a book on human nature. We are convinced, however, that if we are to honor the Bible as such a document, we must expand our perspectives and firmly locate the Bible within the river of time that comes rushing toward us out of the deepest reaches of prehistory.

Although biological scientists some time ago became interested in the study of human nature, they have begun only in recent years to examine the nature and function of religion. This has led to an interdisciplinary approach that takes its cue from the observation that no human culture, regardless of its time in history, has managed to get by without some form of belief system. And this research agrees on one key point: religiosity, most likely a side effect of our predisposition to see actors behind every event, is a universal part of our fundamental psychology. Our innate tendency to attribute just about everything to supernatural powers, adaptive or not, is a basic part of the human condition.

And so neurobiologists began to give Christian nuns and Buddhist monks magnetic resonance imaging scans in a bid to discover which parts of the brain are active during spiritual experiences. Likewise, geneticists have started to search for a predisposition toward faith in our genome, and psychologists are studying the medical benefits of prayer.[5] But this research has yet to come up with truly groundbreaking discoveries. More promising are evolutionary-biological approaches. The belief in God, many evolutionary psychologists believe, promotes cooperation, and religious rituals enhance our sense of community. According to this theory, religion serves as "social glue." It therefore helps to solidify morality.[6]

Missing so far is religious-historical research from an evolutionary perspective.[7] To date, no biblical researcher has conducted a comprehensive analysis aimed at determining whether the Bible can confirm his or her hypothesis. At best only individual biblical passages have been examined from an anthropological perspective. Evolutionary psychologist Steven Pinker refers to the Old Testament's "staggering savagery" in order to illustrate how we have never lived in a world so free of violence as the one we inhabit today. Anthropologists Mary Douglas and Marvin Harris took on the Bible's food taboos, and Robert Wright and John Teehan referred to the Bible in their books on the evolution of religion, but they only examined single episodes.[8] No one has undertaken a systematic reading of the Bible to discover what the Book of Books can tell us about human cultural evolution.

This might have something to do with the fact that scientists aren't all that comfortable working with literary products. What's more, theologians have inadvertently barricaded the Bible against scientists' prying by erecting a protective mountain of texts around it: for nearly every verse there are at least three monographs. Wouldn't this mean that every word of the Bible has already been subjected to close examination? Anyone attempting to scale this peak of theological scholarship may find the heights a bit too dizzying, for the story of the Bible's construction is extremely complicated, and its interpretations are incredibly diverse. And we're not even talking about the issues surrounding dating and translation.

Another obstacle to scholarly examinations of the Bible is the hardened front that has arisen between religion and science, particularly in the United States. It certainly doesn't seem to be the most auspicious time to establish the "word of God" as a subject of scientific investigation. As Richard Dawkins, today's most prominent critic of religion, once said, "To be fair, much of the Bible is not systematically evil but just plain weird, as you would expect of a chaotically cobbled-together anthology of disjointed documents, composed, revised, translated, distorted and 'improved' by hundreds of anonymous authors, editors and copyists, unknown to us and mostly unknown to each other, spanning nine centuries."[9] We suspect that many researchers would agree with this statement. But just because very few sources on the early history of religion have survived, this does not mean that we should simply ignore the Bible, for no other document provides as much insight into the evolution of religion—and of culture. And anyone who takes the time to delve deeper into its pages will soon discover that it is anything but a "chaotically cobbled-together anthology."

"THE WORST MISTAKE IN THE HISTORY OF THE HUMAN RACE"

What, then, led the two of us to believe the Bible might yield other insights than those of interest only to true believers? Quite simply it was the lucky coincidence that we—an evolutionary biologist and a historian—found ourselves drawn together by a shared curiosity about what the Bible has to tell us. Although both agnostics, we have always found ourselves fascinated by the fantastic stories served up in the Old and New Testaments and intrigued by their mysteries. We believed we were holding in our hands a colorful kaleidoscope that reveals human existence in all its splendor and drama. The Bible contains stories of love, death, and the devil, of riches, violence, and slaughter. It raises questions about earthly and heavenly morality and the nature of angels. And it's full of dramatic topics such as incest and sodomy, human sacrifice, and what can best be described as promiscuity-control measures. Holy Scripture, it seems, is certainly not for those with delicate sensibilities. And so we began to read.

Both evolutionary biology and the cultural sciences aided us in our undertaking: we know what happened on earth *before* as well as *after* the Bible was written. This of course made it easier to understand what the Bible is really all about. To get straight to the point, we know how *Homo sapiens* became what he is today—roughly speaking, at least. We know how humans evolved over the last 2 million years and how and to what degree the prehistoric environment shaped the human psyche. That is to say, we can regard our emotions and behaviors as adaptations to a world that has long since disappeared—which explains why many of us don't find life in modern societies all that easy. We can therefore reconstruct the problems the Bible was trying to solve. Armed with this knowledge, we were able to extract many amazing insights from the Bible's pages. At the same time, this procedure allowed us to identify the problems that the Bible itself introduced into our world.

We soon discovered that a systematic reading of the Bible helped us reconstruct how religion as well as culture developed, at least in one particular corner of the world, and therefore the driving forces and the laws underlying their development. We discovered the key to this understanding in the Torah, the five books of Moses. We were taken aback by the countless and diverse calamities with which the Bible confronts its readers from the very outset. God had made humans for Paradise, but his creations eventually had to make their own way east of Eden. More than birthing pains, patriarchy, and living by the sweat of one's brow defined the fate of humankind; people also had to deal with family feuds, murder, and catastrophes. All of these problems are crammed into the short book of Genesis.

But then, embedded in the monumental epic of Moses and the exodus, we came across a whole collection of rules aimed at reining in human conduct (and the Ten Commandments are just the tip of the iceberg). One major goal was soothing God's wrath and thus putting an end to all the calamities. The measures condensed in the Bible's 613 *mitzvoth* are truly impressive in their protoscientific sophistication. People who believe religion is an irrational affair have never taken a close look at the Torah.

Now, it's important to realize that humanity hasn't always lived in a world full of catastrophes in which even brothers like Cain and Abel go for each other's throats. In order to understand this simple truth, we first have to scrutinize the single greatest change in behavior of any species on the planet. We are talking about the adoption of a sedentary way of life and all of the changes that followed in its wake. This was when we ceased to roam the wilderness as hunter-gatherers, as our ancestors had done for hundreds of thousands of years. It was the time we abandoned the small groups in which everyone knew everyone else for a life of coping with membership in large, anonymous societies. It was the time in which existence suddenly became much more stressful and social inequality ballooned.

Jared Diamond has provocatively called this event, which most people admiringly refer to as the Neolithic Revolution, as "the worst mistake in the history of the human race." He dedicates fundamental parts of his classic *Guns, Germs, and Steel* to this transformation, which got under way some 10,000 to 12,000 years ago.[10] Unfortunately, not a lot has been done since then to illuminate the consequences of the momentous changes that began at the start of the Holocene. Only slowly are those evolutionary biologists who trace the principles of cultural evolution entering the arena of pre- and early history. It must be said, however, that historians are not always pleased to find them trespassing into their territory, and perhaps this is why "the worst mistake in the history of the human race" is still waiting for a larger audience to understand it for what it really is: the decisive turning point in human evolution.

Most attention has usually focused on the progress this civilizing step brought with it. There's no doubt that the Neolithic Revolution set the stage for an unprecedented success story. Over the course of the last 10,000 years, *Homo sapiens*'s population has exploded from 4 million to nearly 8 billion individuals, and we have seen revolutionary technological progress. But very little has been said about the collateral damage of this demographic and economic success story. Archaeological excavations have shown that violence became a prominent part of everyday life; people became shorter, suffered more often from starvation, and died younger than their hunter-gatherer forebears. And with the domestication of animals, a number of diseases leaped from livestock to humans. Diseases such as the plague, smallpox, measles, influenza, and cholera emerged and quickly evolved. Tooth decay exploded. At the same time, injustice and repression made themselves at home in the new societies, and women bore the brunt of the suffering. Over the course of the coming millennia, these various scourges would beset humanity like the Four Horsemen of the apocalypse. But there was no going back to the old way of life. The point of no return had long since come and gone.

But still people needed to do something to come to terms with all of the epidemics and other disasters that befell them. The problems were too urgent, too life threatening to ignore. Had our only option been to wait for biological evolution to produce adaptations to these threats, humanity probably wouldn't have survived at all. It was time for our species to exercise its greatest talent—for culture to take over the reins. In order to gain the upper hand over misfortune, people began to look for the causes of all the catastrophes, violence, and epidemics so as to develop ways of protecting themselves from these dangers in the future. Cultural solutions were needed, and soon these efforts would result in a cultural "big bang."

It's important to remember that back then—and by "then" we mean the span of millennia in which first chiefdoms, then states, and finally advanced civilizations such as in Mesopotamia and Egypt developed—there were no independent fields of inquiry like science, medicine, law, and religion. Instead, there existed something of a primordial cultural soup. Only slowly did the individual spheres of knowledge begin to differentiate themselves, each eventually developing its own experts, methodologies, and institutions. But all of them remained deeply influenced by religion, for belief in the rule of supernatural powers was woven into every aspect of life. Any type of misfortune was credited to wrathful spirits and gods.

People tried the most varied strategies to vanquish their adversity. They faced a double challenge. On the one hand, their basic psychological makeup had evolved to deal with completely different problems stemming from completely different ways of life and was not up to the new obstacles people now faced. On the other hand, they had no knowledge whatsoever about the origins of the misery that came with sedentarization. They had absolutely no idea about microbial pathogens and their paths of infection, for example. So people improvised as best they could.

And thus they began to formulate rules and other measures that would eventually develop into entire systems of crisis management aimed at soothing the gods' rage—in hopes of protecting people from diseases and catastrophes, putting an end to all the rampant violence, and promoting cooperation. Much of what we widely attribute to "religion" today began as part of this "cultural-protection system." We shall examine this topic in greater detail in later chapters and also show why religion didn't end there. After all, it provides much more than a means of coping with crises.

But we'd first like to address a possible misunderstanding. We are not trying to say that the Bible directly reflects humanity's sedentarization. This would be difficult to imagine, for several thousand years lie between those events and the appearance of the written text. Much more importantly, sedentism brought with it a number of problems that remained threatening for long periods—some of them in fact still pose a danger to this day. Diseases are a good example: their virulence has fallen over the years, and we have developed medicines to combat them, but we can still catch the flu. The hunter-gatherers of the past were mostly spared all of this misery.

The Bible is much more than a reflection of historical conditions. Indeed, it presents us with the most ambitious strategies for dealing with "the worst mistake in the history of the human race" and all the calamities that followed. It documents congealed human experience from the time of the most decisive turning point in human history. But it also manifests the paths taken that would determine the fates of future generations. In sum, the Bible is humanity's diary, chronicling both the problems our ancestors faced and the solutions they came up with.

If we focus on the Bible's central themes, we can reverse engineer the problems people struggled with for thousands of years. Using this approach, we can work out where gulfs opened up between our innate intuitions and new social and ecological conditions. Here we are thinking about, among other things, how people dealt with the new cultural concepts of property, patriarchy, monogamy, and monotheism, to name but a few. This makes it easy to grasp why our undertaking is anything but backward looking, for an evolutionary reading of the Bible can offer the key to understanding human beings. It can explain a lot of the problems we still face each and every day.

OUR APPROACH

This book follows the Bible's own chronology. We concentrated on the most revealing stories. Because there were so surprisingly many of them, we were also able to offer an overview of the Bible's key stories. The journey through the pages of the Book of Books was at times like an upward-twisting spiral in which we repeatedly encountered the same themes at different levels. These represented questions that nagged ceaselessly at the people of the past—questions in search of new answers. We, too, underwent our own learning process during our journey through the Bible, particularly when it came to our understanding of that fascinating object of study called religion—a term that doesn't even begin to do the phenomenon justice.

Our book pays special attention to the Old Testament, for this part of the Bible is the most direct response to the events that influenced the course of human history like nothing before or since. But our examination also lays the foundation for our understanding of the New Testament, which presents us with a figure whose charisma continues to captivate people to this day: Jesus of Nazareth.

Depending on the religion or denomination, Bible versions vary when it comes to the contents of the Old Testament. We generally follow the order found in the Hebrew Bible, the Tanakh, which consists of three main parts: Torah, Nevi'im, and Ketuvim. The first two parts of our book discuss the Torah, or "instruction"—the perfect introduction to the world of the Bible. We begin in Part I with Genesis, which contains some of the Bible's most famous stories and also presents us with many of the problems that offer a consummate starting point for describing the formation of such a unique religion. Part II of our book focuses on the remaining four books of Moses, in which we also learn of the exodus of the Israelites from Egypt. We will not only discover the cultural masterpiece known as the Mosaic Law but also witness the meteoric career of a middling war-and-weather god and develop a sense of how belief in the one and only God brought with it a degree of religious distress.

In Part III, we turn our attention to the Nevi'im, the "prophets," to which belong the books of Joshua, Judges, Samuel, and Kings as well as the writings of the prophets themselves. Here we encounter not only questions of divine justice and morality but also the touchy subject of heavenly violence and even the understanding that the egoism of despots predictably ends up destroying society's social capital.

Part IV of the book examines the Ketuvim, the "writings," in which we find texts such as the psalms and the book of Job. Here we encounter a new form of religion that is familiar to believers today, for here God appears as a personal counterpart. The Ketuvim also deals with existential questions such as "Where does suffering come from?" and "Why do we have to fear death?" We suspect that the monotheistic God might play a role in these uncertainties: he himself introduced some of the problems with which people often struggle today.

Finally, Part V of this book looks at the New Testament, whose special character gains clarity when viewed against the backdrop of the Hebrew Bible. Cultural evolution produced an astounding marvel that managed to best fulfill humanity's needs. The fact that God got a human face in the form of Jesus of Nazareth played a fundamental role in this process. But the New Testament has a lot more in store, from Mary to the devil, from the resurrection to the apocalypse. The last of the Bible's additions fully satisfies human nature's hunger.

One of the most amazing aspects of our tour through the Bible was almost seeing how the writers were thinking as they wrote. The changes in the Bible reflect their struggles over the centuries to cope with a world full of problems and challenges. Book by book, the authors gradually made their model of how divine forces shape the world more consistent with reality. For more than 1,000 years, the Bible was not a fixed document but an ongoing, urgent mission to finally wrest control over their fate.

Even though we set out to uncover the evolutionary context of the Bible, we often found we had to argue in historical and sometimes theological ways. Anything else would have been impossible, for the Bible grew up in a very special habitat, and ignoring this basic fact would have rendered all of our statements off base. We consistently made an effort to stay close to the original text and to avoid retreating into metaphor or reading interpretations into the Bible that have no basis in its pages. The ease with which we were able to do so reassured us that we were on the right track. When viewed from an evolutionary perspective, all of the stories we examined turned up insightful and comprehensible interpretations in line with historical realities—an outcome few biblical interpretations can offer.

When quoting biblical texts, we drew upon the most famous of all Christian Bibles: the King James Bible. Of course there are more exact translations to be had, but these would do little to remedy the deficit that we couldn't read the original Hebrew, Aramaic, or Greek versions. Fortunately we found there was no need to take a microscopic view of the Bible for what we set out to do in our book. We are interested in the big picture, not individual verses.

To ensure readability, we have generally refrained from indicating the exact source of direct quotes from the Bible, since we usually refer to the book under discussion, and they should therefore be easy to find. Whenever the source was not obvious, we did quote chapter and verse.

Follow the Jewish Tanakh but quote the King James Bible? Yes, our approach is rather eclectic, but the Bible is the holy text of Judaism as well as Christianity. We tend to focus on what these two religions have in common rather than what separates them. After all, cultural evolution is a cumulative process. That's why we generally use the terms "Hebrew Bible" and "Old Testament" interchangeably. We found it important to move away from the view of the relationship between Judaism and Christianity as that of mother and daughter. Current rabbinical Judaism is a child of the same time and environment as Christianity. Both are products of early Judaism at the time of the Second Temple (515 BCE–70 CE), which itself was no clearly delineated religion (the English-speaking world often uses the term "Judaisms" to refer to the various religious systems of the early Jews).[11] Rabbinical Judaism and Christianity therefore developed in close contact with one another, and this is why we believe—along with Harvard religious scholars Kevin Madigan and Jon Levenson—it makes sense to speak of Judaism and Christianity as "siblings."[12]

And this is the source of one of our book's blemishes: our limiting ourselves to the Bible with the Christian New Testament might convey the impression that we view Christianity as an improvement upon Judaism. This, of course, is not our intention. After the destruction of the temple in Jerusalem by the Romans in 70 CE, Judaism and Christianity both continued to develop on their own. Ideally, we should therefore have included the Mishnah and Talmud—the core rabbinical texts—in our examination, but we believe our decision to take on the entire Bible was already hubristic enough.

And while we're on the topic of terminology, by the time of the Old Testament, mainly the time of the kingdoms and the Babylonian captivity (ca. 900–500 BCE), we are generally referring to ancient Israel and the people of Israel or the Israelites. It's important to remember that ancient Israel originally comprised two kingdoms: Israel in the north and Judah in the south. When we refer to the northern kingdom destroyed by the Assyrians in 722 BCE, we are talking about the kingdom of Israel.

There is one more thing we need to make clear. It is not our goal to present a theological or scholarly religious interpretation of the Bible. Of course we owe a great deal of thanks to the findings of theologians, religious scholars, and archaeologists, but we aim to show readers all the other things that can be found in the Bible. We strongly believe that a hidden side of the Bible still exists, one that deserves to be brought into the light. It can help provide answers to mysterious questions such as those relating to God's wrath in the Old Testament and the astonishing phenomenon that even people who do not believe in God find themselves drawn to Jesus.

Perhaps our evolutionary study of the Bible can even offer the faithful clues that will help them distinguish their own authentic needs from beliefs that are the product of hundreds of years of tradition and doctrine. But our real intention is simply this: we want to show readers just what an exciting book the Bible is. It contains countless undiscovered treasures, for we could certainly not uncover them all. It would be a shame if the Bible remained solely in the domain of religion.

Before we finally begin our examination by delving into the story of Adam and Eve, we would like to provide readers with a brief introduction to modern biblical studies and a concise primer on what we mean by cultural evolution and human nature. For all of those who fear that we adhere to a deterministic view of humanity that underestimates the value of culture, we would like to state emphatically that this is by no means the case. Humans don't have one single nature—as we shall see, they actually have three. The Bible—according to our credo (or perhaps we should just say working hypothesis)—can help us to understand them all.

A BRIEF HISTORY OF THE BIBLE

God didn't really write the Bible himself; nor did Moses write his five books. Insights such as these began to emerge in the seventeenth century, although they met with fierce resistance. English philosopher Thomas Hobbes (1588–1679) was one of the first to question whether Moses actually wrote the five books of the Hebrew Torah (Greek: Pentateuch). Then the Amsterdam-born Jewish philosopher Baruch Spinoza (1632–1677) published his *Tractatus Theologico-Politicus* in 1670, in which he listed all of the passages in the Torah that Moses couldn't have possibly written himself, for they told of events that occurred after his death. For his troubles, Spinoza was banned from the synagogue and made the target of two assassination attempts. At least he survived. Less lucky was the Spanish humanist Michael Servet (1509/11–1553), who nearly a century earlier had produced evidence that the Christian Trinity of the Father, Son, and Holy Spirit never actually appears in the New Testament. Servet was burned at the stake.[13]

The first clues as to how the Bible was created were found in the eighteenth century. Independently of each other, German theologian Henning Bernhard Witter (1683–1715) and French scholar Jean Astruc (1684–1766) discovered that in a number of texts in the book of Genesis, God is referred to as Elohim, whereas in other passages he is known as Yahweh.[14] Every reader of the King James Bible can reconstruct this, for the "God" who made heaven and earth in the story of creation suddenly becomes the "LORD God" who ejected Adam and Eve from the Garden of Eden. This indicated that at least two different authors were at work.

These early efforts gave rise to a veritable research tradition that firmly established how the Pentateuch drew upon a number of different sources that had been combined and reworked in a number of editorial phases. Then German biblical scholar Julius Wellhausen (1844–1918) achieved fame with the "documentary hypothesis," which claimed that the Pentateuch was the result of the successive interweaving of four different sources—a theory that continued to have a significant impact well into the twentieth century. The theory is now considered outdated, not least because the existence of the "Yahwist"—a hypothetical author whose texts were believed to have been composed in around 950 BCE—has been refuted. What has remained, however, is the notion of "multiple authorship," although the Bible's creation is now believed to have been so complicated that reconstructing it in detail would be all but impossible.[15]

Nowadays scholars estimate that the Hebrew Bible was composed over the course of approximately eight hundred years (ca. 900–100 BCE). The main work thus took place much later than previously thought. "The historical-critical biblical scholarship has assembled enough evidence to show that the Old Testament books in their present form were clearly influenced by the theology of the Judaism of the Persian and Hellenistic periods" (ca. 540–100 BCE).[16] Of course the authors drew upon older material, but this means that the Bible discusses events in the distant past that were written down and reworked at a much later date. Hence, the Bible cannot offer a dependable picture of historical reality. The Bible actually tells stories.

We do not wish to spend too much time on issues surrounding the dating of the Bible's books—after all, even the experts have a hard time agreeing. William Dever, one of the most important archaeologists focusing on ancient Israel, notes, "The various biblical texts (or 'traditions') cannot in fact be dated very well at all. Even mainstream biblical scholars vary as much as 500 years in dating the material."[17]

A Jumbled Mess: How the Bible Was Created

The following is an overview of what we now believe to be true about the Bible's creation—and here we are focusing on the Hebrew Bible, which Christians refer to as the "Old" or "First Testament." The word of God is actually the work of humans through and through. Most importantly, the Holy Scripture is not a single work but a collection of different writings.[18] We can take references to it as the Book of Books quite literally, for it does indeed encompass an entire library. The number and arrangement of the books vary: the current Hebrew Bible comprises thirty-nine books,[19] including, among others, Genesis, the books of the prophets, and the books of Ruth and Job; in Christian Bibles the texts and books in the Old Testament vary according to denomination.[20] In fact, even the individual books are by no means the homogenous works of single authors, for they have expanded over time to reach their present forms. A number of authors and editors worked on the texts, more often than not creating contradictions and repetitions. In his *The Old Testament: A Literary History*, Zurich professor of theology Konrad Schmid describes such writings as "composite literature."[21]

Oral traditions tend to be much older than their first transcription. The Bible's raw materials are stories of varied origins passed down from generation to generation. Over time, narrative cycles developed around figures such as the patriarch Jacob, for example.[22] As writing blossomed in the kingdom of Israel in the ninth century BCE,[23] scholars began to put these stories to paper, and later they were combined, reinterpreted, and expanded upon. Theologian Christoph Levin compares the process to a snowball, "which begins to pick up layer after layer with each revolution." In the Old Testament, "there is hardly a single textual unit which is not composed of several literary strata," Levin writes.[24] German Nobel Prize laureate for literature Thomas Mann (1875–1955) had something similar in mind when he called the Bible a "chronicle of humanity," nothing but a "mountain of books that has grown together from the rocks of different geological epochs."[25] Because of this, it is just as impossible to "clearly and exclusively" assign a text to a particular time[26] as it is to "clearly and exclusively" assign it a final interpretation. This, too, contributes to the Bible's dazzling diversity.

None of this is all that unusual, of course. Quite a few of the world's great narratives—such as *The Iliad* and *The Nibelungenlied*—took years to grow into great epics or collections of sagas. In the case of the Bible, unusual historical circumstances played an additional role. The Bible was born in not one but two small kingdoms: Judah, with its capital Jerusalem, in the south and Israel, with its capital Samaria, in the north. The problem, however, was that these two kingdoms were located at the intersection of the world's great cultures of the era, which also happened to be humanity's oldest civilizations. To the west lay Egypt, home of the pharaohs and their pyramids. To the east was Mesopotamia, where the empires of the Assyrians, Babylonians, and Persians took their turns at power. The "Holy Land" therefore repeatedly found itself between hammer and anvil. Again and again it was laid to waste and its people led away into captivity. As we will see in later chapters, without these catastrophes the Bible could never have become the book it is today.

We are dealing with an exceptionally productive melting pot. A succession of additional major powers and their cultures—the Hittites, the Philistines, the Phoenicians, the Greeks, and the Romans—also influenced developments in Palestine, as did smaller nations, such as the Aramaeans, the Amorites, the Moabites, and the Edomites, whose names come up again and again in the Old Testament. The competition among these cultural and religious influences was tremendous, and this jumble was to have a huge impact on the Bible's development. Yahweh, the same god who would later champion monotheism, was still a work in progress at the time the Bible was written; he got his start as just one of a whole collection of deities. In all things religious, syncretism was in vogue—consciously or subconsciously the Israelites let neighboring peoples inspire them. Some stories were simply too good to ignore: the tale of Noah's ark had its roots in an older Mesopotamian one, and the episode of the baby Moses floating down the river in a basket of bulrushes was lifted from a legend surrounding the birth of an Assyrian king.[27]

In sum, material that was highly heterogeneous with respect to text type, origin, and religious intention was combined. Much had to be smoothed over, edited, and censored. There was no master plan for the Bible's composition—modern terms such as "bricolage" and "sampling" are certainly appropriate when it comes to describing the story of the Bible's creation. Whereas it is true that Israel's monotheistic God replaced the polytheistic pantheon of the ancient Near East, not all traces of the old belief system were completely obliterated. Opposing positions lingered on, and here and there we still encounter a strange ghost haunting the pages of the Holy Scripture. Everything from magic to witchcraft turns up in the Book of Books.

Many a believer may have to swallow hard on hearing Swiss Old Testament scholars Othmar Keel and Thomas Staubli speak of the Bible as a "river of a hundred voices."[28] Such faithful prefer to believe that "God sat behind His big oak desk in heaven and dictated the words verbatim to a bunch of flawless secretaries," in the words of A. J. Jacobs, who himself wrote an amusing book about his attempt to live one year according to the Mosaic Law.[29] The fact is, however, that the Bible has countless authors—and God most likely wasn't one of them.

This "river of a hundred voices" also offers a beautiful illustration of how the process of cultural evolution works. The secret of *Homo sapiens*'s success is that we understand not only how to create knowledge, ideas, and discoveries but also how to communicate, refine, and—again and again—add to them. This is known as "cumulative cultural evolution." And we shall now turn to this ancient specialty of our species, for this phenomenon transformed humans into creatures with three distinct natures.

CULTURAL EVOLUTION AND OUR THREE NATURES

Unlike purely historical examinations of the Bible, our anthropological approach considers humanity itself as a player on the great stage of cultural evolution. We see humankind as biological beings, animals that only differ from their primate relatives in that they also have cultural evolution at their disposal. Humans are not only capable of making inventions and passing them on to others; they are also able to improve on them, combine them with other ideas, and develop them into increasingly complicated systems. Cultural evolution enables us to quickly adapt to new habitats and even modify these environments to fit our own needs. It emancipates us from "biological" (or "organic," as biologists like to call it) evolution, which leads to changes in our genetic material brought on by our adapting to ecological challenges over the course of countless generations.[30]

We argue that the Bible is so interesting because in its pages we find manifestations of the cultural strategies that helped *Homo sapiens* master problems and crises that can all ultimately be traced back to the greatest change in human behavior the world had ever seen: sedentarization. We now would like to briefly examine that process's impact on humanity and human psychology. It's important to realize that these changes took place in a relatively short period—over the last 12,000 years, to be exact. But what's 12,000 years of cultural evolution compared to the roughly 2 million years of biological evolution in the genus *Homo*? Given this short time frame, it's no wonder that we continue to feel the reverberations of sedentarization to this day.

A Cultural Quantum Leap

Why can we only write this type of book about humans and not some other species? Many will consider this a silly question; after all, only humans have cultures. Over the course of the last few decades, however, we have learned that other species also have their own cultures: they display distinct behavior patterns that vary from place to place and arose when an individual invented a new technique whose usefulness or novelty caused it to be adopted by others and to spread throughout a population. This process continues until some barrier prevents the innovation from spreading any farther. Take orangutans, for example. In the swamp forests of Sumatra, the orangutans on one side of an impassable river have mastered the skill of using sticks to bypass the sharp hairs of the *Neesia* fruit and so access its nutritious seeds. On the other side of the river, however, orangutans have not discovered this trick and must use force to crack open the woody fruits or simply leave them hanging.[31]

The importance of such simple cultural techniques in adapting to local conditions or environmental changes can hardly be overstated, for they can lead to survival advantages that may play out within an individual's lifetime. Biological evolution, on the other hand, always begins with smaller or larger genetic mutations within a single individual, which means it can take countless generations to produce changes at the level of the population. Because most mutations do not lead to actual improvements—and generally more than one favorable mutation is needed to produce viable adaptations—new adaptations clearly do not arise so readily. Given all of these factors, organic evolution is a painfully slow process that hardly ever produces fine-tuned adaptations to local environments, especially when environmental conditions change again before the prolonged adaptation process can achieve optimum results.[32] This means that for many species, there is a limit to how precise the fit between local conditions and behavioral adaptations can be. Cultural adaptation is much faster. Cultural "mutations," new types of behavior or technological innovations, are usually produced to meet a particular need or solve a particular problem.

Great apes are also capable of producing such changes and can therefore be said to have culture. Yet, their cultures differ fundamentally from those of even the simplest human populations. Our ancestors managed to establish the novel process of cumulative cultural evolution. Innovations were refined, corrected, or combined with other ideas. And, again in contrast to the great apes, humans are capable of actively communicating advances to others, meaning they can be taught directly to other people. This is the decisive difference. We have the unique ability to continually improve on and expand our culture. In our reading of the Bible, we will see how certain changes can develop a momentum of their own and lead to even more changes.

The process of cultural evolution led to series of major transitions in our lifeways. The speed at which they took place accelerated rapidly over the course of the last 100,000 years, beginning with the invention of ever more sophisticated weapons, tools, art, and symbolic behavior. Over the past 10,000 years or so, these transformations began to take place at an even faster rate with the arrival of agriculture in its various forms. Predictably, these changes engendered large societies with their kings, cities, and wars.[33] This amazing acceleration is unique and most likely the consequence of a positive feedback loop in which demography plays a significant role: more people mean more inventions, and more inventions mean more people. But every change challenged us to come up with new solutions to the new problems we had ourselves created. Unsurprisingly, however, the increased rate of rearrangement ultimately began to tax our own flexibility and inventiveness.

Viewed from this perspective, sedentarization represented a cultural quantum leap beyond all comparison. Within the shortest period, humans found themselves bombarded with new sets of problems. First and foremost were the diseases brought about by these changes, but hunger and violence resulting from increased competition also represented existential threats. Such rapid and dramatic change meant that genetic mutations did not have sufficient time to produce adaptations. In such situations populations usually go extinct or move on to more favorable regions. Instead, our ancestors developed new survival strategies that they implemented very quickly. Culture alone had to produce the solutions to these problems by building upon clever thinking, lucky guesses, or some combination of the two. People invented all sorts of new tools, developed new rules of behavior and rituals, and built cult sites where they could summon the spirits and—later—the gods. And even though these innovations didn't always work out, culture itself advanced to become the recipe for success. In the end, culture produced the world in which we live today.

Just a Crutch?

Cultural evolution provided us with the tools we needed to live in our new environment, and in doing so it made biological adaptation nearly redundant. We generally see this as progress, but people seldom realize that culture does not always succeed in eliminating our problems entirely (or indeed at all); consider eyeglasses, which do nothing to cure poor vision, even though they do help us to see. Culture only offers tools to help us tackle the new and precarious living conditions we face. Some of these tools are hardly more than crutches, whereas others have become high-tech prostheses with which we can achieve undreamed of heights. But without them we can no longer live in comfort—and if we lose them we can really get ourselves in trouble.

But let us pause briefly to prevent any misunderstandings. We are not promoting cultural pessimism. The last thing we want to do is glorify French philosopher Jean-Jacques Rousseau's (1712–1778) "noble savage," who lived a natural life unspoiled by civilization. Humans are not now, nor have they ever been, truly noble. Our point is merely that we should be aware of how culture has affected humanity. As much as we owe to culture, it really is something of a double-edged sword. Culture produced our unprecedented affluence and the world's immense population, but it is also responsible for a number of the problems we were not designed for. We simply lack the necessary psychological tools to deal with them.

To put it in simple terms, thanks to biological evolution both our bodies and our minds became more or less adapted to our environment, but cultural evolution could never provide us with the same sense of well-being. Quite the opposite, in fact, for it led to a great divergence between our preferences and our actual state of mind (or body).[34] We call these resulting maladjustments "mismatches." Mismatch happens when some adaptation, say a particular behavior, founders because the environment in which it manifests itself is radically different from the one in which it evolved. Many readers will already be familiar with the idea: we move and feed ourselves today quite differently than we did in the times of the hunter-gatherers, when the specific physiological needs of our bodies were formed, and this has had a significant impact on the size of our bellies as well as the incidence of what are fittingly termed "diseases of civilization."[35]

But fewer readers will be aware that we also struggle with such mismatch problems in the sphere of social behavior. When these problems rear their heads, we usually blame them on the poor choices of individuals and do not recognize them as the cultural products they really are. We will encounter many such examples in this book, especially in our examination of Genesis, but for now let us take an example from everyday life. For most of human history we lived in small communities in which everyone knew everyone else. Our fates depended on what everyone else thought about us, so we took great pains to maintain our reputations in front of the group. We automatically viewed any strangers we encountered as a potential threat. With the rise of cities, however, we began to live in large, anonymous societies. Cultural innovations have ensured that we can come to terms with the most serious complications: not all of us have to carry weapons, and we have learned that strangers are generally harmless. The rules of politeness ensure that our behavior is predictable. And then there are the inoculations that protect us from the infectious diseases strangers may carry.

This might sound like a case of successful cultural adaptation, but some things still don't match. To this day we still take great care to protect our reputations, even though urban anonymity means that this is not nearly as important as it once was, and money now plays a much greater role in determining our status. And when packed together in big groups—in a busy department store, for example—people are at great pains to ignore all of the other strangers around them. The fact that some people might have trouble coming to terms with so many strangers is by no means due to their own shortcomings, for these feelings have their origins in a completely natural reaction. Nearly all of today's phobias can be traced back to adaptive—that is to say, sensible—strategies for avoiding danger.

Three Natures

For our evolutionary reading of the Bible, we have developed a simple model of the human mind that can help us describe the effect culture has on people, one that pays heed to the biological as well as the cultural dimensions of our personalities. We differentiate between three kinds of human nature—and all three are found in all of us.

Our *first nature* consists of innate feelings, reactions, and preferences that evolved over the course of hundreds of thousands of years and proved their effectiveness in the daily lives of small groups of hunter-gatherers. Overall, these intuitions enabled people to function fairly smoothly in their social and ecological environments. Because they are anchored in our genetic codes, they require very little inculcation. The products of our first nature include myriad propensities, such as love between parent and child, a sense of fairness and outrage at injustice and inequality, a loathing for incest and infanticide, a fear of strangers and concern for reputation, a feeling of obligation after receiving gifts or assistance, jealousy, revulsion, and—not to forget our sense of religion—the tendency to see supernatural actors at work everywhere. In a nutshell, our first nature makes itself known in the form of intuitions and gut feelings.

The existential problems that arose in the wake of sedentarization were so severe that, as explained above, quick cultural workarounds were of the essence. These solutions in turn led to new habits, conventions, and ways of thinking. These cultural products cannot be inherited; once they have proven themselves, they can only be handed down and learned. Adults actively ensure that children internalize these behaviors during early childhood—to such a great extent that they come to make up our *second nature.* One doesn't have to justify this behavior. If we consider our first nature our "natural nature," this second nature is our "cultural nature." French sociologist Pierre Bourdieu coined the term "habitus" to describe this phenomenon.[36] Our second nature often varies from culture to culture and rarely reaches the same level of taken-for-grantedness—the same emotional depth—as our first nature. Yet our innate feelings still underlie our second nature, which may even co-opt them for its own purposes. This is why people react with revulsion (first nature) to the eating habits (second nature) of other cultures: "How can they bring themselves to eat dogs?" The realm of second nature includes traditions and customs, religion as a cultural product (the kind that is practiced in church, for example), and most of the things that Norbert Elias describes in *The Civilizing Process*: rules of decency, politeness, and good manners.[37] In other words, all of the things to which we might say, "People here just don't do that!" or "It's just the way things are around here!"

Our *third nature* reflects our rational side. Third nature consists of the maxims, practices, and institutions that we follow consciously—due to a targeted analysis of a given situation, for example. This doesn't mean that these rules always make perfect sense, of course. But we usually believe it's a good idea to follow them, for not doing so might get us into trouble. Third-nature products are also internalized to a certain degree, but this usually occurs at a later phase of development—at school or in other institutions, for example. Over time, some successful third-nature products sink to the level of second nature, especially if they are not at odds with our first nature and we are indoctrinated with them in early childhood. Most aspects of our rational nature, though, remain only skin-deep, owing to their cognitive complexity as well as the fact that they often conflict with the innate proclivities of our first nature. The result is reluctance. Examples include all the things we balk at doing even though we know they are in our best interest: eating healthily, exercising, not drinking and driving, and obeying the speed limit, to name just a few. New Year's resolutions, too, are typical products of our third nature. That's why we almost always fail to live up to them. They are merely prudent, after all.

In Praise of Our Gut Feelings

In the prehistoric environment in which human evolution took place, we used our first nature's feelings and intuitions as a compass to navigate daily life. Our second nature supplied us with special habits, techniques, moral prescriptions, and social practices such as rituals, which varied from group to group. Our third nature only came into play in emergency situations when we faced new challenges and tried-and-true mechanisms no longer performed as expected. The cognitive solutions of our commonsense nature could—if they were any good and did not conflict too greatly with our first nature—become habits and conventions and thus, over time, achieve second-nature status. This process of cultural accumulation continuously expands the courses of action we have at our disposal. For most of human history, it took place at a modest pace, for truly new challenges were uncommon in our ancestral environment.

The dramatic shift in human behavior that took place at the beginning of the Holocene opened up to us a whole new world, one for which we simply were not made. Our first nature could no longer keep up with the changes that came in the wake of sedentarization, such as immense population densities and technological advancement. So our third nature took over the helm—and with impressive efficiency. Today hardly any situation we might encounter would induce us to happily hand control back to our first nature. But because our third nature was so adept at ensuring our survival, natural selection had very little opportunity to introduce fundamental changes of its own. And this meant that the mismatch problems that came with these changes were here to stay.

This outcome was inevitable. Because our first nature is anchored in our genetic code, and 12,000 years have not been enough for these changes to modify our psychological preferences, we continue to carry these mismatch problems around with us. When Johann Wolfgang von Goethe's Faust spoke of two souls dwelling in his breast, he was talking about our first nature, which is always happy to make itself known. It protests against what it still considers a new way of life; it niggles at the back of our minds and gets us into moral dilemmas. We have to summon up all our discipline to maintain control, but sometimes we simply don't have the energy to do so. In recent years, experiments have shown that self-discipline is actually something of a limited resource. At some point, when we are tired and weary, we will simply give in to our desires. That is when our first nature takes over and, like Goethe's Faust, cries out to the fleeting temptation, "Stay, thou art so fair!" Unfortunately, as we know all too well, there's a rude awakening in store for us at the end.

Here is a classic example. A married woman falls in love with another man, or a committed husband falls for another woman. Our first nature sighs with pleasure, "Love!" Our second nature calls out, "Fidelity!" And our third nature objects, "Think of the mortgage, the lawyers' fees, the alimony payments!" As a moral dilemma, this is a thoroughly modern problem, since monogamy is a cultural invention that has barely managed to achieve second-nature status (it's also absent from most parts of the Bible). The church's attempts at remedying the situation—"What therefore God hath joined together, let not man put asunder"—have done little to make it all seem more natural.

"Our systems for social interaction did not evolve in the context of vast groups and abstract institutions like states, corporations, unions and social classes," explains anthropologist Pascal Boyer. "Sedentary settlements, large tribes, kingdoms and other such modern institutions are so recent in evolutionary time that we have not yet developed reliable intuitions about them."[38] The result is a latent discontent with civilization, a recurrent feeling of living in an upside-down world. Our second and third natures are cultural products. They have helped us survive, but they don't necessarily make us happy. As workarounds they can only partially fulfill needs that hearken back to a long forgotten world. Ever since we gave up being mobile foragers, life has become a matter of the head instead of the heart.

In our anthropological reading of the Bible, we have developed a greater appreciation for our gut feelings, which have proven something of an effective divining rod in our examinations. Whenever we came across something that just didn't sit well, we knew we had stumbled across another mismatch problem. Our first nature was telling us that something wasn't quite right. In fact, this happened right at the start, with the story of Adam and Eve. The Bible begins with a story that, to be honest, is plain weird. Why did God cast his own creations—and all of their descendants to boot—out of Paradise for a minor misdemeanor? Such cruelty cannot but rile our first-nature repugnance for injustice.

PART I

GENESIS

When Life Became Difficult

Genesis. What a treasure trove of stories! Speaking snakes, mysterious trees in the Garden of Eden, a God who forms the first man from dust, Noah and his ark filled with animals, a tower that soars into the heavens, and intimate glimpses into the private lives of Abraham, his wives Sarah and Hagar, and their numerous descendants. Even respected theologians are convinced that the Bible's "most wonderful stories" are found in the book of Genesis.[1] And they are right. Few other works have managed to capture our imaginations for such a long time. Could this have something to do with the fact that these tales—if you bother to actually read them—are profoundly enigmatic?

The number of calamities encountered at the very beginning of the Good Book will shock the modern reader. Adam and Eve are ejected from Paradise. Cain slays his brother Abel. Humanity endures a catastrophic flood and then, for its audacity in building the Tower of

Babel, is scattered across the globe. The tales of the patriarchs abound in intrigue, murder, and mayhem.

Bereshit, meaning "in the beginning," is the Hebrew name for the first book of the Torah. Its Greek name is Genesis, which translates as "birth," "origin," or "emergence." This name also jibes perfectly with our evolution-inspired perspective. The Bible has identified the true origins of human misery and the resulting—and still omnipresent—feeling of alienation that comes from living in a world for which we were not made.

Some biblical exegetes simply interpret these difficulties as a consequence of the fall from grace. "That's the way the world is," they say. "Our troubles began when people disobeyed God's commandment and were forced to leave the Garden of Eden." Although we do not find this explanation very convincing, we do agree with the surprising conclusion that *Homo sapiens* has not always lived with life's tribulations—the drudgery, the pain of giving birth, the fratricide. Indeed, these problems are new, the result of the behavioral changes that followed in the wake of our ancestors' adoption of farming and livestock breeding. These innovations brought injustice, disease, and massive violence into the world with them.

That's why the Bible is such an inexhaustible and incomparable anthropological resource. By surveying the problems described in its pages, we can assemble a catalog of the afflictions that plague humanity. The biblical narratives tell of a time when our species faced problems so pressing and new that we were unable to evolve effective genetic adaptations to deal with them. People therefore had to develop their own coping strategies, and this led to a cultural big bang whose echoes still resonate today. The Bible reports on what might be the most ambitious, systematic, and sustained attempt to come to terms with all of the sorrow that has confronted humanity for thousands of years. Its pages pack a lot of cultural evolution.

At the end of our journey through Genesis, we will come to see the hand of God in this misery and find an answer as to why the stories of the Book of Books continue to fascinate us to this day. Above all, we will appreciate how religion morphed into a veritable cultural Swiss army knife. Today's readers of the Bible may not all find their way to God, but they will certainly come away with a better understanding of human nature.

1

ADAM AND EVE

The Real Fall of Man

THERE IS NO BETTER BEGINNING THAN THE STORY OF ADAM AND Eve. Let's be honest: Who has never marveled at the fact that God threw both of them out of Paradise for the sake of a little nibble on the sly—and in doing so brought eternal doom down upon all of humanity?

Isn't the Bible's beginning puzzling? Adam and Eve's fate throws up one question after another, and we have an urgent need to find meaning in the story. Throughout the ages people have objected to the fundamental injustice: It's just not right! How could God cast the pair out of the Garden of Eden? Our gut instinct refuses to accept it.

In fact, the Good Book's opening story confirms our suspicion that there is a telling of the Bible that remains to be discovered, a story that has remained hidden from readers.[1] How else can we explain that after 2,000 years there is still no consistent, generally accepted interpretation of the story? Most people are simply not aware of this. Readers believe the true message is the "Fall of Man," which headlines the tragic tale of the world's first pair of lovers. This headline, however, was actually a later addition[2] and a rather tendentious one at that.

This is why we wish to scrutinize the story of Adam and Eve. In this chapter we subject it to a more detailed examination than in our handling of the Bible's later stories in order to illustrate how our anthropologically inspired reading of the Bible works. Our first step is to recall the actual course of events in the Garden of Eden rather than what thousands of years of speculation have made of them. We therefore take a close look at the embellishments surrounding the first two people on earth. This is quite entertaining, for it turns out that nothing limits our imagination when it comes to extracting an at least moderately acceptable interpretation from the word of God. The actual story has long since been buried under the many previous attempts at finding meaning, and this is why, like archaeologists, we have to clear away all of civilization's rubble.

Scholars of religious history traditionally begin by critically analyzing the historical circumstances out of which the Hebrew Bible, the Old Testament, arose. In the second step of our analysis, we use their insights to unlock the brilliant world of the ancient Near East. This will allow us to peer back into the depths of history and identify the ancient substrate from which the story arose—the "before" that refers to our transition from foragers to agriculturalists and pastoralists. As it turns out, the church is right. The story of Adam and Eve really is the story of humanity's original sin. But this original sin is something entirely different from the interpretation the church has maintained over the centuries.

In the third step we introduce our biblical anthropology into the mix. We show just what surprising perspectives emerge when we keep in mind that the Bible's story is actually a tale of the existential difficulties brought about by the radical change in behavior that occurred when *Homo sapiens* adopted a sedentary way of life. When we read it in this way, the Bible suddenly becomes transparent.

THE SEARCH FOR MEANING

Anyone who pays close attention to the first pages of the Old Testament will be struck by just how strangely it all began. Genesis tells of how God first created heaven and earth and then everything else in the world. And on the sixth day he said to himself, "Let us make man in our image, after our likeness." And it came to pass. "So God created man in his own image, in the image of God created he him; male and female created he them." His instructions were clear: "Be fruitful, and multiply, and replenish the earth, and subdue it."

Then something peculiar happened. After God looked again upon his work, declared it "very good," and allowed himself a day of rest, what did he do next? He started over. He took earth from the fields, formed Adam from it, and breathed life into him. Then God "planted a garden eastward in Eden" and placed Adam in it. And because God immediately realized that it was "not good that man is alone," he thought to himself, "I will make him an help meet for him." From earth he once again formed the beasts in the field and the birds in the sky and presented them to Adam, who was allowed to name them. "But for Adam there was not found an help meet for him." At that point God came up with the idea of the rib.

He put Adam into a deep sleep, took one of his ribs, and from it formed a woman. "This is now bone of my bones, and flesh of my flesh," Adam realized immediately. "And they were both naked, the man and his wife, and were not ashamed." God had every reason to be satisfied. But he didn't reckon with the snake.

We all remember the rest of the story: Eve is tempted by the snake to defy God's commandment and taste the fruit of the tree of knowledge of good and evil. Adam does nothing to stop her. From that point on, they are both ashamed of their nakedness, and God casts them out of Paradise. We have lived east of Eden ever since.

Generations of believers, deeming every letter of Genesis holy (and not knowing that it combines two creation stories), have puzzled over the details. Why did God create humans twice? What happened to the first people? It's no wonder that the rumor arose that Eve was actually Adam's second wife. The first had run off, and that is why God resorted to the trick with the rib.

Not surprisingly, the story of Paradise has given rise to endless questions—and brought forth new stories. Seldom are they of a pious nature. Why did God first show Adam the animals when looking for a helpmate? Theologians like German Old Testament scholar Erhard Blum speak of a "trial and error process."[3] The Babylonian Talmud really gives free rein to the imagination: Adam supposedly tried to mate with one female animal after another.[4]

We can understand all this intense interest. The creation account seeks to explain the curious course of our history. We want to know all the details—it's about our own fate, after all. There are no limits when it comes to conjuring up new interpretations, and even the smallest detail is worthy of our attention. Did Adam and Eve have navels although they were not natural-born? Most painters seem to believe they did. Did Eve have one rib more than Adam because she was made from one of his? No, men and women have the same number of ribs. Does that mean that Adam originally had thirteen ribs? Maybe: around 10 percent of people do. And the snake that was damned to crawl around on its belly after it tempted Eve—haven't snakes always done that? No, some paintings depict the snake sauntering around on two legs; some even show it riding a camel.[5]

All Because of an Apple That Wasn't

Not only are the details contested, but even the core question of the story's message remains unresolved. Although mountains of learned interpretations have piled up over the years, there is still no convincing answer to the most important question of all: What does the Bible want to tell us through the story of Adam and Eve? Why did God punish them? It almost seems as if every rabbi or theologian has his or her own opinion on the subject.

At this point a number of believers will shake their heads in bewilderment and say, "There's no question about it! It's right there in the chapter headline of every Bible: man's disobedience." But isn't this a strange tale? All of humanity is held collectively responsible and punished over hundreds of generations—all because of a single apple? In fact, the Bible does not even mention an apple, speaking simply of a "fruit." Only in late antiquity did scholars begin to refer to it as an apple because the Latin word for apple, *mālum*, was similar to the Latin *malum*, meaning "misdeed," "evil," or "calamity."[6]

More importantly for our analysis, the story of the Garden of Eden makes no mention of sin,[7] a term that does not appear until the story of Cain and Abel. And if you consider how mild God's punishment was in the face of a capital offense—it was, after all, humanity's first case of murder, and Cain was merely exiled—the inexorable penalty meted out to our first parents is quite disproportionate. Why was God unable to forgive the theft of a single piece of fruit?

Could it be that he wanted to draw attention away from his own guilt? After all, God alone bears the responsibility for his creations. He had just formed the actors in this story, including the snake that tempted Eve. If one of his creations messes up at the first opportunity and his creatures don't know how to behave, then that's a problem for him. Isn't he to be held accountable?

Other questions arise once we begin to reflect on these issues. What was so bad about what the first couple actually did? Eve only wanted to learn, and what's wrong with that? "The fact that humankind strives for knowledge about good and evil, that it wishes to be wise," notes theologian Rainer Albertz, "is not the 'original sin' in all of the rest of the Old Testament, nor is it anywhere else in the Near East of antiquity."[8] Albertz goes one better and asks whether the humankind "that God actually wanted was meant to be dumb and incapable."[9] In the past such speculation might have got him burned at the stake.

Some Adam and Eve advocates doubt the two were even capable of guilt. At the moment of committing their crime, they existed in a state prior to the knowledge of good and evil. They could not have known that they were doing something wrong, for they had no idea what evil was. As they were incapable of guilt, it follows that they could not be punished. In fact, theologians have described their behavior as something more akin to "muddle-headedness and naïve childishness."[10] As intellectually challenged as they were when it came to morality, they deserved some leniency. Or at least a recognition of mitigating circumstances.

But maybe that is what they actually received. The punishment that God gave them is strangely inconsistent. God had threatened Adam, "In the day that thou eatest thereof thou shalt surely die." Yet this did not come to pass, for believe it or not, Adam lived to see his 930th birthday, surviving for many centuries after the fall. Was the serpent right when he recommended they taste of the fruit and told Eve, "Ye shall not surely die"? The exegetes of early Judaism sought to save God's honor by means of numerical acrobatics. They claimed that Adam and Eve did indeed die on the day they ate from the tree of knowledge, but one of God's days is equal to nearly 1,000 human years.[11] This is also a side story, but it shows how God's inconsistency also managed to confound the scholars of antiquity.[12]

What was God's punishment according to Genesis? He cast Adam and Eve out of the Garden of Eden. As he did, he told the woman, "In sorrow thou shalt bring forth children; and thy desire shall be to thy husband, and he shall rule over thee." Then he told the man, "In the sweat of thy face shalt thou eat bread, till thou return unto the ground; for out of it wast thou taken: for dust thou art, and unto dust shalt thou return." Yet then God showed himself to be surprisingly considerate when he fashioned them coats out of skins and clothed them himself. Did God have a bad conscience? This first story in the Bible truly is quite peculiar.

It Must Be True!

Even though we could continue along this line of reasoning for hours, let us try to keep things brief, lest the theologians among our readers grow bored. We are, after all, not the first to point out these inconsistencies. Still, the fact remains that new questions arise out of Genesis's every paragraph. God's message was not clear at all, something that parents understand all too well: "Daddy, why is the snake bad? Why didn't God make him good?" These questions are of interest not only to children.

Homo sapiens is the "storytelling animal," says Jonathan Gottschall, a literary scholar inspired by evolutionary biology.[13] Anthropologist Pascal Boyer also describes that our minds are "narrative" or "literary": "Minds strive to represent events in their environment, however trivial, in terms of causal *stories,* sequences where each event is the result of some other event and paves the way for what is to follow." And we pay close attention to whether these stories are consistent.[14] The story has to be logical even when it deals with the fantastic.

As it happens, we all do have an internal compulsion to seek out coherence aimed at "cognitive dissonance reduction."[15] We have to lend meaning to the things that we do and the things that happen to us. Our very survival has always depended on this: someone who could not interpret the signs in his environment wouldn't make it all that far in life. That is why our "storytelling mind" is downright allergic to uncertainties and true coincidences. And if our mind is unable to come up with a meaningful interpretation, then it has absolutely no qualms about making one up.[16] We will encounter this over and over again in this book.

Confirmation bias—the error cognitive psychologists refer to as the "father of all fallacies"[17]—also has a role to play here. We humans seek out information and prefer to grasp at those bits that confirm our preconceived ideas. Information that conflicts with these ideas we often do not even notice. Confirmation bias is responsible for the fact that fundamental assumptions seldom get called into question. When Adam and

Eve were cast out of the Garden of Eden, there *must* have been a good reason for it.

And if we are dealing with a story that claims to explain why life has become such an ordeal, then our internal speculation engine switches into high gear. We simply have to identify the mistake that led to this state of affairs, for we certainly don't want to make it again. The clergy might claim that the Lord works in mysterious ways, but this does not stop people from trying to decipher them.

Taming Unruly Texts

Even theologians speak of a "plausibility gap" at the beginning of Genesis, by which they mean that the world "apparently is not as God's plan of creation intended."[18] Does this not lead to the heretical question of how firmly God is in control of his handiwork? A subversive tendency resides at the heart of our passion for making up stories, something brought home by the interpretations we have mentioned above. They are seldom all that pious: Adam carries on with the animals! Adam's first wife ran off on him! And Eve has even been accused of sleeping with the devil, leading to her conception of the fratricidal Cain. You cannot even complain there is ill intent behind all of this speculation: it reflects a desperate attempt to make logical sense of the biblical story.

A great deal of effort has therefore gone into searching for religiously acceptable interpretations. The Jewish philosopher Philo (15/10 BCE–40 CE) is considered the first to make use of allegorical exegesis borrowed from the Greeks. From then on, whenever they came across something too "naïve, anthropomorphic, or Oriental—when, for example, God regrets having created humanity or, after having first created a paradise for man, immediately drives them out of it for a childish mistake, or when he either out of anger and/or fear of competition destroys their nice big tower and confounds their language—then the philosophical scholars claimed that all of this was figurative in meaning."[19] Following this formula, we can extract appropriate meaning from just about any unruly text.[20]

Let us now examine the most important attempts at making sense of this story. They are all expressions of our craving for coherence, and they also represent the rubble we must first clear away if we hope to reach the Bible's substrate. It is actually a fairly enjoyable job: the rubble makes for rousing reading.

The Devil in the Serpent

We have already noted how the story of the Garden of Eden makes no mention of sin. It is therefore remarkable that the notion of original sin derives from such a sin-free story, a sin of such intensity that it affected all of Adam and Eve's children as well as their children's children. The idea of original sin does not originate in Genesis, and throughout the rest of the Old Testament, it is never invoked to explain the existence of death, disease, or suffering.[21] Not until the book of Sirach (ca. 175 BCE), which Jews and Protestants consider apocryphal, does the fatal sentence first appear: "Sin began with a woman, and thanks to her we must all die." Even after that, the idea took some time to gain traction. In the Syriac Apocalypse of Baruch one reads, "And what will be said of the first Eve who obeyed the serpent, so that this whole multitude is going to corruption? And countless are those whom the fire devours."[22] On the Christian side of things, Paul[23] (ca. 5–64 CE) and, most importantly, the church father Augustine of Hippo[24] (354–430 CE) are responsible for propagating the idea of original sin.

We also know by now that the devil appeared—or, to be more precise, entered into the snake—much later.[25] This interpretation, even if not predominant, is still common today: "Man, tempted by the devil, let his trust in his Creator die in his heart and, abusing his freedom, disobeyed God's command. This is what man's first sin consisted of," according to the current catechism of the Catholic Church.[26] But Genesis itself makes no mention whatsoever of Satan or Beelzebub or fallen angels or the envy of the demons.

For this reason, industrious biblical scholars have devised alternative interpretations. Here we take a look at three of them. The first is a very modern reading in which hubris lies at the heart of the first couple's transgression. According to this theory, Eve decided to take fate into her own hands as if she herself were God. The desire for independence thus acquired the "stigma of godlessness."[27] The second exegesis is all about sex.[28] As early as the Middle Ages, scholars speculated that the couple—naked and without shame—were tempted not by the tree's sweet fruit but by the sweet allure of intercourse: "God did not mean to say: *ne edatis*, but *ne coeatis*," according to Kurt Flasch.[29] The consequences of their sinful behavior were agony in childbirth for the woman and hard work to feed the family for the man.

More intellectually demanding is a third interpretation, which views the story of the Garden of Eden as a metaphor for adolescence. The apple symbolizes maturity, and the bite represents a coming of age, as well as, in the words of Konrad Schmid, professor of Old Testament studies in Zurich, the consequence that "adult human life guided by the 'knowledge of good and evil' . . . necessarily moves away from God."[30] Schmid's German colleague Erhard Blum believes the story manifests "man's self-inflicted exit from his blessed immaturity."[31]

In all these cases authors have foisted their interpretations on the Bible to transform this stubborn material into something more manageable. And in every case doing so requires moving away from the original text, either by adding material that isn't there—you can keep on looking, but you won't find the devil hiding in Eden—or interpreting it allegorically. In 2,500 years of exegesis, no definitive theological interpretation of the story of Adam and Eve has emerged. Compared with this, the religious historians have done better.

THE GLITTERING ORIENT

Religious history provides two major frameworks for interpreting these stories. First, whether they arose in Egypt, Mesopotamia, or Greece, stories of the gods explained how the cosmos and humans were created and why the conditions of life were as they were and not different. Such stories are known as etiologies.[32] In this framework, the story of the Garden of Eden offers explanations for why our life on earth consists first and foremost of hard work. But it also describes how the world became patriarchal, why the Sabbath became holy, and why God no longer walks among us.

The second perspective uses the story of the loss of Paradise to explain the realities of life among the Israelites. In 722 BCE, the Assyrians destroyed the northern kingdom of Israel, and in 587 BCE the Babylonians captured the southern kingdom of Judah—including its capital, Jerusalem—and exiled the upper classes to Babylon. According to the priesthood, God had made use of the foreign armies to punish his own people for their disobedience. Just as Adam and Eve were cast out of Paradise for their waywardness, the kingdoms of Israel and Judah were lost to the Assyrian and Babylonian oppressors because their peoples had ignored God's commandments. The losses of the Promised Land and the Garden of Eden—according to this historical theological interpretation—were God's punishments for a lack of loyalty.[33] The Diaspora thus began with Adam and Eve.

Few believers are comfortable with such interpretations: God's word is eternal, after all, not bound to a particular time. Yet only a historical analysis can explain why the story of Adam and Eve continues to capture the imagination today. But this analysis should not limit itself to the world of the Bible alone. Indeed, if we broaden our scope to cover another one or two millennia and also encompass the broader region stretching from the Nile to the Tigris and Euphrates, then the analysis becomes more powerful and the Bible more sensible, largely because much of the ancient Orient is buried in the story of Genesis.

There is nothing remarkable about that, for Palestine was situated at the crossroads of the ancient high cultures. It was a transit zone for goods, ideas, and stories, whether they came from Egypt or Mesopotamia. Most of the episodes appearing in the Old Testament were written down and reworked either during or after the Babylonian exile—and all of this took place against the backdrop of contemporary Oriental religions. The authors and editors were well versed in the stories of their rivals' gods, and they had absolutely no qualms about using them for inspiration—or even copying a few stories in their entirety.[34]

Biblical scholars have uncovered a number of ancient stories that fed into the story of Genesis. For our research, the most interesting of these originated in Mesopotamia. The *Enuma Elish*, dating from the second millennium BCE, recounts the creation of the heavenly bodies and humans much as Genesis does.[35] The myth of Atrahasis (ca. 1800 BCE) even describes a great flood, sent to destroy the humans, who had come to trouble the gods. But Atrahasis, warned by the god Enki, built an ark, brought animals aboard, and survived.[36] This flood narrative also found its way into the most renowned epic of the era: the saga of the feats of the Sumerian king Gilgamesh.[37] Here, too, we find motifs familiar to readers of the Bible. For example, Gilgamesh finds the herb of life, only to have it stolen by a snake.[38] Gilgamesh also encounters a second snake, which hides in the roots of the huluppu tree planted in a holy garden.[39] Finally, in 2014 biblical scholars Marjo Korpel and Johannes de Moor reported that they had discovered a tale from the Canaanite city of Ugarit, now located in modern-day Syria, that could have served as a model for the authors of the story of Adam and Eve. In the text, dating from the thirteenth century BCE, the god Horon takes on the form of a snake and, by means of a bite, transforms the tree of life located in a "vineyard of the gods" into a tree of death.[40]

Whether vilified as plagiarism or celebrated as "bricolage" or "sampling" in the humanities, this form of adaptation is one of humankind's oldest cultural techniques. In cultural evolution "endless cross-fertilization" is only to be expected.[41] Motifs, characters, and ideas wander from culture to culture. They are crossed with one another to create new stories.

He Was Not Alone

Not only is Genesis full of motifs taken from older stories, but these stories also played out against the backdrop of the ancient Middle Eastern cosmos. It is a colorful world full of life, a polytheistic world full of gods and spirits—a concrete setting unfettered by abstract ideas. It is hardly ever explicitly moralistic and often not even moral.

The final editors of the Bible tried to eliminate this Oriental legacy in favor of their new idea of monotheism, and they sought to adapt the older stories to fit the new religion. This is particularly apparent in the story of the flood. Whereas there were three gods and a goddess at work in the Mesopotamian original, in the Bible we only meet God, and he plays all of these roles. This fact explains why his actions are not all that "coherent."[42]

The creation story downsizes divine diversity: whereas the sun and moon were powerful gods in neighboring cultures, the God of the Bible demotes them to mere "lights."[43] The primordial ocean, once the formidable goddess Tiamat, is now nothing more than just water. And whereas the Oriental gods had to slaughter one of their own in order to obtain the blood needed to awaken the people they had formed out of clay,[44] in the Bible it took only the divine breath of life. Here it is easy to see what Max Weber meant when he spoke of the "disenchantment of the world."[45]

This process of disenchantment was never carried to its full conclusion, however. How could it be? Otherwise, we would have had to do without the story of Adam and Eve. For the story to be retold, one could not do without the talking serpent. It is not an evil intruder but a native inhabitant of the Garden of Eden.[46] And what is true for the snake is also true for God himself. At the beginning of creation he remains a representative of the older polytheistic world. He is neither abstract nor discarnate, neither omniscient nor omnipresent. In the evenings he goes for a walk in his garden and cannot find his creatures. He has to call out for Adam:

"Where art thou?" And to whom is he speaking when he says, "Let us make man in our own image"? Religious scholars see this as evidence that he was once accompanied by a divine entourage.[47] As Robert Wright wrote in his *Evolution of God*, "Apparently God himself didn't start life as a monotheist."[48]

Today we lack the sensorium required to detect the remnants of the ancient cosmos of the gods. Some of them are obscure, such as when Eve is dubbed "the mother of all living," a title derived from a very special lady indeed: Asherah.[49] Once upon a time she was most likely God's wife (more on this later). Other clues are less obvious, because we no longer perceive them as such divine remnants. We believe, for example, that God posted an angel armed with a sword to guard the entry to the Garden of Eden, but in reality there were two guards, who were much more exotic: the cherub is a winged creature, a hybrid of man and animal, and the flaming sword is a solidified flash of lightning.[50]

Not a Question of Morality

Once we are aware of this backdrop, Genesis really begins to sparkle. The story of Adam and Eve has its origins in an old polytheistic world swarming with supernatural beings. And they often got in each other's way—even a snake could thwart a god's plan. This dazzling array of powers made it impossible to differentiate between good and evil.

All of this also makes it clear that the biblical story of the Garden of Eden contains one prominent object that does not belong there: the tree of knowledge of good and evil from which Adam and Eve were said to have eaten. It stood, according to Scripture, next to the tree of life in the center of the garden. Yet its strangely long-winded name is enough to suggest that the tree was not indigenous.[51] And why, going against narrative economy, does the Bible mention two trees? Only one of them is important when it comes to the story. Exegetes have repeatedly pointed out that Eve, having been approached by the serpent, only seems to know of one tree.[52]

In fact the tree of life was an old, familiar fixture in the ancient Middle East. The Canaanites[53] believed it to be the seat of the goddess, and in *Gilgamesh* we already saw the connection between snakes and plants or trees of eternal life. Never before seen in the ancient Middle East, however, was a tree of knowledge of good and evil.[54] Some biblical scholars therefore surmise that the tree was a later addition.[55] We have to agree: the black-and-white logic of good and evil postulated by the tree of knowledge simply doesn't fit into the old, scintillating world of the Eden episode. Such strict morality only appears when monotheism arrives on the scene: if there is only one God, you are either with him or against him; you are either good or evil. *Tertium non datur*—there is no third possibility.[56] The polytheistic cosmos lacks such rigid dualisms. No godly decision is irrevocable, for another god might easily upend everything. People find this attractive. If they don't agree with one god's way of thinking, they can turn to one of his rivals. Adam and Eve do this as well when they follow the advice of the serpent instead of listening to God. The realization that the tree of knowledge is a later addition has far-reaching consequences. Whatever the story of Adam and Eve was about, it was *not* about the knowledge of good and evil or humanity's acquisition of morality.

Why We Really Got Expelled

So far we have exhaustively documented what the story is not about. Here is what we think its actual subject really is: the Garden of Eden story tells the tale of a worsening existence. It is about a cultural step: in the beginning we lived in a world of abundance, and this gave way to the involuntary adoption of a life of tillage and toil. We must follow this trail as it leads directly to a point in history that is deemed a decisive turning point in human cultural evolution: the adoption of a sedentary way of life.

Variations on this theme appear in a number of stories of the ancient Middle East. First primordial man is created, and in a second step he is civilized. A Sumerian literary work with the curious title *The Debate Between Grain and Sheep* begins by describing how humans knew nothing of culture and lived like the animals: "The people of those days did not know about eating bread. They did not know about wearing clothes; they went about with naked limbs in the Land. Like sheep they ate grass with their mouths and drank water from the ditches." The gods then decided to civilize people so that they could provide food for them. According to the myth of Atrahasis, humans were solely created to nourish the gods and later were decimated by plagues, because the gods felt they made too much noise.[57]

Probably the closest religious-historical parallel to the story of the Garden of Eden appears in an early Babylonian version of the epic of *Gilgamesh*—in the biography of Gilgamesh's friend Enkidu, to be precise. After Enkidu was formed from clay, he first lived like a wild man together with the animals. It was not the gods, however, who civilized him, but rather a harlot skilled in all the womanly arts of seduction. Although all of the animals left Enkidu and his ability to run was compromised, his intellect expanded, and for the first time he understood her words. The harlot commented on his transformation as follows: "You are wise, Enkidu, and now you have become like a god." That sounds familiar—"and you shall be as gods," or so reads the serpent's well-known promise to Adam and Eve. The similarities are striking. In the *Gilgamesh* epic, the civilizing act has equally negative repercussions: Enkidu loses his innocence and his feeling of security among the animals. Now that he is a civilized human being, his thirst for action and glory will harm the animals and offend the gods. As a result he brings the gods' wrath upon himself and suffers a tragic end.[58]

Henrik Pfeiffer describes these stories so typical of the ancient world of the Middle East as "two-staged anthropogonies."[59] Man's transformation from an evidently wild creature into a cultural being led to a life of backbreaking labor. "In these texts work is always described as drudgery," Old Testament specialists Othmar Keel and Silvia Schroer explain.[60] A key point in all of them, however, is the understanding of this step as the result of a transgression. "The existence that arose as a result of a transgression against God's commandment . . . is clearly seen as negative and must become humanity's doom."[61] The new way of life is a burden. And, astonishingly, such interpretations show that these two-stage anthropogonies are actually surprisingly close to historical reality. And with that it is time for us to get to work on the evolutionary interpretation.

THE TRULY ASTONISHING

We now turn to the prehistoric world, when humans roamed the earth in small groups of hunter-gatherers. During this time our first nature was formed, that innate set of emotions, psychological needs, and moral intuitions necessary for us to master life in groups of manageable size. We shall see that the hunter-gatherer epoch provides the necessary backdrop for understanding biblical phenomena.

They Were Not Ashamed

For most of our history, we were nomadic hunter-gatherers. A tiny number of scattered people still live like that today, and we can reconstruct how we would have lived based on knowledge of these people's lifeways. Hunter-gatherers resided in small, multifamily bands comprising some thirty to fifty members that were always on the move. They formed loose networks with neighboring bands, even though they were not in constant contact. Hunter-gatherers required large swathes of territory. Only occasionally did all members of these "macrobands" come together to arrange marriages or to service their alliances.

Within the individual groups, people lived in close personal contact with one another. At least in the warmer regions of the world, they had few clothes and mainly covered only their genitals. They also had very little in the way of belongings. Distinct hierarchies were just as absent as significant concentrations of power. Social differentiation was minor, determined by individual abilities or prominent personality traits. Resources, game in particular, were shared. Generosity improved hunters' reputations. In short, for the longest time human existence was generally egalitarian and democratic. This, etched into our psyches, determines the perception and interpretation of our social surroundings to this day.[62]

Cooperation was everything. Interdependence among group members was the foundation of communal living. Decisions were made as a group, often after long discussions. An individual's reputation was of great importance. Could one rely on his knowledge and experience? Was he dependable and ready to help? Reputation was the capital of prehistory. As humans in those days had no stored provisions at their disposal, foragers had to invest in social relationships, and these included relationships with other bands of the same community. Whoever had proven that he was prepared to support the other members of the community could count on their support in times of need. Cooperation was a form of life insurance.

Accordingly, harmony was of paramount importance. Every transgression against community life was noted and, if necessary, punished. In serious cases this might lead to expulsion from the group or even the death of the troublemaker.[63] But these were rare occurrences, for no one stood a chance of surviving on his own, and this meant everyone was at pains to be a good group member. The group was also capable of forgiving misconduct if the culprit showed the proper remorse. In the wilderness, every woman and every man counted. There was no place for rigorism.

The emotions of our first nature governed how we lived together. Over hundreds of thousands of years of evolution, innate preferences and moral intuitions ensured that hunter-gatherers functioned in a way the others could count on. Particular traditions such as customs or rituals had long since become a part of their second nature, and they rarely found themselves confronting a situation that forced a third-nature solution to the problem. We can safely assume that the state of grace that Adam and Eve enjoyed in the beginning of the story—"And they were both naked, the man and his wife, and were not ashamed"—closely matched life in prehistoric times.

To avoid any misunderstandings, we do not wish here to romanticize the prehistoric world before humans adopted sedentary ways of life. We only wish to establish that the life of the hunter-gatherer did possess some "paradisiacal" traits—at least compared to the living conditions that came next. We are interested in the fact that humans lived in an environment to which they had genetically adapted over the course of hundreds of thousands of years. The "mismatch," as it is often called, did not yet exist. Before sedentarization, there was no gulf between the "old" psyche and the "new" environment—the source of doom and gloom. Still, that does not mean the life of the hunter-gatherer was a picnic, so to speak.

Paradise Lost

With the waning of the last ice age around 15,000 years ago, many parts of the area known as the Fertile Crescent, situated between the Nile to the west and the Tigris and Euphrates to the east, were transformed into a virtual land of milk and honey. Herds of antelope, gazelles, horses, and wild cattle populated the extensive grasslands. In many places, hunter-gatherers who previously had constantly roamed around found they no longer had to do so. They established permanent camps and began to enjoy all of this bounty. In a way their lives were not so far from that in the Garden of Eden—but their luck would not hold for long.

Prehistorians continue to debate whether the end to these paradisiacal conditions around 12,000 years ago stemmed from climate change or human activity. Regardless, there is clear evidence that overhunting caused animal populations to collapse. If they wanted to avoid a similar fate, the foragers had to come up with something new. A return to nomadic life was no longer an option in many of these places, as densities had become too high and neighboring communities no longer tolerated trespassers on their lands. Another problem surely was the loss, after generations of sedentary living, of a great deal of the knowledge needed to survive in the wilderness. People desperately tried to come up with new strategies for survival.

We can be sure that no one in those days ever proclaimed, "Eureka! Let's be farmers!" The early days of agriculture were a haphazard affair. Humans had always collected berries, nuts, and wild grains, and seeds would have continually fallen on the ground close to their settlements. This would have led to the seeding of whatever plants they had brought back. At some point, humans began to make systematic use of this discovery. The domestication of animals such as goats and sheep also took place during this time.

This new life was hard, however, which Genesis sums up in a nutshell. That Adam and Eve have to make a living east of Eden by toiling in the fields and pulling weeds is a sign of punishment. "Cursed is the ground for thy sake!" God was pretty clear when it came to explaining to Adam and Eve their new fate. "In sorrow shalt thou eat of it all the days of thy life. Thorns also and thistles shall it bring forth to thee; and thou shalt eat the herb of the field. In the sweat of thy face shalt thou eat bread."

The new way of life really did appear to be a curse, and over the centuries things got progressively worse. Evidence of this surfaces in prehistoric skeleton finds. It seems that sedentary life didn't agree with humans: people were no longer as tall as their hunter-gatherer ancestors, they suffered from hunger[64] and disease, and they died younger. They had to toil in the fields. Without decent seeds, fertilizer, and effective irrigation techniques, they were lucky to bring in a good harvest. Droughts and floods had a much greater impact than in the past, for people were now closely bound to a particular place. Sam Bowles, an economist and anthropologist at the Santa Fe Institute, has calculated that the earliest agriculturalists had to invest a great deal more time to obtain the same amount of calories as the hunter-gatherers of yore enjoyed.[65] This means that the latter had more time for maintaining social relationships. Viewed from this perspective, the life of their ancestors must have seemed like paradise to the earth's earliest farmers.

Is the Bible Right After All?

The biggest question is, how could the authors of the Old Testament have remembered all this? The Old Testament was not set to paper until the first millennium BCE—thousands of years after sedentarization had taken hold. There is a tremendous gap between the two events.

The idea of a "collective memory" has firmly established itself in the human sciences.[66] This memory comprises personal "communicative memory," which includes no more than three generations, and a "cultural memory" that reaches back further into the past. The latter preserves memories in the form of myths, rituals and beliefs, stories, songs, and sayings. There has never been any proof that events could be retained in the cultural memory over thousands of years before the invention of writing. Nevertheless, many biblical stories sound like distant echoes of the Neolithic Revolution, particularly when the Bible falls back on older traditions. As we shall see in subsequent chapters, the Bible repeatedly focuses on precisely the problems that developed during this time.

There are, of course, alternative explanations. Lost paradises, golden ages, and lands of milk and honey appear in myths and fairytales from all over the globe. The idea that "nature provides its full blessings of its own accord and without human assistance" appears to be a persistent fantasy throughout human history.[67] Such stories are often interpreted as backward-looking utopian tales whose true purpose is a critique of the present. Perhaps.

We might also be dealing with an encounter between contrasting experiences. In some places, agriculturalists must have had sporadic contact or even trade with nearby groups of hunter-gatherers. Travelers spread rumors of unclothed people who knew no shame and appeared to receive nature's fullest blessings. Or perhaps the nomadic herdsman lifestyle of neighboring pastoralists struck the agriculturalists as primitive?

Distant echo, backward-looking utopia, or contrasting experience—we do not wish to exclude any of these explanations. Ultimately, however, disentangling these sources is of merely historical interest, because such an analysis cannot explain why so many of the Bible's stories remain so fascinating to this very day. But that is precisely what our anthropological argument can do! It doesn't really matter if people were aware that their ancestors had not always had to deal with the problems that came with sedentism, for many of the problems are just as persistent now as they were thousands of years ago. It still feels as wrong to our first nature that we have to toil in the field by the sweat of our brow as it did on the very first day.

In terms of evolution, the few millennia since these problems initially appeared simply represent too short a period for human psychology to adapt. Had such adaptations evolved, our psyches would no longer find such living conditions problematic; they would seem normal—the problems would simply have ceased to exist. And no one would have ever found it necessary to build a gripping story around these issues, for stories never center on the obvious. Instead they are time-tested simulations that we use to reflect upon problems and develop solutions. Jonathan Gottschall describes the telling of stories as a powerful "virtual reality" technology that enables us to take a theoretical approach to challenges, determine possible actions, and test their acceptability. "Just as flight simulators allow pilots to train safely," Gottschall notes, "stories safely train us for the big challenges of the social world."[68] People everywhere have always fallen back on the medium of the story to deal with life's miseries, and mismatch problems were the seeds from which these stories sprouted. The best of them survived because they offered explanations that people felt were plausible.

This is what our evolutionary reading of the Bible has to offer: by seeking to expose the core of these stories, we can recognize the problems humans faced after the greatest change in behavior that *Homo sapiens* had ever experienced. It allows us to identify those challenges that evolution saddled us with. These are not the idiosyncratic problems of particular individuals: to a greater or lesser extent, they continue to trouble us all. This is why the Bible still has something to say even to those who don't believe it is God's word.

SCANDAL IN PARADISE

With all of these considerations in mind, we now return to Adam and Eve. As we shall see, their story reflects more than our desire to comprehend why life for most of us consists of so much drudgery, for it also manifests two other important problems of our new existence: the invention of property and the oppression of women.

The Invention of Property

The first and only rule that God had issued in Eden was simple: "Of every tree of the garden thou mayest freely eat: But of the tree of the knowledge of good and evil, thou shalt not eat of it." It was God's property, and because Adam and Eve did not follow this simple rule, they were ejected from Paradise. Indeed, the invention of private property is the most consequential outcome of the adoption of sedentism, and respecting someone else's property is the first commandment of the settled world.

It is hard for us today to imagine the significance of this change. Property rights are so familiar that we take them for granted; indeed, we treat them as natural rights. Nomadic hunter-gatherers, however, only owned a few everyday objects, such as a spear and butchering knife. Game, large fish, and honey, however, they shared, and they even celebrated this act of sharing. Anyone who tried to keep all of the meat for himself would suffer a loss of reputation—and if it happened repeatedly, the perpetrator would face more severe sanctions. The land itself belonged to the group, the tribe. Everyone knew exactly where the neighboring tribe's territory began, but within the tribe's own stretch of land, every member enjoyed the same rights of use. Anyone who claimed, "That there is my tree. You may not eat of its fruit," would have been ridiculed.

With the arrival of sedentism, all of this inevitably changed. Agriculture demanded that certain things could no longer be shared. How can one have a good harvest if everyone helps himself to the fruits of the field? From this point on, the farmer would claim, "This is my land! Those are my plants and my stores!" Others were no longer allowed to take from them. But establishing this new concept of property was by no means easy. It required an enormous intellectual effort to convince members of a community that certain things now belonged to a single individual or family. Why should this land or this tree suddenly be off-limits to everyone else? Everyone would have made use of it in the past! What right did an individual have to call it his own?

Ethnologist Frank Marlowe observed among the Hadza of today's Tanzania how one individual hunter-gatherer once began to tend a crop but soon abandoned the idea when other members of the group helped themselves to the produce. Today we might call their actions shameless—and, indeed, they never would have thought for a moment that they were in the wrong.[69]

This should not come as a surprise, however. As there was no property worth mentioning in the world of the hunter-gatherer, our recognition of it never became anchored in our first nature. American psychologists Jonathan Haidt and Craig Joseph compiled a list of five universal moral modules: "suffering (it's good to help and not harm others), reciprocity (from this comes a sense of fairness), hierarchy (respect for elders and those in legitimate authority), coalitionary bonding (loyalty to your group) and purity (praising cleanliness and shunning contamination and carnal behavior)."[70] Conspicuous in its absence is a moral module that drives us to respect the property of others. It does not exist because 12,000 years turned out not to be enough for evolution to anchor it in our genes.

In order to effectively protect their belongings, people had to come up with new ideas—a classic task for our third nature, our ability to reason. What are needed are cultural rules that help to establish the new idea. As we have seen when it comes to the Bible, prohibitions coupled with a threat were quite popular in the past, something along the lines of "If you eat that, something bad is going to happen to you." Among horticulturists (foragers who also plant gardens), spirits are responsible for protecting the garden in its owner's absence. In Polynesia, ethnologists have observed how owners utter a taboo over their fruit or vegetable garden and leave it up to the gods to punish the thief.[71]

The Bible shows us that verbal measures are not that effective, however. Eve herself was no longer sure exactly what God's commandment meant after the snake had whispered in her ear. Strong institutions are better than mere words. In order to prevent humanity from stealing from the tree of life, God transformed Paradise into a *hortus conclusus*, an enclosed garden, and even posted sentries outside for good measure.

It took a few generations for the new rules concerning private property to establish themselves. The more acceptance of these conventions grew, the more they became a part of our upbringing, and this is how respect for other people's property became a part of our second nature. Its position is tenuous, however: "property" is a concept that we have to teach children ("Give her back her toys! They're not yours!"). And even adults bear sympathies for robbers such as Robin Hood who steal from the rich and give to the poor.

In light of this perspective, we have to ask ourselves just what the actual scandal in Paradise was. The fact that Adam and Eve failed to heed God's first commandment was, as we have shown, certainly not it—at least not to nonreligious sensibilities. One does not have to be a hunter-gatherer to find it hard to believe that the fruit of a particular tree is taboo. Pick the fruit before someone else does—that is our first nature.

The real scandal therefore lies elsewhere. When humans adopted a sedentary lifestyle, a fundamental rule of human coexistence was cast aside, an everyday norm developed over the course of hundreds of thousands of years: food must be shared; selfishness is shameful. For ages this had been obvious. The new concept of property subverted the prehistoric reliance on solidarity. What was commonly owned—nature's food supply—suddenly became monopolized. That is the real scandal! We have to imagine how an everyday, even necessary activity—the gathering of fruit—was not merely forbidden but criminalized. The scandal reverberates to this very day. If we had found it reasonable that God punished Adam and Eve for picking the forbidden fruit, their story would not have captivated us for so long.

Materially Rich, Socially Poor

Sedentarization brought about a number of processes that sparked radical changes in human societies. Earlier we touched on how hunter-gatherers, unable to store provisions, invested in social relationships to ensure mutual support in times of emergency. Cooperation was everything; solidarity served as a form of life insurance. Suddenly, the world was turned on its head: the privatization of resources made the farmer far less dependent on his neighbors. People no longer relied on the critical support of other families, and they have neglected these less crucial social relationships ever since.

Thus begins our journey down a one-way street leading to a world in which life is ever richer in terms of material goods but also socially and emotionally impoverished.[72] Relationships with people outside the family become less and less important. What's more, property has to be protected, with violence if necessary. And because one's own relatives offer the best allies—blood is thicker than water, after all—sons begin to stay at home with their fathers. That means that women now have to be brought in from elsewhere. These developing patriarchies transform women into tradable goods, into chattel.

Wherever the new stockpiling economy is successful, population numbers begin to soar. Competition is everywhere, social distinctions grow larger, hierarchies are born, and a privileged class arises. The discontent of men with neither property nor women of their own begins to swell. This can only lead to one thing: more violence. In this respect, philosopher Jean-Jacques Rousseau was probably right when he wrote,

> The first man, who, after enclosing a piece of ground, took it into his head to say, 'This is mine,' and found people simple enough to believe him, was the true founder of civil society. How many crimes, how many

> wars, how many murders, how many misfortunes and horrors, would that man have saved the human species, who pulling up the stakes or filling up the ditches should have cried to his fellow: Be sure not to listen to this imposter; you are lost, if you forget that the fruits of the earth belong equally to us all, and the earth itself to nobody![73]

We will pay close attention to the Bible and uncover the wide variety of problems called into being by the new property-based way of life. Women were its first victims, and this conclusion brings us to the second major problem of this novel lifestyle.

All About Eve

Poor Eve. Just look at all she was made to suffer. God cursed her with the words, "I will greatly multiply thy sorrow and thy conception; in sorrow thou shalt bring forth children." If that were not enough, he added, "And thy desire shall be to thy husband, and he shall rule over thee." The biblical exegetes upped the ante with their interpretation of Eve's role in committing the original sin, which brought bitter suffering down upon everyone who came after her.

This bias against women is quite a recent phenomenon. The relationship between the sexes was actually much more balanced among hunter-gatherers. For sure men were the dominant sex, but women were able to return to their families at any time or even to switch partners if their current husband behaved too high-handedly. Pair bonds at that time were not necessarily exclusive, and in practice the notion that a woman was bound to one man for her entire life was alien. Sometimes women even maintained parallel relationships. Ethnological observations among the Aché people, who lived as hunter-gatherers in the highlands of Paraguay until quite recently, have shown that over the course of her life, one woman had an average of twelve husbands.[74] Engaging in sexual relations with several men was in a woman's best interest, allowing her to establish a network of potential fathers for each of her children.

The fact that women knew how to best use their charms was a natural part of their sexual freedom. In most hunter-gatherer societies, women leave their breasts uncovered. Women really were naked and knew no shame, just like the Bible tells us. Sexual freedom was effective in egalitarian groups because it brought all members a bit closer together by creating an invisible "network of love" (even if jealousy was just as big an issue then as it is now). This was all over as soon as this group's way of life fell apart and every man started worrying about his property and demanding absolute faithfulness from his wives.

As mentioned, the new concept of ownership led to a situation in which sons remained with their fathers. If it had been the other way around and sons had left the farm, then not only would the clan lose able-bodied males who could be depended on as fighters thanks to a shared genetic heritage, but families would have to bring strangers into their homes as husbands for their daughters. There was always the risk that these outsiders would side with their original male kin should a conflict arise.

So the daughters were married off into other families. They served to help forge alliances or were simply viewed as tradable goods. The new families, however, treated the new wives with suspicion. As these were arranged marriages and thus not generally love matches, the women did not necessarily harbor strong feelings of solidarity with their new families. Only after they had born children to the son of the household did things change, for then a shared genetic interest had taken hold that bound the entire family together.

In this new state of affairs, patriarchs did all they could to prevent their wives from sleeping with other men. And in those areas where wealth and power prospered, men turned to polygyny, that is, marriage to more than one woman. This is commonplace in the stories of the Old Testament; nearly everyone from Cain's descendant Lamech to Abraham and Solomon had more than one wife. When women become the property of men, their power has to be reined in, and a large share of this power lies in their sexual attractiveness. The Bible underscores this point. Adam and Eve have to reach for the fig leaf, and God himself fashions them clothes. Although Genesis lets them both wear clothes, in agrarian societies, the women in particular must dress more modestly than they did in the days of the hunter-gatherer.

Fidelity in the Age of Bestiality

The patriarchal world raises a woman's fidelity to her husband to the level of commandment. That is one moral of the story of the Garden of Eden: "And thy desire shall be to thy husband." Yet a woman's sexual obligation to only one man is a cultural rule, not a biological one. Events of the recent past have also helped demonstrate that fidelity as a moral issue is subject to cycles. After the birth control pill ushered in sexual liberation and promiscuous lifestyles in the Western world, fidelity experienced a comeback in the 1980s. The reason was AIDS. This sexually transmitted immune deficiency changed sexual habits in Western societies. In order to prevent infection, one had to reduce his or her number of sexual partners.

When humans took up farming, sexually transmitted diseases played a key role in establishing fidelity. Microbiologist Dorothy Crawford explains in *Deadly Companions* how the domestication of animals enabled animal microbes to infect humans, whether due to proximity, shared shelter, or ingestion of milk or meat.[75] The humans had no immunity, no genetically anchored resistance to the new threats. The microbes would have a devastating effect in these new societies.

Crawford did not mention one significant factor in this context: bestiality, or fornication with animals. In his chapter in *Guns, Germs, and Steel* on the diseases faced by the first agriculturalists, Jared Diamond tells a modern anecdote about a woman who hit her husband over the head with a metal bottle after being told that his mysterious illness was due to "repeated intercourse with sheep on a recent visit to the family farm."[76] Opportunities for hunter-gatherers to practice bestiality were rather limited but also, given their general sexual laxity, unnecessary. Agriculture put tame animals close at hand while simultaneously creating a new social context in which many men had no opportunity to satisfy their desires. The Neolithic Revolution therefore also marks the birth of bestiality, which did a great deal to increase the number of sexually transmitted diseases. The connection between sex with animals and the signs of disease on intimate parts of the body was easily grasped. Indeed, in the Old Testament, bestiality is repeatedly denounced as a capital offense.[77]

Curtailing female promiscuity served two patriarchal ends: preventing illegitimate children and avoiding sexually transmitted diseases. The ones with the greatest interest in keeping things under control were of course those men with more than one wife. Since they had more sexual partners than most, they were most at risk of becoming infected.[78] Accordingly, the elite were particularly dedicated to keeping their women "pure" and restricting their sexuality.

We see the rise of a cult centered on a woman's virginity before marriage in patriarchal societies, and thus in the Bible as well, for another reason.[79] A woman who had had no previous sexual contact was considered disease-free.

At a time when women were traded, this significantly increased their market value. Foragers, who rarely suffered from sexually transmitted diseases, would have met with incomprehension the idea that a woman should enter into marriage as a virgin. Indeed, incidents of what we describe as "premarital sex" are still high among today's remaining hunter-gatherers.[80]

Why Adam's First Wife Left Him

And the women, did they put up with all of this? We simply do not know, for we have no texts describing their reactions. The Bible was written by men in pursuit of men's interests. Less than 10 percent of the names mentioned in the Old Testament belong to women, for example.[81] But since we do wish to follow up on every lead, we cannot leave out the story of Lilith, rumored as she is to have been Adam's first wife. Strictly speaking, the story of Lilith doesn't really belong here, for she does not appear in the context of the Garden of Eden until nearly 1,500 years after the Old Testament was put to paper. But let us first tell her story.

According to Genesis, God created man and woman in his own image only to later create woman again from Adam's rib. One popular interpretation is that Eve was actually Adam's second wife. The first had run off! The rumor that Adam was previously married to Lilith appears in around 1000 CE in the *Alphabet of Ben Sira*: "Soon they began to quarrel with each other. She said to him: I will not lie underneath, and he said: I will not lie underneath but above, for you are meant to lie underneath and I to lie above. She said to him: We are both equal, because we are both created from the earth. But they didn't listen to each other." Lilith flew into the air and disappeared into the desert.[82]

Religious scholars suspect that Lilith is actually a reinterpretation of a demon that lives in the tree of Inanna, the goddess of heaven, as recounted in the Sumerian epic *Gilgamesh, Enkidu, and the Underworld*. In the Hebrew Bible she stirs up trouble in the book of Isaiah. She also appears in rabbinical literature: when not eating children, she visits sleeping men to rob them of their semen. She owes much of her current fame to German poet Johann Wolfgang von Goethe (1749–1832), whose Faust sees her on Walpurgis Night: "That is Lilith," warns Mephistopheles, who knows more about black magic than anyone else, "Adam's first wife. Beware of her fair hair." This she-devil is the primordial mother of all femmes fatales as well as something of a pillar saint for feminists.

Why do we bring up Lilith? Because her story appears to offer yet another proof that we modern people do not feel comfortable with the official version of the Bible. The submissive Eve who bears her fate without complaint was supposed to be Adam's first wife? Our gut tells us that something is not quite right here, and our gut instincts rarely lie when it comes to our innate intuitions. The concept of Eve cultivated by the Christian Church in which she takes up a subordinate position to her husband after their expulsion is clearly male wishful thinking. Lilith, in contrast, is the personification of the threat to male dominance. Her powers of seduction are tremendous. We will stumble across such primordial female powers again in our reading of the Bible.

Lilith's tale reveals the perfidious nature of everything that God did to Eve and every woman who came after her. Man is not only to be his wife's master; she will be the one who lusts for him: "Thy desire shall be to thy husband." The message: the woman is the source of sexual desire, for her flesh is weak. This is why women need strong male leadership. Over the centuries, the story of Adam and Eve has been used to show what happens if the man slackens his control: "We were cast out of Paradise," men have preached from the pulpits for centuries, "because a man listened to his wife." The story of Adam and Eve is therefore also a story of misogyny.

The Pain of Childbirth

Let us return to God's verdict over Eve: "I will greatly multiply thy sorrow and thy conception; in sorrow thou shalt bring forth children." This is traditionally interpreted as an etiology, an explanation for the existence of birth pangs. This is quite exciting, for it would also indicate that people did not appear to have accepted this as matter of course, something like the pain you feel when you fall out of a tree, for example. That sort of pain does not require explanation, which is why it is never mentioned in the Book of Books.

No one has ever noticed, however, that the Bible is actually right when it comes to birth pangs, for this is indeed a phenomenon in need of an explanation. They are not normal, so to speak, for they are the product of a relatively recent development—a mismatch. With the arrival of agriculture, harsher living conditions and the increased number of diseases caused women's bodies to become smaller. The size of the children in the womb, however, particularly the baby's head, remained pretty much the same. This made for a more difficult journey through the birth canal that increased pain as well as risk during childbirth. Paleodemographic analysis has shown that early agrarians were subject to much higher rates of perinatal mortality of both mother and child (stillbirth and newborn death) than hunter-gatherers.[83] On top of this, the reduced mobility of agriculturalists may have also made giving birth more difficult.

In light of this information, could it be that Genesis presents us with a shadow of a memory from the time in which women were once spared this misery? Because birth pangs are a hardship that women have only had to face for the last few thousand years, they have not had time to develop an appropriate psychological response. The pains signaled a life-threatening danger—a danger against which hardly anything could be done. And here we see the effect of the admittedly crude story of Eve's curse. In a way, it is even a soothing story, for at least it suggests that the pain is normal: "Every woman must pay the price for Eve's transgression in the Garden of Eden." This may do nothing to relieve the pain, but it helps to prevent panic, making for an easier birth. In fact, the "curse" was a common term for birth pangs until the middle of the nineteenth century, when chloroform was first introduced.[84]

Why It's Good to Blame God

So what is the Bible actually trying to tell us? Its etiological stories, most of which are found in Genesis, seek to explain the various aspects of the world that people found strange. One could say that these scintillating stories formed themselves around these mismatch problems, much like oysters form pearls around threatening irritants. These stories aim at restoring normality.

Stories that tell of why God did this or that a certain way can also be described as "reifications." Sociologists Peter Berger and Thomas Luckmann define reification as "the apprehension of the products of human activity as *if* they were something else than human products—such as facts of nature, results of cosmic laws, or manifestations of divine will."[85] Reification means that the phenomena responsible for a given problem are believed to be eternal and immutable. So when the story of the Garden of Eden presents women's subordination or the rise of property as matter of course because they are the will of God, we can justifiably regard this as ideology, for here both deception and the legitimization of power play an important role.

There is no question that the Bible's stories have been abused over the course of human history in order to justify domination, violence, and repression. However, the claim that stories such as the one about Adam and Eve were produced by a group of men merely to cement the rule of the patriarchy—as critics of the church are prone to do—overreaches. Our psychological dispositions cause us to see the influence of spirits, gods, or demons everywhere we look. Therefore, people back then were seeking an explanation for why God had done the things he had. Perhaps the way our mind delivers explanations inadvertently ends up legitimizing the status quo as God's will. Needless to say, those who benefited saw no reason to object to this logic. Such explanations foster social harmony because any discussion is cut short by referring to God as the architect of the state of affairs. They also prevent people from worrying too much about problems that cannot be solved anyway, such as birth pain.

FAREWELL TO PARADISE

The story of Adam and Eve reveals three important themes: the wonder at why life involves so much drudgery, the difficulty in coming to terms with the concept of property, and the peculiar fact that women are meant to be subordinate to men. These are three of the most urgent problems with which humans have had to cope ever since they began living in sedentary societies—the true original sin. In this respect, things haven't changed all that much in 10,000 years. As we will see, these problems (and more) underlie many of the early parts of the Bible. Genesis does not deal with what we, from our modern perspective, see as the supposedly "eternal" questions of salvation or the meaning of life. In reality the people back then worried about the calamities confronting them when they were plunged into a new way of life: injustice, violence, oppression, illness, and misogyny. With neither science nor philosophy, medicine nor political ethics, people found in religion a coping strategy—and an extremely successful one. But although religion's monocausal solution ("God") did indeed help people come to terms with these problems, it did not eliminate them; they remained as pernicious as ever—and precisely this ensured that the biblical stories have continued to move us ever since.

Now it is time to take our biblical anthropology out of the Garden of Eden. Many readers of the Bible find it shocking that things went awry so quickly and that brother murdered brother right after Adam and Eve left Paradise. We are not surprised, however. It had to happen that way.

2

CAIN AND ABEL

The Birth of Violence

CALAMITY MAY HAVE GOTTEN ITS START IN THE GARDEN OF EDEN, but things really take off after the fall. Cain strikes Abel dead, the flood takes millions of human lives, and even the Tower of Babel ends in failure. The book of Genesis stumbles from one catastrophe to the next, and all the while God, as the "master of disaster," is at the center of things.

Isn't this all rather unusual? If a supreme deity creates the world, then shouldn't this lead to a world in which brothers don't try to kill each other and no one has to live in fear of God's wrath? The prophet Isaiah promises that the wolves shall dwell alongside the lambs and the lion will eat straw. Instead, chaos reigns from the moment Adam and Eve are cast out of Paradise.

We find ourselves compelled to interpret the fact that these scenes of terror begin the Bible as evidence of their basis in reality. No believer would ever burden God with such a heavy load of his or her own accord. Evidently the Bible's authors could not afford to ignore actual events at the heart of these episodes.

As we shall see in the following pages, there is indeed a close correspondence between biblical (often referred to as primeval) history and actual prehistory, the time when people stopped foraging as hunter-gatherers and began to live in permanent settlements. We will examine closely the core problems of this new way of life. The fact that murder—and a murder among brothers at that—is the first incident worth mentioning after the ejection of Adam and Eve is the inevitable result of these new conditions.

MURDER AMONG BROTHERS

It doesn't take long to relate their story: Cain, the farmer, made an offering to God of the fruits of his field. His younger brother Abel, the herdsman, made a sacrificial offering of the firstlings of his flock and their fat. "And the LORD had respect unto Abel and to his offering: But unto Cain and to his offering he had not respect," whereupon Cain flew into a rage, and although God had sought to appease him, he lured his brother into the field and slew him. For his deed, Cain was cursed to live as a fugitive. But God also set a mark upon Cain to protect him from others. "Vengeance shall be taken on him sevenfold," should anyone harm him. "And Cain went out from the presence of the LORD, and dwelt in the land of Nod, on the east of Eden." Later he would found his own successful lineage.

This raises a pressing question: Does God not bear at least a part of the blame for the bloody deed? Both brothers had sought to honor him. If he had accepted both sacrifices, nothing would have happened. Why did he have to snub Cain, who was the firstborn, after all? Other questions arise as soon as we begin to question God's motives. Why does God first favor Abel only to allow him to be killed later? And why is he so lenient in his punishment of Cain?

It's easy to understand why biblical exegesis has struggled so hard with this story.[1] As God is beyond all criticism, many rabbis and theologians have posited that something must have been wrong with Cain's offering. Some authors have suggested he must have been greedy. Others have introduced Satan into the mix. According to their readings, the devil was Cain's true father, not Adam, and this made the elder brother's offering inacceptable.[2]

Today's scholars speculate that the story illustrates how God shows mercy even unto sinners,[3] but this explanation is not all that convincing, given that only two pages later he drowns all of humanity in a great flood—and for much lesser transgressions than fratricide. In the pages of the Pentateuch, mercy isn't among God's great strengths. Some scholars interpret the story as the Bible's anthropological statement of man's true nature, as proof that we are all Cain and Abel, meaning that each of us is both capable of and vulnerable to evil.[4] But this raises an uncomfortable question: Why did God make us so?

Some Bible scholars believe that the story of Cain and Abel is actually an etiological one about a nomadic people known as the Kenites.[5] Others see it as an explanation for why the sacrifice of firstborn animals is so important.[6] But does that make it necessary to introduce a story of blood and thunder right at the outset? Theologian Benedikt Hensel's theory is more interesting. He claims to have discovered a narrative motif: in Genesis, the "promised lineage of 'Israel' is often perpetuated by the younger brother and not by the otherwise significant first-begotten." Here the Bible's authors wished to show that the status of God's chosen is a conscious decision and not simply the automatic privilege of the elder child. Behind all of this, Hensel speculates, is a statement of self-validity on the part of the people of Israel, who were seeking to assert themselves "in the midst of a number of larger, older, and more culturally dominating nations." In other words, they were trying to come to terms with their "big brothers, the nations surrounding them."[7] In Abel's case, this identification doesn't augur well for Israel: Abel might have been God's favorite, but he was killed anyway. As a result, the Israelites had to trace their ancestry back to a third-born, Seth.[8] Even if we do not wish to follow Hensel's metaphorical interpretation, the observation at the heart of his conclusion is important. There is something special about younger sons (more on this later).

Any valid interpretation of this biblical case of fratricide must find plausible answers to four questions: Why does God favor one brother over the other? Why does he choose the second son, Abel? Why is Cain driven to murder? And why does God show the murderer leniency? Here are the answers we came up with.

Why Does God Favor Only the One?

Why is Abel the sole recipient of God's grace? Because God is forced to make a choice. The law of the new world east of Eden demands it: there can be only one. In light of the reproductive success resulting from agriculture, fertile land was soon in short supply. A previously unknown problem appeared within this context: how to pass on property to children. Here we have another example of a mismatch problem for which there was no adequate solution. If the land were divided among several heirs, eventually the fields and pastures would be too small to sustain a viable existence. One solution took hold in many places: one son would receive everything. The patriarchs had a great interest in this development, for if all property remained in the hands of a single heir, he would remain reproductively successful, provided he could keep the family strong enough to compete with neighbors. But this model has an inevitable drawback: the heir's siblings must go empty-handed, creating a whole class of propertyless people in one fell swoop.[9] Nevertheless, this state of affairs rapidly became the status quo—and the God of the Hebrew Bible was there to lend his support.

As is always the case when confronted with a mismatch, our first nature tends to rebel: privilege for the few clashes with our innate hunter-gatherer preferences based on egalitarianism and cooperation. We find it deeply unjust when some individuals are inordinately favored over others—unless we ourselves are the ones to profit, of course.

Why Does God Choose the Wrong One?

God is acting in line with the new custom in that he chooses one brother, but he then violates that same custom by favoring the wrong one. The law of the land is the legal institution known as primogeniture, according to which the firstborn inherits everything—or at least the lion's share. As the older brother, Cain was the rightful heir. This is clearly reflected in the division of labor. Cain was the farmer, the future landowner. Abel, however, performed the dirty work of the shepherd looking after the animals that would one day belong to his brother.

Landowners faced a difficult challenge in handing down property from generation to generation with as little conflict as possible. To help us better imagine the importance of this issue, let us look at the question the other way around: How could a father dispossess the majority of his children of their inheritance? He might act as he saw fit and choose the most capable of his sons or the one to whom he felt closest. But this could lead to a tricky situation, for his preference might change over time or be contested by his other sons. And if the patriarch met with sudden death without first having chosen an heir, a vicious struggle among the brothers for the inheritance would be the predictable result.

Clearly, deciding ad hoc upon an heir in each individual case would never bring about peace. Each selection had to be well founded and witnessed—and remain contestable. It's not hard to imagine the family conflicts that would arise in these situations, especially if the patriarch decided that the eldest son would not inherit, for the eldest was normally the strongest and would thus believe he was in the right. If the father died early, the eldest son would be the first to achieve adulthood. If we follow this logic, we arrive at the institution of primogeniture—the firstborn son as heir—as a solid solution to this problem. A previously subjective decision could now claim legitimacy in objective fact. Later in the Torah, God even decrees that the firstborn son, even one born of an unloved wife, is to be recognized.[10] This makes it all the more surprising that God goes against his own maxims at the beginning of Genesis and refuses to accept the firstborn son's offerings.

God himself doesn't seem to have been all that comfortable with his choice either. How else can we explain that he failed to protect his protégé, Abel, who was beyond reproach in every way? As noted in the previous chapter, the God of Genesis is not yet an all-powerful being but a deity much like those we see in a polytheistic world. He even discusses the affair with Cain in person. In God's defense, it could be said that even Zeus, lord of the Greek pantheon, made mistakes, and even a god as powerful as Apollo was unable to protect his favorite Hector from the wrath of Achilles (who was in turn assisted by the goddess Athena). Does the biblical Cain and Abel narrative draw upon earlier stories in which various gods played a hand in events? As American theologian John Byron has discovered, the biblical Cain and Abel story contains a number of syntax errors, gaps in the text, and inconsistencies. All of these strongly suggest that it was subject to intense revision over the years. Researchers also disagree about when the story originated.[11] Any conclusions about origins necessarily remain speculative. But we also have better arguments in hand.

If we view the story from the perspective of cultural evolution, God's unusual decisions shed light on what is really going on. God was dealing with highly contested matters, for the new rights of inheritance were met with fierce opposition. The first to offer protest against this state of affairs would be our first nature. At the risk of repeating ourselves, we should stress that hunter-gatherers were organized into egalitarian groups and had no property to speak of. They bequeathed their most important possessions—knowledge of hunting techniques and the best places to collect food—to all of their children. Cooperation was at the center of everything. The notion that resources could be privatized and willed to a single heir would have seemed absurd.

Furthermore, the letter of the law might not offer a satisfactory solution to the question of inheritance: What if the firstborn son was a loser? The Hebrew Bible is full of ne'er-do-well firstborn sons. Esau sold his birthright for a bowl of lentils, Reuben slept with his father's wife, and David's firstborn son, Amnon, raped his half sister Tamar. Might it be that God had his reasons for rejecting Cain?

Finally, we should not forget that some of primogeniture's most formidable opponents actually had a great deal of sex appeal. Many readers of the Bible may not realize that the wealthy families of biblical times often practiced polygyny. One man married several wives, often taking new ones as he became older and more powerful. Hence, the man's later wives were generally younger than those he had married earlier. Each would do everything in her power to ensure that her son was made heir in order to ensure his—and thus her—genes the best possible chance at reproduction. Polygyny can therefore explain why younger sons are favored in the Bible. Young women bewitched men with their loveliness. The best known of these is Bathsheba. King David, who spied her at her bath, was so taken with her beauty that he sent her husband, who served in David's army, to his death so that he could marry her. And it was Bathsheba's son who—despite the existence of older brothers—took up his father's throne. His name? Solomon.

Why Does Cain Slay His Own Brother?

God's favoring of Abel stood the entire order of precedence on its head. Cain was merely defending his legitimate interests, and he resorted to violence in order to underline his rightful claim. The fact that he carried matters to the extreme is proof of just what was at stake. We find his actions shocking, but conflict among siblings is nothing new when it comes to evolutionary history. Sibling rivalry is a well-known theme in the biological sciences. It is common for young animals to compete for the limited resources provided by their parents. Among various species of birds, for example, only the strongest in the nest will survive to adulthood.

Indeed, sibling rivalry among humans is nothing unusual—if one child feels he or she is being treated unfairly, the resulting fuss can be quite dramatic. But as "cooperative breeders," humans have managed to rein in the problem: older siblings help raise their younger brothers and sisters, for example. This promotes altruism over rivalry.[12]

This also makes good genetic sense. Siblings share a large portion of their genetic makeup. Accordingly, brothers were the most dependable partners in hunter-gatherer societies, and they knew how to make the most of this familiarity. They usually understood each other so well, for example, that there was no need for words during the hunt, and this meant that together they could bring back even better kills. This, in turn, had a positive effect on their reputations and chances for reproductive success. Kill your own brother? You might as well cut off your own arm. This is the scandal of the story of Cain and Abel. The thoroughly unjust distribution of resources in the new world drove one brother to murder another.

Why Does Cain Get Away with It?

Why is God's punishment so mild? According to the law later revealed by Moses, murder was punishable by death. Furthermore, Cain's actions represented an attack on God's authority. Exile to a life of wandering, then, was a surprisingly merciful punishment. In later legends, the fratricidal brother gets what he deserves: the blind Lamech is said to have shot Cain with an arrow while hunting, having mistaken him for an animal.[13]

As we have said before, we are dealing with a transitional situation. The new laws based on formal legal reasoning met with considerable resistance during this time. These third-nature rules had not yet become a part of humans' second nature—and it is questionable that they ever did. In the story of Cain and Abel, God appears just as undecided as the story itself. Disastrous as they were for Adam and Eve's descendants, these events seem to portend the future to come.

For most of human history, a murderer would never have come away so lightly as Cain did. While physical strength is of primary importance to our great ape cousins, the chimpanzees and gorillas, it no longer played a decisive role in the lives of hunter-gatherers. Their weapons allowed them to kill physically stronger men from a distance. The primacy of physical strength gave way to the power of cooperation. Anthropologist Chris Boehm believes that egalitarian groups at some point began to kill those individuals who tended toward excessive violence, thereby systematically excluding them from the gene pool.[14] Despots were not to be tolerated.

In the new world order, however, the clans—patrilineal families—had to remain solidly united in order to defend their property. Here the egoistic impulses of humans' first nature offer an advantage. Once held under tight control by the group, the patriarchs now had free rein—and they had to do all they could to keep social competitors at bay. A heavy hand was needed to maintain clan cohesion within, and violence was used to counter any external threat from without.

This is why the story of Cain and Abel is such a perfect fit. In the new order, those clans that relied on aggression were particularly successful. Cain had to follow God's commandment and vacate the field—in the truest sense of the word—but he is protected by a divine mark: "Whosoever slayeth Cain, vengeance shall be taken on him sevenfold." After his banishment, Cain makes a rather successful career for himself as the "father of all cultures."[15] His son Enoch founded the first city. Lamech, one of his descendants, married two women and beguiled them with the following: "Hear my voice; ye wives of Lamech, hearken unto my speech: for I have slain a man to my wounding, and a young man to my hurt. If Cain shall be avenged sevenfold, truly Lamech seventy and sevenfold."

Evolution does not pursue a particular goal. A nearly obsolete behavioral drive from our deep past—a striving for dominance and monopolization—suddenly became the way of the future in the new world. The alphas got their second chance; their tendency toward ruthless violence turned out to be an advantage when it came to adapting to the new environment and the web of obligations that used to keep them in check no longer existed. As the Bible tells us, Cain's descendants would prove extremely talented when it came to monopolizing property, women, and violence. The alpha males represent the root of tyranny and despotism. They would also become the protagonists who took humanity to the next level of civilization—the era of chiefdoms and the first states. Their wars would plunge the people of the world into misery. And at the risk of sounding repetitive, we can say the Bible sums it all up in a nutshell.

FAR AWAY FROM PARADISE

The story of Cain and Abel is so valuable for the peephole view it offers of the social chaos that reigned in those days. The new society based on property unleashed competition, inequality, and violence on the world. People simply weren't prepared for it. In evolutionary terms, this all happened very quickly, making it impossible for natural selection to adapt our emotions. The new rules went against the old, innate feelings and impulses. "Might makes right" made a successful comeback. Old and new principles inevitably clashed, like in a classic Greek tragedy. The result of this mismatch was a great deal of confusion.

The Bible introduces the story of Cain and Abel in the right place. Conflict between brothers follows in the wake of sedentarization with almost lawlike inevitability. Fraternal conflict affects the very core of the family. The old familial connections are torn apart, and families risk getting scattered by a variety of centrifugal forces. Cain was bundled off to the land of Nod, far away from Paradise. His heinous act was therefore not the consequence of some personal deficit—his intemperate nature—but rather triggered by the mismatched social situation. We will encounter similar events again and again in the Bible.

3

SONS OF MAN, SONS OF GOD

Dubious Relatives

A MOVE STRAIGHT FROM FRATRICIDE TO THE FLOOD MIGHT BE A TAD too depressing, so it's a good thing that between these two tales of calamity we find two brief passages in need of examination. Both have received little attention over the years—a mistake from a biblical-anthropological point of view. Continuing the theme of the previous chapter, both are about sons in the age of patriarchy: first the sons of man and then the sons of God. One group is intended to quell the chaos; the other only creates more of it.

WE ARE FAMILY

In the fifth chapter of Genesis we encounter the Bible's first genealogy. Few other Bible passages come closer to thwarting the good intentions of those wising to read the Good Book in its entirety. From Adam to Noah we follow the family tree in sentences that are not exactly easy on the eyes: "And Seth lived an hundred and five years, and begat Enos: And Seth lived after he begat Enos eight hundred and seven years, and begat sons and daughters: And all the days of Seth were nine hundred and twelve years: and he died." This is a pretty hard slog alright, particularly when we bear in mind that the tale of the flood is followed by the Table of Nations, which describes how sons of Noah went on to found individual nations of their own. Then we are back to verses that read like the following: "The sons of Japheth; Gomer, and Magog, and Madai, and Javan, and Tubal, and Meshech, and Tiras. And the sons of Gomer; Ashkenaz, and Riphath, and Togarmah. And the sons of Javan; Elishah, and . . . "

But we do the genealogies injustice, for they represent a tried-and-true means of combating chaos in ancient times. Whether in Egypt, Greece, or Mesopotamia, everyone had a soft spot for determining the family trees of gods, kings, and noble families in an attempt to find out how everyone was connected to everyone else.[1] There's no need to shake our heads in frustration, for these lists are actually the forebears of today's social media, such as Facebook—at least in genealogical terms.

In a sense, all of Genesis represents "a book whose plot is genealogy."[2] It is an uninterrupted rundown of generation after generation from Adam and Eve through Jacob and his sons, from which the twelve tribes of Israel descend. And it just keeps on going from there. The Hebrew Bible follows the generations from the moment God placed humans on earth until the time of the biblical authors. Later on, the evangelist Luke will do the same and trace Jesus's pedigree all the way back to Adam. It is best to see these genealogies, bridging as they do this whole period, as the ancient equivalent of pressing the fast-forward button. We breeze through the centuries in rapid succession until something spectacular happens again. What counts is maintaining continuity.

From the perspective of evolutionary biology, this comes as no surprise. Each society had to establish social networks in order to counter the centrifugal forces discussed in the previous chapter. Everything possible was done to ensure that the selfish interests of individuals did not end up plunging the community into chaos.

In earlier times among the small groups of hunter-gatherers, three mechanisms ensured altruistic behavior: kinship ("If I help my relatives, I am helping ensure that my genes will be passed on"), reciprocity ("You scratch my back, and I'll scratch yours"), and indirect reciprocity ("I'll help those who I know will help others, and that will boost my reputation"). But how do you foster cooperation in societies in which people are not related and often will never meet again and therefore will not have the opportunity to return favors, not even indirectly?

This is where the pedigrees come in. It's not just in traditional societies that people hold conversations in order to find out if they may somehow be related to one another.[3] Genealogies are like thick ropes woven throughout social space to which the finer points of clan relationships are attached. This is how social networks are created. Even if the relationships between two given families are indirect at best, this is enough to conclude that we are all members of one big family. Shared genealogy thus equates to shared cultural identity.[4]

In this manner, the concept of the hunter-gatherer community was kept alive in this new era in which anonymity was a basic fact of life. What happened next is fascinating. People began to tap into one of the primordial sources of human altruism. When genealogies construct fictive kinships, we are dealing with what anthropologists call the "cultural manipulation of kin psychology" aimed at enabling solidarity within larger societies.[5] The Bible thus simulates a large family and replaces anonymity with a familiarity that comes from a real—or, more commonly, concocted—shared past. Fictive kinship remains an integral component of religion to this day. You can see it in any church in which complete strangers address each other as "brother" and "sister" and pray to "God our Father" and "Mother Mary."[6]

AND THE WOMEN?

Another crucial aspect of these genealogies is their neglect of women—mentioned only as "daughters"—and the emphasis on patrilineal inheritance. This is a negation of our first nature, in which mothers are just as important as fathers (if not more so). The genealogies are there to sing the praises of primogeniture. In reality, things were rather different. Indeed, later in Genesis, women such as Sarah, Rebecca, and Rachel feature importantly when it comes to determining inheritance. It's no overstatement to claim, "While the system was run by men, the women were needed to make it work."[7]

In his examination of the Egyptian king lists, Egyptologist Jan Assmann determined that these documents served not to underscore the importance of history but instead to show its triviality. They confirm continuity: nothing has changed over the millennia. Thus their purpose is to "halt history."[8] The same is true of the genealogies in Genesis. The biblical authors fast-forward through history using long-winded pedigree lists to prove that nothing has changed since the beginning of time. Readers are to think that inheritance has always passed from father to first son. In this way, questioning the status quo is ruled out of order. The pages of Genesis not only create normality but also cement it.

BIBLICAL AGES

As the Good Book's genealogies tell us, Adam lived to 930 years of age, Jered reached 962, and Methuselah made it to a ripe 992. The true ages provided by the Bible offer a lot of room for speculation, however: "In biblical genealogies, the most ancient ancestors are credited with extremely long life spans," claims microbiologist David Clark. He asks, "Is this just hyperbole attached to the memory of heroes, or is it a half-forgotten memory of days gone by when life expectancy was indeed much longer?"[9] Not only were hunter-gatherers healthier, but their maximum age was well above that of the farmers, for a number of deadly diseases only began to plague humans after the advent of agriculture.[10]

Here we should point out that there is no dearth of alternative explanations: many have suggested that the Bible simply confused years for months (meaning Methuselah actually only would have made it to eighty-three). A nice idea, but this would also mean that Seth sired his first son at the tender age of eight. Or did the Bible take inspiration from Mesopotamia with its tales of mythical kings who ruled for thousands of years? That is certainly possible, but an even simpler and nonexclusive explanation is that the ages appearing in the genealogies were stretched so as to create a chronology that made the reconsecration of Jerusalem's temple in 164 BCE coincide perfectly with the four thousandth anniversary of creation.[11]

WHO WAS THE MOTHER?

The next story in the Bible—that of the sons of God—is so brief that we can cite the passage from Genesis 6 in its entirety:

> And it came to pass, when men began to multiply on the face of the earth, and daughters were born unto them, That the sons of God saw the daughters of men that they were fair; and they took them wives of all which they chose. And the LORD said, My spirit shall not always strive with man, for that he also is flesh: yet his days shall be an hundred and twenty years. There were giants in the earth in those days; and also after that, when the sons of God came in unto the daughters of men, and they bare children to them, the same became mighty men which were of old, men of renown.

The men took the women as they chose; this is the reality that Genesis alleges. Women were property, and it was not unusual for them to be traded or sold. Powerful men simply took any woman who pleased them. Obviously, this could not work in the long run, for it gave rise to jealousy, was a potential source of conflict, and heightened the spread of sexually transmitted diseases (not to mention the women's suffering, but the Bible tends to ignore that anyway).

Here, we can make various linkages: we also find people getting mixed up with the gods in the *Gilgamesh* epic and the Greek pantheon, for example.[12] Zeus, the Olympian, fathered Heracles by Alcmene, wife of Amphitryon, and seduced Leda, queen of the Spartans in the guise of a swan. One issue from this coupling was Polydeuces, who himself, together with his brother Castor, would go on to take the woman who caught his fancy: Peter Paul Rubens's baroque masterpiece *The Rape of the Daughters of Leucippus* captures this story on canvas. But what are God's sons doing in a book that is all about the one true God? Who was their mother?

Most likely we are here dealing with an early product of globalization. Palestine was the meeting place of a number of cultures, and, particularly when it came to religious matters, syncretism was the order of the day. People were aware of the myths from other cultures.[13] In the Hellenistic era, under the rule of the Seleucid king Antiochus IV (215–164 BCE), even the cult of Zeus was introduced in Jerusalem.[14] And as one of prehistory's clear goals was to explain the origin of the known world, everyday knowledge of great floods or the architectural megalomania of the Babylonian empire got woven into the Bible's stories. This helped to increase their credibility. We can thus consider the "mighty men which were of old, men of renown" to be such an inclusion. Whether Gilgamesh, Achilles, or Heracles, heroes of neighboring cultural centers were integrated into the biblical narrative in just such a fashion and helped support Genesis's aspiration to offer a universal history of the world.

But note that the above theory is not incompatible with the idea that these sons of God represent a memory of encounters with neighboring hunter-gatherers—something that must have happened independently in numerous places and at various times. Anthropologists have found that contemporary hunter-gatherers were, on average, half a head taller than sedentary peoples,[15] and so they may have been the source of the stories about giants.

Popular commentaries have tried to reassure us of course. "The sons of God are not the physical 'sons' of God, but heavenly beings of a divine nature like those we encounter as gods in the beliefs of all peoples."[16] In many passages, however, as mentioned in chapter 1, the world of the Hebrew Bible is polytheistic, meaning there were many gods. The God of monotheism who rules alone over time and space is a later invention. Nevertheless, most people today think only of the one true God when they read the Bible, which makes much of it difficult to understand.

We can only understand all of the Bible's inconsistencies if we realize that many of the stories comprising it arose in an era swarming with divine beings. What's more, these stories were not cleansed of their polytheistic traces until some time during or after the Babylonian exile. And not all of these traces were fully erased. We have already mentioned the speaking snake in Eden and the winged cherub who guards Paradise; angels will also appear later. All are relicts of the dazzling old divine cosmos, the surviving remnants of a "downsized pantheon."[17] The sons of God also belong in that category. In later times, during the Hellenistic era, they were often interpreted as fallen angels and so provided the devil and his crew with a respectable biblical pedigree.[18]

The thing to remember is that the God we encounter at the beginning of the Bible comes from a world in which many gods existed. All of them had their strengths and weaknesses, their favorites and adversaries, and none of them could claim to be the sole keeper of the truth. This is one reason why the God we encounter in the Bible sometimes seems inconsistent and temperamental. He is simply the product of his time. The polytheistic gods were a lot like us and not abstract principles.

The momentous conclusion is that religion, just like any other institution, is subject to change over time: it evolves. God was not always as big as he later became. He had to start small, much smaller than the God we know from the Bible. In the next chapter we will address what made him so powerful.

4

THE FLOOD

God's Wrath

"AND GOD SAW THAT THE WICKEDNESS OF MAN WAS GREAT IN THE earth, and that every imagination of the thoughts of his heart was only evil continually. And it repented the LORD that he had made man on the earth, and it grieved him at his heart." The Bible doesn't kick up much of a fuss when it comes to God's decision to unleash the flood upon the earth. We, on the other hand, would really like to know why he did so. What had people done to drive a desperate God to let loose forty days and forty nights of rain until every single creature on earth had drowned? Without a doubt, the story of the flood is one of the most disturbing in the entire Bible.

Of course someone had to survive. Noah. God warned him about the flood and instructed him to build an ark and to take his family and a pair of every species of animal with him on board. This, however, changes nothing about the basic facts. If the flood did indeed take place as described in the Bible, there would have been around 20 million dead, according to evolutionary psychologist Steven Pinker,[1] not to mention all the animals also doomed to perish for our sins.

Throughout Genesis it's pretty clear that after the first couple was tossed out of Eden, things took a turn for the disastrous for humanity, but no other story is as explicit as that of the flood when it comes to the consequences of bad behavior. Humanity could expect nothing less than God's fury. Whereas Zeus sent down bolts of lightning, God sent disasters. And with few exceptions, no one could expect the slightest mercy.

This raises a number of questions that, astonishingly, have received little attention to date. Why do people actually believe that God wished to punish them? Today people speak a great deal about God's love and compassion, but we see next to nothing of it in the Bible's early pages. Why did people back then assume that their behavior must have been so egregious that God found he had to chastise and torment them and even wipe them off the face of the earth altogether? If we can answer these questions, we will have identified fundamental features of human psychology and figured out what makes humans religious. In this way, the story of the great flood can help us gain more insight into the cultural evolution of religion.

"BIG BROTHER IN THE SKY"

We find it hard to accept the claim made by some scholars that the great flood was not an expression of God's wrath.[2] What emotion other than extreme anger could have impelled him to eradicate humankind? Wrath was arguably the Old Testament God's most conspicuous feature—and theologians are embarrassed by this fact. As recently as 2009, Bible scholar Jörg Jeremias complained, "So far there is no satisfactory exegetical monograph regarding God's wrath." It is an "unpopular subject for the majority of theologians that they would like to consign to the darkest background in order to avoid having to confront it."[3] But remaining in permanent denial about God's wrath is pretty hard work because there are no fewer than 390 documented instances of and 130 verbal references to it in the Old Testament.[4]

Today's believers likewise tend to find all of this fury hard to take. A vengeful, violent God doesn't fit at all with what people nowadays expect from religion. God is supposed to comfort us, take away our fear of death, help us deal with all of life's unforeseen crises, and make us feel safe. He is supposed to be a benevolent father, not a mass murderer with a short fuse.

While God's traditional clientele might have had their difficulties with the Lord's choleric nature, the divine tendency to punish is a matter of great interest in other circles. Evolutionary psychologists and sociobiologists like to debate why God is so easily enraged. After all, the notion of a heavenly taskmaster is not found just in the Judeo-Christian religious tradition. Gods everywhere sent down plagues in order to punish the peoples of the earth. In recent years, a number of scholars have begun to argue that we are dealing with a key aspect of religion. According to their theories, the idea that supernatural beings keep an eye on events on earth and punish misconduct arose in order to encourage moral behavior.[5]

Based on their findings, these scholars have formulated what is known as the "supernatural punishment hypothesis."[6] "Our evolutionary ancestors required a fictitious moral watcher to tame their animalistic impulses," explains evolutionary psychologist Jesse Bering.[7] Indeed, a number of experiments have shown that people who feel they are being observed and fear that bad behavior might damage their reputations or result in sanctions tend to behave better. An experiment carried out by a research team led by Melissa Bateson from the University of Newcastle provides a particularly impressive example. In their departmental coffee room, they had an honor box to pay for their drinks. As soon as images of human eyes were hung above this box, readiness to pay increased dramatically, suggesting that people subconsciously felt they were being watched.[8] Jesse Bering's own experiments also point to the same conclusions. Children given tasks to complete alone in a room were less likely to cheat if told beforehand that an invisible princess named Alice was in the room observing them.[9]

Of course, the original function of this tendency was not necessarily supernatural: we humans deeply care about our reputation, and sensitivity to being watched is a great motivator to make doubly sure that we do not unwittingly damage it. But equally obviously, once such sensitivity exists, it is grist for the mill of moralizing deities out to punish us for our misdeeds.

The effects of feeling observed by a transcendent moral watcher can be quite enduring, as Richard Sosis and Eric Bressler's research has shown. Their analysis of the lifespans of two hundred nineteenth-century American communes clearly indicates that the religious communes lasted longer than the secular ones.[10] The decisive factor contributing toward their social stability was, first and foremost, their successful "instrumentalization of God in order to morally punish group members," says sociobiologist Eckart Voland. "The notion of divine punishment is the foundation of the simple moral idea behind a social contract. It is a thoroughly worldly idea. If one does not do anything bad one won't be punished." Little wonder, says Voland, "that when comparing cultures, punitive gods crop up more often when groups depend on cooperation."[11] Dominic Johnson and Jesse

Bering put it more provocatively: to prevent people from engaging in antisocial behavior, the divine "stick" is much more effective than the "carrot."[12]

Does this mean that we need God, our heavenly taskmaster, in order to guarantee civilized coexistence? Are monitoring and punishment his true functions? That would mean that the all-seeing eye floating in the heavens—we know it as a companion of the Egyptian god Ra, an emblem of the Christian Church, or even a symbol on the American one-dollar bill—truly is the perfect symbol for God, our "Big Brother in the Sky."[13] This would imply that people only behave decently because they fear God's big stick. It also means that we live in a kind of "Panopticon," the perfect prison dreamed up by British philosopher Jeremy Bentham (1748–1832) in which a ring-shaped prison block with cells opening onto the center is situated around a watchtower, allowing a single guard to keep an eye on all the prisoners at once—and intervene every time someone breaches the rules. Michel Foucault helped Bentham's Panopticon find fame when he wrote *Discipline and Punish: The Birth of the Prison.*[14] Could it be that the earth is not the vale of tears we imagined but one great prison with God as the ultimate warden?

WHAT IS RELIGION?

All this talk of God as the punisher of rule breakers and guarantor of civilized life raises the question of what religion really is. One cannot say there has ever been a conclusive answer. Quite the opposite, in fact, for there is certainly no lack of definitions. In *The World Until Yesterday*, Jared Diamond lists sixteen of them before adding a seventeenth of his own. We quote him here, because Diamond's own attempt reveals the level of complexity we are dealing with: "Religion is a set of traits distinguishing a human social group sharing those traits from other groups not sharing those traits in identical form. Included among those shared traits is always one or more, often all three, out of three traits: supernatural explanation, defusing anxiety about uncontrollable dangers through ritual, and offering comfort for life's pains and the prospect of death. Religions other than early ones became co-opted to promote standardized organization, political obedience, tolerance of strangers belonging to one's own religion, and justification of wars against groups holding other religions."[15]

This definition illustrates quite clearly that religion, as we understand it today, actually results from a long and meandering historical process. It is by no means an eternal and unchanging entity. Rather, it is a product of cumulative cultural evolution, a complex amalgam of different elements, a "cultural package" of beliefs and practices.[16]

As religion's areas of responsibility have shifted over the course of the millennia, so have its functions—to a massive extent, in fact. Much too often, we project our modern understanding of what religion is onto the past. But in today's world, God's job description has become radically different from the one he had at the beginning of his divine career—and certainly from when he tried to drown humanity.

We must fight this projection urge, because everything we know points to massive changes in this concept over time. We therefore begin with the basis of religiousness, the biological substrate and part of our first nature. The conventional wisdom is correct: *religiousness*, the belief in supernatural beings, has been with *Homo sapiens* for eons and can be understood as a human universal. Many scientists consider it a byproduct of some of the brain's cognitive functions that serve "to deduce cause, agency, and intent, to anticipate dangers, and thereby to formulate causal explanations of predictive value that helped us to survive."[17] Even today, children have "a tendency to over-attribute reason and purpose to aspects of the natural world."[18] They are convinced that someone created everything—from animals to the sun and the stars—for a particular purpose; somebody is behind every action, scheming in the background. We actually have to learn *not* to believe in supernatural beings.[19] We will return to this point later. Here it suffices to say that religiousness is an innate trait, although obviously its individual manifestations vary from person to person and are influenced by socialization and social environment.

In prehistoric times, this religiousness first manifested itself as a belief in a natural environment full of ghosts, ancestors, and animal spirits. In the nineteenth century, social anthropologist Edward Tylor introduced the term "animism" to describe this phenomenon. As studies of hunter-gatherer groups existing today have shown, these peoples have an entire cosmos of beings to explain "why it snows, why wind blows, why clouds obscure the moon, why thunder crashes, why dreams contain dead people and so on."[20]

Incantations and rituals, prayers of thanks, and protective spells are cultural products. Shamans arose as experts for dealing with the spirits. And that brings us to *religion* itself, the "special domain of concepts and activities" with more or less set religious teachings and specialists.[21] Religiousness may be innate, but religion is not; it is the result of a process of institutionalization. It must be internalized during early socialization in order to become a seemingly natural, unquestionable part of our personalities—our second nature. Religion as an institutionalized system is therefore a part of culture, meaning that it is also subject to historical transformation. That is why we find such a wide variety of religions throughout the world.

Often, however, we overlook the fact that religion was not originally responsible for questions of morality. The spirits and ancestors were not the guardians of our moral behavior—they simply were not needed for this purpose. When people lived in small, manageable groups, they could not conceal most of their misdeeds from the others. "If you stole a man's digging stick, where would you hide it? And what would be the point of having it if you couldn't use it?" explains Robert Wright. "And, anyway, is it worth the risk of getting caught—incurring the wrath of its owner, his family, and closest friends, and incurring the ongoing suspicion of everyone else?" The fact that people lived with pretty much the same group over their entire lifetimes was a great incentive to behave properly. "Hunter-gatherers aren't paragons of honesty and probity, but departures from these ideals are detected often enough that they don't become a rampant problem. Social order can be preserved without deploying the power of religion."[22]

If we take this historical perspective into account, the "supernatural punishment hypothesis"—the idea that God made a name for himself as a divine taskmaster charged with keeping our "animal instincts" under control—loses some of its appeal. Belief in the punitive God is not an adaptation, not a genetically based, universal human trait, but a late-appearing product of cultural evolution.[23] Fittingly, the concept of the gods as agents having at their disposal all of the information necessary for them to judge human activity also developed gradually.[24] The spirits of the hunter-gatherer communities know a lot about a surprising number of things, but they are by no means omniscient. Sometimes even the opposite is true. "Many spirits are really stupid," says Pascal Boyer. They can be easily fooled.[25] Not really the best qualifications for a judge of moral probity.

Moreover, dastardly deeds often go unpunished, and the culprits get away with them. Otherwise no one would ever be tempted to do anything forbidden. A differentiated afterworld, a place where everyone meets his deserved fate, appeared much later in cultural evolution—at least where the Bible is concerned. We will not encounter this realm until chapter 16.

Big Gods Arrive on the Scene

A decisive step in the cultural evolution of religion occurred when the spirits became gods. They gained a great deal of power and an enormous expansion of their responsibilities. What had caused this quantum career leap? And why were they suddenly interested in human morality and prepared to punish bad behavior with their wrath?

Usually the answer to this question refers to the useful function of such beliefs and invokes French sociologist Émile Durkheim (1858–1917), who places the social aspect of religious ritual in the foreground. He defines a religion as "a unified system of beliefs and practices relative to sacred things, that is to say, things set apart and forbidden—beliefs and practices which unite into one single moral community called a Church, all those who adhere to them."[26] Religion therefore facilitates cooperation.

And this cooperation was a dire necessity, for the new "big gods" were indeed the long-term consequence of humanity's adoption of a sedentary lifestyle during the millennia in which communities transformed into societies, when "communities" became "tribes" and then "chiefdoms" before finally evolving into "states." Religion is therefore traditionally viewed as a solution to a core problem of evolutionary biology: the free rider problem, which became painfully prominent under the new living conditions. How could a society in which everyone no longer knew everyone else maintain altruistic behavior against the selfish interests of individuals? Religion solves this problem by serving as "social glue." It inspires a sense of community in large societies and provides rules and mechanisms that strengthen social cohesion while hampering free riders.

This social glue is the product of shared beliefs and common rituals. And this is where that feeling of being observed by a supernatural agency begins to gain importance. "Watched people are nice people," as psychologist Ara Norenzayan succinctly puts it.[27] His colleague Jonathan Haidt goes into greater detail: "Creating gods who can see everything, and who hate cheaters and oath breakers, turns out to be a good way to reduce cheating and oath breaking."[28] Societies that venerate morally driven gods are thus able to reduce conflict and significantly improve their members' ability to cooperate.

This reconstruction sounds convincing and is widely accepted, but a problem remains. No one questions religion's role as social glue, but what drove the spirits to become gods in the first place? Why did the spirits, who were not above a practical joke now and then, suddenly morph into supernatural moralizers? How did they introduce such a spurt of innovation into the cultural system of religion? We have the impression that, to date, scholarly discussion of religion's evolution has often been too quick to see the practical use of a religion based on a moralizing god as the cause of its emergence.

Norenzayan does not make this mistake. He believes that the cultural evolution that went into overdrive with the arrival of the Neolithic Revolution allowed for the rise of a few cultural "mutants"—a handful of "watchful Big Gods with interventionist inclinations." Those groups fortunate enough to follow them were able to cooperate more effectively and could expand faster at the expense of other groups. This, in turn, helped push the gods up the career ladder.[29]

But how plausible is this assumption? Did the moralizing gods appear on the heavenly stage by accident? Probably not, for we are dealing with societies that faced the danger of being torn apart from within. This would have made it extremely difficult for the gods promoting cooperation to establish themselves if there was no real need for them.

The following might offer a complementary and perhaps even alternative explanation: the "chiefdoms" and first states have been described as "kleptocracies,"[30] as "protection rackets" with "powerful Mafiosi."[31] A few individuals enriched themselves at the expense of the larger society by skimming off the surpluses. This would suggest that the concentration of power among the spirits developed in parallel with the concentration of power among people. One ensured the rule of the other, so to speak. With the big men came the big gods—on earth as in heaven.

But even this is not fully convincing, for why the sudden appearance of gods interested in people's good conduct? The functions of this moral monitoring, such as the legitimization of power, certainly helped increase the gods' chances of success, but they did not help the gods make their grand entry in the first place. Here the Bible suggests a different trajectory—one that is fundamental to our understanding of human culture. No other part of the Good Book illustrates so clearly how these big gods emerged as the story in which one of them, in a full-tilt rage, sends down torrents of water to wipe out humanity. Let us now immerse ourselves in the flood.

HOW GUILT CAME INTO THE WORLD

It is truly astounding how quickly the God of the Bible began to regret "that he had made man on the earth." Of course he had already suffered a few disappointments with Adam and Eve's disobedience and Cain's act of fratricide, but it is odd that his creations, whom he had only recently described as "very good," could so quickly cause him so much distress. He is practically bursting with rage: "I will destroy man whom I have created from the face of the earth; both man, and beast, and the creeping thing, and the fowls of the air; for it repenteth me that I have made them." But isn't God's pressing of the reset button tantamount to an admission of complete failure?

Every child knows what happens next. God lets it rain for forty days, and just in case this is not enough, he opens up the springs below the earth. Only Noah is worthy of God's mercy. He is allowed to build an arc (which, according to the measurements given in the Bible, could never have stayed afloat), take his family with him on board, and bring a pair of every species of animal (even if the details in the Bible vary a bit). Every other living thing perishes in the flood. Artists from Michelangelo to Gustave Doré have captured the events in paintings or frescos full of devotion that depict the remaining people desperately trying to climb up trees or onto high rocks.

The story still bothers sensitive souls. This episode caused Anna, then seven-year-old daughter of one of this book's authors, to stop reading her children's Bible. She simply couldn't understand this God. How could he let all of those people drown? Children! Babies! Guinea pigs! They were all innocent. Why did God create all of these things if he was just going to turn around and destroy them?

Those who believe the Bible represents more than just a collection of antique texts—that is, the word of God—have looked for a motive that could clear God of his reputation as the epitome of irascibility and capriciousness. Who wants to pray to a choleric? Blame therefore often falls on the sons of God, who were amusing themselves with the women of the earth just before the flood. Or else—once again—it is the women's fault, for they had seduced the angels and thus provoked the calamity.[32]

Some exegetes have maintained that the flood never really happened. In the New Testament, for instance, Peter interprets it metaphorically as the archetype of the rite of baptism.[33] In both instances, water is used to wash away humanity's evil. Once again the Bible fails to provide us with an unambiguous message.

The Flood Was Real

As we have stated previously, no one would simply foist such a story on his own god. Something similar must have actually happened. "Floods are among the most common and most consequential of the Earth's natural disasters," according to geographer Jürgen Herget. "They have always occurred, in every natural environment."[34] We know of more than 250 stories of floods from all over the world.[35] The advanced civilizations arose along the great rivers: the Nile, Tigris and Euphrates, Ganges, Yellow River, and Yangtze. Such rivers were especially susceptible to flooding—and their densely populated centers made them particularly vulnerable.

For decades, archaeologists and geologists have been hunting for past megafloods that could have served as the original prototype for the flood. "One considered the confirmed floods in and around Ur (fourth millennium BC), Shurrupak (ca. 2800 BC), or Kish (ca. 2600 BC). Today the Black Sea flood (ca. 5600 BC or earlier) is a widely debated possibility."[36] Scholars even speculate that the Garden of Eden now lies on the floor of the Persian Gulf.[37] The experts discuss meteor strikes, volcanic eruptions, and tsunamis as possible causes for these floods, in addition to the usual variability in rainfall.

As mentioned elsewhere, the biblical account of the flood is based on Mesopotamian texts. The Bible's authors helped themselves to the *Gilgamesh* and *Atrahasis* epics; in other words, Genesis's story of the flood was plagiarized,[38] but an interesting twist was added. Where it takes only the one God to get the job done in the Hebrew Bible, there was an entire horde of gods in the original stories.

In the oldest version of the story, the myth of Atrahasis, humanity was not doomed to destruction for sinful behavior. The real reason was overpopulation. People just made too much racket. Initially, the gods sought to reduce their number by sending down droughts and famines. Only after these proved ineffective did they resort to their ultimate weapon: the flood.

The abundance of flood stories allows us to conclude that there really was a flood, maybe even several, but certainly not one as large as the flood of the Bible. The occurrence of a great flood was a historical fact deeply engraved in the collective memory of the people of the Near East, and the Bible could not simply have chosen to ignore it.

Why Is It Our Fault?

So there really was a flood, a violent event in great need of explanation. Every attempt to clarify why the gods sent it down to earth was a rationalization after the fact—a desperate effort to make sense of a senseless natural disaster. But what led people to believe that a god was responsible for natural disasters? And why did they believe that they themselves were the reason for divine displeasure?

German philosopher Friedrich Nietzsche (1844–1900)—who in his *Twilight of the Idols* claimed to "philosophize with a hammer"—did indeed hit the nail on the head when he diagnosed humanity's "error of false causality" going back to the "oldest psychology," in which "every occurrence was an action, every action was the result of a will; the world, according to it, became a plurality of acting agents; an acting agent (a 'subject') was insinuated into every occurrence."[39] Nietzsche was not the first to notice this, however. About a century earlier, Scottish philosopher David Hume (1711–1776) had noted a similar phenomenon. In *The Natural History of Religion* he wrote, "We find human faces in the moon, armies in the clouds; and by a natural propensity, if not corrected by experience and reflection, ascribe malice and good will to everything that hurts or pleases us."[40]

In recent years, numerous scholars have subjected this "oldest psychology" to detailed analysis. The modern formulation of Hume's and Nietzsche's animistic psychology is the Hyperactive Agency Detection Device (HADD), a concept developed by researchers, including Stewart Guthrie, Justin Barrett, and Pascal Boyer, that is a part of every single one of us.[41] Boyer explains why this cognitive detection system is hyperactive and produces so many false positives: "Our evolutionary heritage is that of organisms that must deal with both predators and prey. In either situation, it is far more advantageous to over-detect agency than to under-detect it. The cost of false positives (seeing agents where there are none) is minimal, if we can abandon these misguided intuitions quickly. By contrast, the cost of not detecting agents when they are actually around (either predator or prey) could be very high."[42] Or, to put it simply, it's better to run away from a supposed predator one time too many than one time too few.

Admittedly, our brains have a hard time accepting that they include a cognitive system that can cause false alarms, but HADD provides us with valuable services—at least three, in fact. We have already mentioned the first of these: the one that helps us to survive. Anyone seeing ghosts might be subject to ridicule, but he or she is more likely to survive than a pure rationalist.

HADD's second function is to ensure that we react quickly enough in a worst-case scenario. It trains us to examine the clues and recognize patterns of action, but it also keeps us thinking about who might represent a threat—or who is responsible for a disaster. This includes scrutinizing our fellow human beings and asking ourselves, Does someone wish to do me harm?

Third, HADD keeps us calm, which Friedrich Nietzsche also understood:

> To trace back something unknown to something known, relieves, quiets, and satisfies, besides giving us a sensation of power. There is danger, disquiet, and solicitude associated with the unknown—the primary instinct aims at *doing away with* these painful conditions. First principle: any explanation whatsoever is better than none. Since, after all, it is only a question of wanting to get rid of depressing ideas, people are not especially careful about the means of getting rid of them: the first conception, by which the unknown declares itself to be something known, is so pleasing that it is "taken as true."[43]

"Any explanation whatsoever is better than none." This amazing realization has also established itself in modern psychology. Psychologist Alison Gopnik describes it as an inherent "theory formation system" that craves causal knowledge and—if we find an answer—rewards us with a feeling of immense satisfaction. Just like Nietzsche, Gopnik concludes that the explanations don't even have to be correct to make us feel good.[44]

HADD also causes us to look for agents everywhere. When unusual things take place, we answer the question of why in social terms. In the past, we believed pretty much everything—even the things we now describe as a part of "inanimate nature"—to have a spirit of its own. Ethnological descriptions of modern-day hunter-gatherer groups show that natural occurrences are experienced as "social drama," as the interaction of various actors. "When the Klamath," a hunter-gatherer people in what is now Oregon, "saw clouds obscuring the moon, it could mean that Muash, the south wind, was trying to kill the moon—and in fact might succeed, though the moon seems always to have gotten resurrected in the end."[45]

To briefly summarize, evolution made us in such a way that it is easier for us to believe the world around us is inhabited by living beings, even if these are not always visible, than to see it as the product of the abstract laws of nature. We see the world in social terms. We ascribe beliefs and motives to everyone (a tendency known as theory of mind) and think that everyone acts to achieve a goal. That is of course nothing special, although at present it is unclear how unusual humans are among animals in this respect. But amazingly, we also apply exactly this same reasoning to natural events. Somehow natural selection did not find it necessary to equip us with a sophisticated objective approach to the world. In fact, people have to learn that complex astronomical or meteorological mechanisms are responsible for what we see in the skies. Children tend toward animistic ways of thinking when they try to understand these phenomena. We have known this since the 1920s, when developmental psychologist Jean Piaget (1896–1980) showed that children attribute intentions to both living and nonliving entities. To give another example, whatever hurts them does so on purpose: if a child bumps into a rock, the rock is bad, and her hand hurts because the wall hit it.[46] This is also why so many religious explanations intuitively make sense to us, whereas we must work quite hard if we want to understand scientific explanations. Philosopher Robert McCauley got to the heart of the matter when he titled his book *Why Religion Is Natural and Science Is Not.*[47]

If something unusual happened, people started looking for potential agents. Was it an animal? A person? But what about events that no person or animal could have caused? A flood, for example. Other agents were required—and they were readily at hand.

Do Not Speak Ill of the Dead

Having seen how humans interpret the world, we can no longer find it surprising that our natural psychology does not make a clean distinction between the physical and spiritual worlds. The belief that humans have souls that can leave the body reaches back into the darkest depths of prehistory. Sometimes our spirits even seem to leave our bodies and wander the far corners of the earth when we are awake. Imagine what they do when we are asleep! They experience one adventure after the other.

Death has an even stronger influence on us. We believe there must be something more to life than what we can see and touch. One moment a person is alive—laughing, dancing, running around—and then suddenly it's all over. The body lies unmoving, not even a flinch, its warmth fading. What remains is the cold, stiff husk of a person. What happened? Might there be something else as well that slips away with a dying person's final breath? Let's call it the spirit or the soul. And doesn't our own experience tell us that something does indeed live on after we die? The dead visit us in our dreams, for example. If the dead are soulless bodies, then what do their bodyless souls do?[48] They must live on as spirits. "A belief in the immortality (in some form or another) of the dead occurs in all cultures as does the worship (again, in some form or another) of ancestors."[49]

But there is no hereafter: the realm of the dead is not yet completely separate from the world of the living. There's nothing supernatural about spirits. Yes, they are transcendent, of course, but the results of their actions can be seen in the here and now, meaning that they exist in the same world as we do. We can interact with them and even influence them to our own ends.[50] After all, we know a lot about them, for in a way they are just like us. And one day our own souls will leave our bodies.

Our ancestors participate in our lives; they protect us or plague us—just like the living.[51] To placate them, we have to do the same things hunter-gatherers learned to do with the living: we speak well of them and offer them food. In these actions we see the origins of praising and offerings. We can see the impact of these actions in the story of the flood. After the waters subsided, Noah sacrificed cattle and birds upon a burning altar: "And the LORD smelled a sweet savor; and the LORD said in his heart, I will not again curse the ground any more for man's sake."

It is advisable to remain in the spirits' good graces. Hunter-gatherers might have no inhibitions when it comes to berating a spirit now and then, should it get up to mischief, but they have to be sure of the protection of another spirit before doing so. You don't want to fool around with the denizens of the other side, after all.[52] This is an ancient, innate bit of knowledge: the dead can kill.

It's also a well-known fact that corpses not only attract beasts of prey but are also a possible source of infection once they begin to decay.[53] And this is where things start to get really exciting: even back when people had absolutely no idea about microbes or other means of infection, they had made the unsettling connection that dead bodies were "dangerous sources of unseen and barely describable danger."[54] The dead can make us ill or even kill us without doing anything visible. Once this deeply disturbing realization had etched itself into the human mind, it developed a powerful momentum of its own. The spirits of the dead became capable of anything, even without recourse to a physical body. It is not surprising that our ancestors are the usual suspects when it comes to all types of mishaps. This attribution underpins the importance of providing the dead with a proper funeral and only speaking well of them. *De mortuis nil nisi bene*—say nothing but good of the dead. Otherwise they might start looking for revenge.

The scarier the stories about spirits told around the campfire, the more likely they were to be retold and embellished. This is how an entire cosmos of supernatural beings came into existence. And so we lived in an animate world and coped with it as best we could, a world in which mishaps were isolated incidents and generally only affected individual persons. In extreme cases people could seek out the assistance of a shaman, those experts at communicating with the spirits. Otherwise the set of feelings and intuitions arising from our first nature were enough to show us how to interact with the spirit world—after all, we used the same feelings and intuitions to guide us in our dealings with the living. This worked for millennia, but the Neolithic Revolution changed the way the world functioned. And the spirits began their career.

Dramatic Events Require Dramatic Explanations

Let us again examine this new world. As our reading of the Bible has shown, sedentary life east of Eden had one very important characteristic: disaster. The calamity began with the end of the Pleistocene, as the animals hunted by humans began to disappear, forcing us to give up a way of life that had sustained us for thousands of years. Sedentary humans found themselves fighting for survival. Droughts and other natural disasters hit them particularly hard, as they could no longer just move on as they had before. The old social structures crumbled; the new ones fostered competition and engendered violence. Even within the family peace no longer reigned.

The strangest—and deadliest—changes were the epidemics that suddenly appeared, seemingly out of nowhere, and decimated human communities. The first farmers made it easy for the new microbes. Hunter-gatherers, who lived in small groups and rarely stayed in one place for long, were almost never exposed to infectious diseases. On the off chance that they did encounter a calamitous new virus, it would wipe out the entire group, immediately destroying its own basis for existence. Once people became farmers, increasing numbers began living side by side, and their communities grew from ten to a hundred times larger than those of the foragers.[55] Farming also created frequent connections through trade and so generated a reservoir for viruses and bacteria. There were always enough sick people spread among a number of villages to secure the pathogens' existence, and the situation only worsened once cities began to develop.

Hygienic conditions had also undergone a transformation. Because people did not properly dispose of their waste, human excrement contaminated their drinking water, paving the way for continuous reinfection. It was the ideal habitat for germs and parasites of any kind. Analyses of coprolites—semifossilized feces—have shown that the incidence of roundworms and whipworms increased dramatically among the early farmers.[56]

Though the rise of agriculture represented a "bonanza"[57] for microbes and worms, the new epidemics came as a complete shock. No one knew how or why unknown illnesses occasionally and suddenly visited them. The epidemics were an unpredictable and deadly mystery, and they remained so for the greater part of history. They were also a source of intense evolutionary pressure. Anything that lessened their impact directly increased fitness. "If a virulent plague rages through society the obvious response is to stop it by whatever means possible."[58] Our first nature, however, does not offer us any effective way of coping with such hardship. Such a situation demanded more of our third nature than ever before. To devise new defensive measures to combat disease, we had to shift our ability to rationalize into high gear.

We have already described the means by which humans look for explanations. Our HADDs scream within our brains, "Agents must be at work!" And who would have been the prime suspects for people who knew nothing of viruses, bacteria, or climatology? Of course it had to be the same agents who had always been responsible: the spirits! They had played a part in everything—from everyday problems to human illness—for thousands of years. This time, however, the spirits had come up with something much greater than anything they had wrought in the past.

Our ancestors were not to remain unaffected by such changes. Whereas individual spirits might have tormented one or two people in the past, they were now in the business of wiping out entire villages or even tribes. And here we see the role played by our "proportionality bias"—our tendency to match the power of the cause to the size of the event. A horrible epidemic, for example, must be the work of a powerful being. As Pascal Boyer reminds us, "We would like the explanation of dramatic things to be equally dramatic."[59]

We still see this at work today. Our proportionality bias fans the flames of conspiracy theories, for example. We simply cannot accept that the death of John F. Kennedy, the most powerful man in the world, was the work of the lone loser Lee Harvey Oswald. The mafia must have played a part! The CIA! The Soviets! Proportionality bias prompts us to look for big causes behind big events, a tendency that psychologists have recreated in a number of experiments.[60] When humanity found itself confronting all these new evils, this mechanism enabled our ancestors to make a mighty leap up the career ladder. The spirits became gods.

The Invention of Sin

And these gods were angry, a disposition that the God of the Old Testament definitely did not have a monopoly on. Throughout the entire ancient Near East, people lived in terror of the gods' wrath,[61] and the same was true in ancient Greece. Another famous book of antiquity, Homer's *The Iliad,* begins with Apollo sending a plague down on the Greeks besieging Troy because they had humiliated his priest. Gods were universally held responsible for disease, droughts, earthquakes, and military defeats. Catastrophes were their way of punishing humanity.[62]

Walter Burkert, a German classicist, points out that "an excessive elaboration of the principle of causality" underlies this notion of punishing gods. Calamities are personalized. They reflect "the wrath of a superior being punishing his subjects. . . . The concepts of gods behaving this way must have been established long before our documentation begins." Burkert maintains that this behavioral program is "universal and aboriginal and typical of the human mind and human behavior in general."[63] We couldn't agree more. This is the "oldest psychology" discussed by Hume and Nietzsche, the HADD referred to by today's anthropologists.

Burkert reconstructs the individual processes in this detection program: people experience calamity, and this leads them to ask, "Why? Why now? Why us?" Answers are provided by a "special mediator who claims superhuman knowledge: a seer, priest, or interpreter of dreams." He offers a "diagnosis" and identifies the cause of the disaster. He also assigns blame, "identifying what wrong was committed and by whom, and whether recently or long ago." This is followed by "appropriate acts of atonement, measures both ritual and practical to escape from evil and to find salvation." These usually take the form of sacrifices and the removal of the calamity's cause.[64]

One could divide this process of analysis into the steps used by modern medicine: listing the symptoms, arriving at a diagnosis, and providing treatment. But we would like to add another element, one that actually became the decisive factor in the construction of the Torah: prevention. Survival critically depended on figuring out how to forestall future occurrences. People began to go beyond diagnosing the immediate cause of the gods' wrath and shifted their focus toward determining the behaviors that angered the gods in order to avoid punishment in the future. And so a new concept arose that summarized all those forms of "wrongful" conduct: sin. The rise of the gods, then, begot the fallibility of man. After all, punishment assumes guilt.

Already the Babylonians had developed divination—refined methods of interpreting the course of heavenly bodies, the flight of birds, or the entrails of sacrificed animals—to determine what enrages the gods. The seers' findings were archived in order to evaluate the efficacy of the calamity-avoidance system and develop improvements.[65] Elsewhere, priests drew up "catalogues of sins" for use in interviewing those affected by illness or other misfortune to quickly identify their transgressions. As sociologist Max Weber explains, society's interest in identifying sins grew "with the increasing signs of divine wrath."[66] They could leave no stone unturned when it came to mollifying the angry gods, especially since the latter seemed hell-bent on inflicting collective punishment on the entire community.

These developments represented an enormous catalyst for the institutionalization of religion. New rules were continuously devised and integrated into complex systems, rituals were tested, and new sacred sites were built. Increasing numbers of priests had to try new means of disciplining "sinful" individuals—or eliminating them entirely—in order to placate the gods. Religion achieved a previously unknown level of institutionalization. A new kind of morality, established as a kind of disaster-avoidance system, had to be maintained at all costs, and the priests became its custodians. Part II of this book documents how much this whole enterprise shaped the Bible.

For the first time in human history, a form of morality developed whose maxims were the product of our rational third nature, which sought to deduce the new rules through observation of reality. Of course, hunter-gatherers also had their own forms of morality, but they were rooted in our first nature. Everyone knew what was and wasn't allowed, for these rules grew largely out of our innate emotions. If a problem arose or someone got out of line, the issue was discussed in depth. Every single aspect of the case was debated—sometimes for days on end—until the group came up with a solution. In extreme cases the group might decide that it was necessary to kill one of its members, but usually the punishment was nowhere nearly that drastic. In such hunter-gatherer groups, everyone was dependent on everyone else, and the group's survival required a certain level of harmony. People came to arrangements. In these societies, uncompromising forms of morality requiring strict adherence to rigid rules would have brought about the group's demise.

In this new agricultural world, such a set of strict rules did establish itself, but its uncompromising nature was not only a response to growing societies monitored by experts whose prestige and survival depended on keeping divine rage in check. The uncompromising new morality stemmed primarily from the fact that the gods seemed to go in for collective punishment: illnesses, floods, and droughts affect everyone, after all. The Bible contains many stories in which entire peoples are punished for the misdeeds of a single individual. This idea also surfaces in pagan religions. In ancient Greece, for instance, people feared that the "godlessness of a single citizen could unleash the wrath from heaven over an entire city."[67] Since no one wanted to suffer for the sins of others, everyone now had to keep a close eye on everyone else's behavior!

We cannot overestimate the consequences of these developments. With the proliferation of diseases seemingly connected to sexuality, hygiene, or eating habits, any behavior linked to these activities became increasingly suspect. The Torah is full of rules that correspond to these areas, with bitter consequences for those desires of our first nature directed toward satisfying our old, genetically anchored drives. Many of these were suddenly declared sinful, meaning that our "everyday desires and passions"[68] were now considered a curse. The rejection of all things carnal and corporeal harbored by many religions today originated in this way.

This rigid cultural system was a success, however. Some of the efforts actually improved hygiene and so helped to stop the spread of disease. Increased monitoring of moral behavior and adherence to the rules also actually contributed a great deal to social cohesion. Furthermore, this system of control was attractive to the shamans and priests, as it greatly boosted their power.[69] Finally, it offered the elites the opportunity to define the rules of the community and legitimize their authority.[70]

Religion as Cultural Calamity Control

These assertions are of such crucial importance that we would like to summarize them here. Ecological catastrophes, epidemics, and social tensions were the midwives of the great gods. Religion and morality—in the modern sense of the words—manifested themselves in the form of a cultural-protection system that sought to explain why calamities occurred and how to prevent them and, incidentally, served to justify the social order. This new system was needed because our innate intuitions had failed to keep up with the new realities of sedentary life. The fundamental assumption was that humanity had enraged the gods, and now we had to do everything possible to placate them. Never again should the gods have a reason to collectively punish individual misdeeds with plagues or catastrophes.

There is no question that the new gods' success could be put down to their effect on social morality and cooperation. According to our theory, however, social problems were not the primary threat early societies faced; disease was the real existential danger, and nothing spurs innovation more than it. Over the course of history, bacteria, viruses, and fungi have killed more people than wars or famines.[71] It is no wonder, therefore, that religion, an unprecedented cultural achievement, was called upon to combat these scourges. But the success of a religion depends on how well it protects its followers. As microbiologist David Clark says, "For most of history, infectious disease was the major killer, so how religion responded to disease was a crucial aspect."[72] "In fact disease is most intimately linked to religion in most civilizations," confirms classic scholar Walter Burkert.[73] The gods were indeed widely regarded as the real disease agents.

Catastrophes and diseases not only propelled the gods along their career paths but also served as their most powerful propagandists. Every new epidemic, every flood, and every famine helped reinforce the notion of supernatural agency, and thanks to the proportionality bias, the gods' authority grew with every new disaster. Anyone capable of unleashing a gigantic flood, like the God of the Old Testament, must be very powerful indeed! This rationale would remain a historical constant; every catastrophe fanned the flames of religious zeal.

Recent studies confirm that people in harsh environments are more prone to believe in all-powerful gods with clear moral demands[74] and that "people are more religious when living in regions that are more frequently razed by natural disasters."[75] The reason for this belief in "acts of God"[76] is not, as supposed by traditional interpretations, a greater need for solace. Instead, where disasters are common and unpredictable, people will ask more often and with greater urgency why a disaster struck and who was behind it. And it is only natural to suspect that human wrongdoing invited the misfortune of godly interference.

In the event of recurrent catastrophes, religion as a coping strategy always operates under extreme pressure. The disasters stoke the furnace of cultural evolution. Each new misfortune proves that there is a constant need to improve the system, with more detailed rules, more precise instructions. Especially the priests were in a tight spot; it is not uncommon for them to be killed if their actions prove ineffective.[77] Virulent diseases create virulent religion.

After this excursion, let us return to the social-glue argument favored by many as the raison d'être for religion in the larger, denser agriculture-based societies that arose after humans abandoned foraging. We hope to have shown that our new cultural-protection hypothesis provides a pithier explanation for how simple spirits evolved into big gods to meet these dangers. Their role as social glue is real, of course, but it emerged later—as an adventitious advantage, so to speak.

"YOU WILL SURVIVE"

We are now ready to take up the story of the flood again, for in it we can observe the power of the rational function of religion. Once God was sure that he had destroyed all that he had set out to, he plugged the fountains of the great deep and turned off the shower of heaven. When the dove released by Noah returned with an olive branch in its beak, "Noah knew that the waters were abated from the earth." The survivors returned to land, and with them came "every living thing . . . of all flesh, both of fowl, and of cattle, and of every creeping thing that creepeth upon the earth." Noah built an altar and made an offering to God, who then appeared and spoke soothing words: "I [will not] again smite any more every thing living, as I have done. While the earth remaineth, seedtime and harvest, and cold and heat, and summer and winter, and day and night shall not cease."

What happens next is important for our argument. God makes a covenant, a contract with Noah and his sons and all the animals: "And I will establish my covenant with you; neither shall all flesh be cut off any more by the waters of a flood; neither shall there any more be a flood to destroy the earth."

This passage summarizes in a nutshell what religion can accomplish as a rationalizing force. It renders God and all his catastrophes predictable. The most high himself is now bound by a contract. No longer able to high-handedly let his wrath run rampant, he must keep his emotions in check and prove himself a dependable negotiating partner. The priests can now tell their followers that the worst is behind them. The certainty that there will not be another flood makes it possible to plan for the future. And the fact that a catastrophic flood is, as Nassim Taleb would describe it, a "black swan,"[78] an extremely rare event, means that the priesthood is only proven wrong in the rarest of cases.

But that is not all. God sends a sign to the people to remind them of the covenant: the rainbow. "And it shall come to pass, when I bring a cloud over the earth, that the bow shall be seen in the cloud," God says to Noah. "And I will remember my covenant, which is between me and you and every living creature of all flesh; and the waters shall no more become a flood to destroy all flesh."

Friedrich Nietzsche pointed out that religion's attributing all disaster to known forces has a "comforting, liberating and reassuring" effect. "The new factor, that which has not been experienced and which is unfamiliar, is excluded from the sphere of causes."[79] Here the evolutionary value of religion is clear, just as it was when we analyzed Eve's curse of painful childbirth. The rainbow is a virtual "Don't Panic!" sign intended to take away our fear of even the worst storms. Its message is "You will survive." Religion helps people keep calm and reassures them that sound minds are looking for the solutions that can save us.

One Pulled Through

We began with the flood—and look where we washed up! But before we move on to the Tower of Babel, we'd like to say a few more words about

Noah. One man and his family survived the catastrophe, and from this family a new humanity would arise. In terms of evolutionary history, we all have ancestors who survived all the catastrophes and epidemics that have ever taken place, at least long enough to reproduce.[80] That is not the result of a conscious divine choice—as it was for Noah—but an accident of natural selection, more or less. This thought makes us all a bit uneasy.[81] But that is why stories like the one about the flood also remain so popular: one man pulled through, and he deserved it. That is what the Bible tells us.

The fact is, circumstances forced the biblical authors' hand. They simply had to turn Noah into a virtuous man by force of logic—an example of the coherence craving innate to human thought. If there is only one God, and he is responsible for everything, this implies that anyone singled out from his collective punishment and allowed to survive must have been good. Otherwise God would not have spared him. From the idea of one God arises the moralistic binary construct of good and evil, something we shall examine in greater detail later on.

The priests pounced on the survivor and declared him a saint: "Noah was just a man and perfect in his generations, and Noah walked with God." They had no other choice, for otherwise they would have been proven wrong—and not only because the people behaved just as badly after the flood as they had before. Later in the story we get the impression that Noah probably didn't deserve the divine salvation he received: no sooner had he survived the flood than he began to drink himself silly and "was uncovered within his tent." What happened next is a source of mystery that has led to a great deal of speculation. "And Ham, the father of Canaan, saw the nakedness of his father, and told his two brethren without. And Shem and Japheth took a garment, and laid it upon both their shoulders, and went backward, and covered the nakedness of their father; and their faces were backward, and they saw not their father's nakedness. And Noah awoke from his wine, and knew what his younger son had done unto him. And he said, Cursed be Canaan; a servant of servants shall he be unto his brethren."

Why did Noah curse his grandson Canaan? He hadn't done anything wrong, unlike his father, Ham, who made Noah's embarrassing nakedness public. It remains a mystery, but here we are only interested in one aspect of the story. Even after the flood, family life remained just as chaotic as during the times of Cain and Abel. People had not bettered themselves. Not even the Bible can cover up this naked fact.

5

THE TOWER OF BABEL

Death Traps

THE STORY OF THE TOWER OF BABEL RANKS AMONG THE BIBLE'S classic tales. The renowned painting by Pieter Bruegel the Elder, in which the mountain-like tower reaches into the clouds, illustrates the hubris of boundless belief in progress. The unfinished construction project of the Tower of Babel is the archetype of human delusions of grandeur.

That's a big success for such a small tale. The whole story is only a few lines long. Humans begin to multiply after the flood and soon move into Mesopotamia. In those days, everyone "was of one language, and of one speech." The people said to themselves, "Go to, let us build us a city and a tower, whose top may reach unto heaven; and let us make us a name, lest we be scattered abroad upon the face of the whole earth." This set God's alarm bells ringing. "And the LORD said, Behold, the people is one, and they have all one language; and this they begin to do: and now nothing will be restrained from them, which they have imagined to do. Go to, let us go down, and there confound their language, that they may not understand one another's speech." No sooner said than done. "So the LORD scattered them abroad from thence upon the face of all the earth: and they left off to build the city. Therefore is the name of it called Babel; because the LORD did there confound the language of all the earth: and from thence did the LORD scatter them abroad upon the face of all the earth."

Remarkably, God came down personally to sort things out. In this tale he still represents an older world of the gods and has not yet become the shapeless deity floating high above us whom we know from later passages in the Bible. Also striking—although it shouldn't really surprise readers who have made it this far—is the absence of a universally accepted interpretation of this story.

This text, subjected to multiple revisions, is often interpreted as an etiological story meant to explain why the peoples of the world live in scattered groups and speak different languages. Or it is understood as a tale of divine punishment for human arrogance. Some scholars have also read it as a criticism of the megalomania of the Assyrians and Babylonians, whose ziggurats, those towering terrace-stepped temples, were legendary in ancient times.[1]

CITIES WITHOUT STREETS

The story of the Tower of Babel is proof of just what good historians the Bible's authors were. The sequence of events they describe from the biblical primeval history up until after this tale reflects the real historical development of humans from small groups of hunter-gatherers to dwellers in the big cities of advanced civilizations. Cities represent a quantum leap in terms of cultural development. Whereas foragers roamed over wide expanses of wilderness, with the rise of the cities people lived by the thousands within a confined space that they themselves created. This presented a serious challenge to their first nature. In the hunter-gatherer past, people only associated with people they had known all or most of their lives. Now they were constantly surrounded by strangers.

We should not forget that, like property, the city was a new cultural invention. Prior to its emergence, people had absolutely no concept of what a city could even be. The settlement of Çatal Höyük, located on the high plains of central Anatolia in today's Turkey, is one of the first known examples of a large human settlement. Archaeologists have identified fourteen layers dating from 7400 to 6200 BCE. This protocity of about 2,500 inhabitants was built in an "agglutinating" fashion, meaning that the houses were built right up against one another, like cells in a wasp's nest. There were no streets, no alleys, no passageways! Getting from one house to another in Çatal Höyük required climbing over the rooftops of the other buildings. One can easily imagine that this led to a variety of problems besides transportation and hygiene.[2]

Archaeologists have identified a typical pattern. Many of the protocities were abandoned after a boom phase, in a cycle observed throughout the Levant and later also in Europe.[3] Remarkably, these early settlements lack evidence of functional and social differentiation. The buildings are nearly all of the same size, for example. Could it be that the old hunter-gatherer spirit was still at work, preventing the "keeping up with the Joneses" phenomenon from taking hold in these protocities? It seems that no one was allowed to rise above anyone else.

Hierarchies are necessary in order to keep such a large community together, but anthropologist Ian Kuijt suspects that people's egalitarian mentalities had not yet changed that much. This would have led to tremendous social tensions that possibly brought about the demise of these first large settlements.[4] "It's an intriguing hypothesis," says science author Annalee Newitz, "especially when you consider that when cities re-emerge in the 4,000s BCE, they have rigid social hierarchies with kings, shamans, and slaves."[5]

There is some evidence that religion produced the social know-how required for urban life to develop into the form that arose in around 4000 BCE in Mesopotamian Uruk. Transformed into gods, the spirits could now be mobilized to help legitimize the new power structures. They watched over the people's morality, and the festivals and rituals organized in their honor provided the social glue that held the community together. But it took some time for the tensions to subside—the new relationships first had to become part of humankind's second nature. Only then did enormous entities such as cities become possible; only then do we see the appearance of monumental structures.[6]

Viewed against this backdrop, it is somewhat ironic that the Bible's Tower of Babel story describes the exact opposite process: God doesn't bind the people together but sabotages their collaborative joint effort and scatters them in all directions. This suggests that another force might have brought about the cities' collapse.

THE INVISIBLE LEGIONS

Readers of Genesis will quickly notice that God has no love for cities. Just a few pages after the Tower of Babel story, he passes judgment on the cities of Sodom and Gomorrah. But exactly what is his problem with cities? The question is easily answered once we know that the biblical story of the Tower of Babel is no isolated case. Stories of tower building gone awry are found among a number of other peoples. The fact that the biblical Tower of Babel tale draws upon earlier Mesopotamian narratives will no longer come as a surprise. The story originally had to do "with neither the motif of the dispersion of the peoples nor the confusion of tongues,"[7] which means that its etiological aspect must be secondary. The important question therefore remains unanswered: Why did the project fail?

We are still familiar with unfinished buildings today. The architectural and logistical challenges posed by monumental construction projects are tremendous. It's no wonder that during the earliest phases of civilization—when societies were still experimenting with ways to make large groups of strangers work together effectively—quite a few construction projects never reached completion. Even in the absence of structural engineering hurdles, many megaprojects were never completed. Witness various unfinished pyramids resting in the sands of Egypt.

What lies behind many of these architectural failures as well as the collapse of a number of cities? In *Germs, Genes, & Civilization*, David Clark succinctly identifies the culprit: "Sooner or later, some pestilence or plague will strike the emerging city."[8] Jared Diamond explains, "If the rise of farming was thus a bonanza for our microbes, the rise of cities was a greater one, as still more densely packed human populations festered under even worse sanitation conditions."[9] For the first time in history, cities provided a reservoir large enough to provide a permanent home to a number of diseases. Whereas the surviving inhabitants of a city could develop resistance over time, newcomers were at risk, for they had never come into contact with the local pathogens.[10] As a consequence, cities developed into virtual death traps, for more people died in them than were born. The cities could only grow as a constant stream of people from the countryside moved in and compensated for those losses. And the new inhabitants often brought new diseases with them.[11]

The large-scale building projects of ancient times constituted an even more potent source of infection. Here workers from all over the region, many of them prisoners of war, slaves, and deportees, mingled. They labored under terrible sanitary and dietary conditions: an ideal habitat for all sorts of germs. The suffering caused by epidemics under these conditions can scarcely be imagined. No sooner had an epidemic appeared than hundreds of workers perished. Illnesses had always been deemed punishments meted by the gods—and here people could see the level of destruction of which God was capable when enraged. His power must have been immeasurable. People simply did not have the slightest clue about "the unseen and unsung legions of microbes" that wreaked havoc among urban inhabitants.[12] But one thing they did understand: God hated cities.

One self-evident explanation for the divine rage readily emerged. The delusions of grandeur that led the elite to erect enormous palaces, temples, and fortifications that appeared to reach into the heavens represented a direct challenge to the gods. The story of the Tower of Babel thus relates and interprets events that occurred over and over again in ancient times.

The last thing we wish to do in this book is present monocausal, exclusive explanations for the phenomena we describe. The final version of every biblical story allows for any number of interpretations. We have already explained how one must view their growth like that of a snowball rolling downhill. The story of the Tower of Babel enables us to observe this process: around the historical core—probably the failure to complete a large-scale project due to an epidemic—new layers were deposited. This explains how elements such as confusion of the tongues and the scattering of the peoples came to be a part of the story.

For the first time in history, the early states that built the cities were organized along the lines of multitribal alliances. Inhabitants no longer lived together with members of their own clan or at least as common descendants of a shared mythical forebear. People had to learn to get along with others from all sorts of backgrounds who oftentimes spoke different languages. Growing social divisions made this even harder—and often led to a rending of the social fabric. The biblical motifs of the confusion of the tongues and the scattering of the peoples are thus a perfect fit.

TAKING THE PROBLEMS SERIOUSLY

With the Tower of Babel we arrive at the end of biblical primeval history, which has taken us from the Garden of Eden to Babel, from existence as nomadic hunter-gatherers to life in the first cities. The anthropological-historical reading of Genesis was surprisingly easy. Primeval history documents the many occasions when humans confronted problems threatening to their very existence with unexpected and fateful vehemence. People could only assume that some agent—namely God—was responsible for all these catastrophes. One thing was sure: they urgently needed a solution to these problems. If people wanted to avert catastrophes, then they must go to great lengths to avoid arousing God's wrath.

In our reading of the Bible, we got by with a lot less effort than many an exegete. This should please at least the scholastics among our readers: Occam's razor, also known as the principle of parsimony, states that when dealing with a number of explanations, one should prefer the sparest among them. This we have attempted to do by taking the problems described in the Bible literally instead of attempting to explain them away as allegories.

We are now approaching Genesis's final cycle of stories, those of Abraham, Sarah, Hagar, and their many descendants. Entertainment is ensured: the problems we shall discuss are certainly no less important—just a bit more human.

6

PATRIARCHS AND MATRIARCHS

Family Feud

REMEMBER THE STORY OF CAIN AND ABEL? WE INTERPRETED THE murder of brother by brother as a clear indication of the immense difficulties people faced in their new environment. It showed how competition for property affected the very heart of the family. We have to admit that we formulated this thesis on the narrow empirical basis of a single family—although, according to the Bible, this family represented 100 percent of the world's population in those days. Fortunately, we can also draw on another family saga in Genesis that follows the fate of a single clan over the course of four generations.

We are talking about the patriarchs, the *Erzväter* (founding fathers) of ancient Israel: Abraham, Isaac, and Jacob. These days, German commentators commonly refer to the *Erzeltern* (founding parents), and the "matriarchs," Sarah, Rebecca, Leah, and Rachel, often accompany the patriarchs.[1] This is understandable, for these women played a central role in the family saga. But that the three men had at least four other wives as well is usually forgotten. The Bible mentions Hagar, Keturah, Bilhah, and Zilpah, and it quite conceivably left out an extra wife or two.

Traditionally, the narrative of the patriarchs is interpreted as "God's promise to the Patriarchs that their descendants would become a great people and possess their own land." We generally have to read between the lines of the popular commentaries to glean that there is more to this story: "God's saving plan," one popular commentary explains, "cannot be undone by humanity's lies and deception."[2] This represents a sign of progress after the flood: God has learned to control his rage and have patience with his creations—except, of course, the denizens of Sodom and Gomorrah, the cities he decides to destroy in a shower of fire and brimstone. But in this chapter we are less interested in God's actions and focus instead on humanity's shocking capacity for "lies and deception." In the end we will even come to understand why monogamy really isn't such a bad idea.

THE PATRIARCHS: SEASONS ONE TO FOUR

We are dealing with a family saga fit for a TV soap opera in the grand tradition of *Dallas. The Patriarchs* is a perfect example of what wealth can do to a family. Sons sleep with their fathers' wives; men couple with their daughters-in-law and even get their own daughters pregnant. Women plot and scheme. Brothers hatch bloody conspiracies of revenge. There's easily enough material here for four seasons. A brief overview:

Abraham is the star of season one. Because he follows God's call, Jews, Christians, and Muslims all revere him as the progenitor of monotheism. Abraham's father was from Ur in Chaldea, and he himself leaves Harran in northern Mesopotamia at God's bidding to move to Canaan. His nephew Lot accompanies him. Both are "very rich in cattle, in silver, and in gold." They live in tents and are always on the move.

Because she is infertile, Sarah advises her husband, "I pray thee, go in unto my maid; it may be that I may obtain children by her." Abraham doesn't need to be told twice. He takes Hagar as a wife, who soon gives birth to Ishmael. Years later, God promises Sarah that she, too, shall give birth to a child. She laughs, "After I am waxed old shall I have pleasure, my lord being old also?" Nevertheless, a son is born unto her. His name: Isaac.

This marks the end of family harmony. Sarah demands that Abraham send her rival away: "Cast out this bondwoman and her son: for the son of this bondwoman shall not be heir with my son, even with Isaac." Although it displeases Abraham, God advises him to fulfill her wish, and he sends his second wife, Hagar, and their son into the desert. They would have died of thirst had an angel not appeared at the very last second. Muslims consider Ishmael the father of the Arabs.

Isaac is the lead actor in the second season of *The Patriarchs.* He almost didn't survive the first season, however, in which God commands Abraham, "Take now thy son, thine only son Isaac, whom thou lovest, and get thee into the land of Moriah; and offer him there for a burnt offering upon one of the mountains which I will tell thee of." Abraham does as he is told. Only when he pulls out his knife and is about to kill Isaac does an angel stop him. God is assured of Abraham's loyalty.

After Sarah dies, Abraham marries once again. His new wife, Keturah, gives birth to six more sons, but Isaac nevertheless inherits everything. Isaac marries Rebecca, a relative from Harran. She gives birth to the twins Jacob and Esau. Isaac's wealth continues to grow, and this leads to disputes with his neighbors. Throughout this season, Isaac remains a rather flat figure, and it would seem that his wife Rebecca is the true star of the show. Apparently not all of this season's episodes have survived the past few thousand years.[3]

Things are completely different in the third season. The older of the twins, Esau, achieves infamy for selling his birthright for a plate of lentils. Rebecca helps the younger twin, Jacob, obtain his blind father's blessing by deception. In deadly fear of Esau, Jacob flees to his uncle Laban. Here the swindler himself is swindled. As recompense for seven years of compulsory labor (in return for the right to marry his daughter), Laban gives Jacob his older daughter Leah as a wife—not Rachel, the daughter Jacob truly desires. Jacob is allowed to marry her, too, but only after another seven years. Things don't go so well between the two wives, and they enter into something of a birth-giving competition. They even bring their maids in on the action. In the end, Jacob becomes the father of twelve sons born by four different women. There is no mention of daughters, with the exception of Dinah, who later falls victim to rape. Her brothers avenge the crime with a great deal of bloodshed. Jacob dreams of a heavenly ladder with angels climbing up and up and gets into a wrestling match with God. Jacob too becomes wealthy.

At the center of the fourth season of *The Patriarchs* is Joseph. He is the eleventh of Jacob's sons and his father's favorite, which drives his other brothers into a rage. They plot to kill him but instead have him sold off to Egypt, where he has a number of adventures. The wife of Potiphar, his owner, tries to seduce him, but Joseph rejects her advances. She then falsely accuses him of attempted rape and has him thrown in prison. His ability to interpret dreams leads to a new career, however. After he accurately prophesies seven rich and seven lean years, the Pharaoh appoints him vice-pharaoh and gives him the daughter of the sun priest in marriage. When Joseph's brothers, driven by hunger, arrive in Egypt, they are forgiven. The final season of *The Patriarchs* concludes happily—something quite unusual for a biblical production.

The Problem with the Camels

The stories of the patriarchs are so vibrant that many readers are tempted to "believe the narrated textual world represents a description of a living world filled with historical personalities."[4] These tales describe a rural world based on livestock farming. The clans move around the Syrian-Palestinian region, living in tents, not cities. Nothing other than clues provided in the texts indicates when these events supposedly took place. Based on biblical chronology, Abraham was long believed to have lived in around 2300 to 2200 BCE.

Scholars now know this estimate is untenable. The nomadic routes of the patriarchs are not at all realistic, and cities that did not yet exist appear in the narrative. And camels are traded as status symbols, when camels did not arrive in Palestine until sometime after 1000 BCE. In short, the stories of the patriarchs must be much younger creations. They were simply projected back upon the past.[5]

The succession of generations from Abraham to Isaac to Jacob to Joseph has also been proven a fiction. The four were originally unrelated, and each was the main figure of a separate narrative. Some of these stories came from the northern kingdom of Israel; others, from the southern kingdom of Judah. With the demise of the northern kingdom after its defeat at the hands of the Assyrians in 721 BCE, large numbers of refugees streamed south into Judah, bringing their stories with them. In order to better integrate the newcomers, the biblical authors blended the tales together into a larger work, according to professor of Old Testament studies Barbara Schmitz. The patriarchs and their wives "as a *single* family" were promoted "as the *common* ancestors of the entire people."[6]

The stories took on a new significance after the south fell in 587 BCE and the elites were packed off to Babylon. The alleged wandering of the "forefathers" from Ur in Chaldea into an unknown country promised to them by God reflected the journey that the exiles themselves would have to take: the way "out of Babylon (= Ur in Chaldea)" back to the old homeland, one now foreign to second- and third-generation exiles in Babylon and Persia.[7] Abraham thus advocates a return from exile. Against this background, it makes sense that God chose Abraham, forged a covenant with him and his descendants, and promised him a homeland and numerous progeny. But these details are mainly of religious-historical interest. Our true subject here is the "lies and deception."

FOUR CASES OF LIES AND DECEPTION

Literary historians' determination that the various patriarchs' family stories originated in different locations and times is highly significant. It means that we're not dealing with the whims of a single biblical editor. Rather, we are witnessing the effects of structural problems. Domestic bliss was in short supply just about everywhere in biblical times, at least among the elite.

The root cause of all these conflicts is well known: the introduction of agriculture brought with it ownership of land, animals, and food stores—a completely novel concept in the evolutionary history of our species. In many locations this new way of life was so successful that people managed to accumulate surpluses. As a result, some individual men became rich and could afford two or more wives. Of course, as in real life, only some men in a particular community in ancient Israel achieved the status of patriarch.[8] Polygyny brought these wealthy men immense advantages in terms of reproduction; more women meant more children, after all. But more women also always meant more problems. Polygyny, then, is the ultimate source of the constant "lies and deception" that plague the patriarchs' families.

We now turn to the conflicts that arose from these new property relations and are largely responsible for *The Patriarchs'* soap-operatic character. These conflicts happen at four different levels. Next we will scrutinize the strategies with which cultural evolution experimented in order to protect the clans from destroying themselves.

Competition Among Wives

Polygyny was already practiced among hunter-gatherers, albeit rarely, as a woman could share a husband with another woman. This was by all accounts a voluntary arrangement, which brought more advantages to a woman than having a less skillful hunter to herself or no husband at all. Under the new living conditions, however, women could no longer make these decisions independently. Quite the contrary, in fact: oftentimes they were treated no better than animals and were even traded like objects. Marriage for love became the exception, at least among the rich. The institution's primary function was to help solidify alliances.

Patrilocality—in which women move in with their husband's family—ensures that women enjoy fewer rights. Whereas in their birth families they are considered "short-term members and thus excluded from inheritance," in their new families they are viewed as outsiders due to their "foreign blood."[9] Biologist William Hamilton formulated the reasons for this development: the closer the kin relationships among organisms, thus the more genetic material they share, the more they behave altruistically toward one another. This is the fundamental maxim of kin selection.

Hamilton's rule can also explain why conflict between different wives in polygynous families is inevitable. These women are usually not related, which means they have no motivation to show altruism to one another. In fact, the opposite is true, for they compete directly with one another for their husband's limited resources. Each will vie to become his favorite, because that is the best means of improving her own situation and putting her sons in pole position in the race for inheritance.

As discussed in the chapter on Cain and Abel, the institution of primogeniture was established to prevent exactly this kind of conflict from erupting. It ensured that the oldest son was designated the heir. This makes good sense from the patriarchs' point of view, for in principle the oldest son is most likely to have enough authority to hold the clan together. The patriarch's first wife is also supportive of this institution, for neither she nor her son need fear the children of her husband's other wives.

But time and again in the Bible, primogeniture fails to fully establish itself. Be it Ishmael (sent out into the desert), Esau (tricked out of his inheritance for a plate of lentils), or Reuben (disinherited for an affair with one of his father's wives), the patriarchs' firstborn sons never enjoy their fathers' blessings. The actual heirs—Isaac, Jacob, and Joseph—were all later-born sons. We have already mentioned that the Torah would make up for this deficit and bluntly decree that the firstborn son be made heir even when born of an unloved wife.[10]

Primogeniture stands in direct opposition to the vital interests of all other family members, who are set to inherit nothing. Why should the mothers look on passively when another son to whom they aren't even related receives all of the father's favor? The Bible is quite clear that women are among the protagonists behind struggles to distribute inherited wealth, a characterization substantiated by ethnological studies showing how polygyny leads to competition among "cowives." In polygynous households, the danger of violence, abuse, child neglect, and even murder is significantly higher, because unrelated people in direct competition with one another for a single, nonsharable resource all live together under one roof.[11]

Let us return to the families of the patriarchs with this knowledge in mind. The fierce and ruthless behavior of the women no longer comes as a surprise. Problems arise as soon as Sarah's slave Hagar becomes pregnant with Abraham's child. The conflict escalates when Sarah also gives birth to a son. Her position as Abraham's first wife (she was also his half sister) enables her to ensure that the firstborn Ishmael, the rightful heir, is sent into the desert together with his mother. Had the angel not intervened, this would have been a case of cold-blooded murder.

Rebecca, on the other hand, is busy pulling strings in her son Jacob's struggle for inheritance. This is surprising, for what could be her interest in depriving her own firstborn son, Esau, of his inheritance? After all, he shares as much genetic material with her as Jacob. In doing so she creates an enemy within her own household. This does not add up, but since not all of the stories about Isaac made it into the Bible,[12] we are likely dealing with an abridged version of events. Rebecca's behavior, combined with the fact that Esau and Jacob are entirely different in terms of behavior, stature, and hirsuteness, supports the idea that Isaac once had a second, older wife: Esau's mother. After all, it would beggar belief if the rich patriarch Isaac only had one wife. This makes all of Rebecca's scheming to establish Jacob as the family heir much easier to understand. Obviously, this interpretation remains speculative, even if highly plausible.

Jacob married the sisters Leah and Rachel, but whereas Leah gave Jacob four sons, Rachel remained childless: "And when Rachel saw that she bare Jacob no children, Rachel envied her sister; and said unto Jacob, Give me children, or else I die. And Jacob's anger was kindled against Rachel: and he said, Am I in God's stead, who hath withheld from thee the fruit of the womb? And she said, Behold my maid Bilhah, go in unto her; and she shall bear upon my knees, that I may also have children by her." After Bilhah gave birth to two sons, Rachel was triumphant: "With great wrestlings have I wrestled with my sister, and I have prevailed." Leah, not to be outdone, also recruited a slave to serve as birthing machine. By the end of the story, Jacob has twelve sons born of four different women.

We must understand one thing about Rachel's behavior. In the world of the Old Testament, barrenness was grounds for divorce. Even worse, "childlessness was the equivalent of an early death for a woman."[13] Through their sons, women cultivated influence to ensure they were well taken care of. Thus, in order to gain power and secure a better life for herself in old age, a woman must have a son. In a pinch, she might have to resort to a maid as a surrogate, as did Rachel.

Primogeniture, as logical and practical as it might seem at first glance, has a built-in fundamental flaw. The mothers of the youngest sons are also the most attractive—thanks to their youthfulness—and they are able to use this advantage to bend the patriarch's ear. If this means they must make use of intrigue, lies, and deception to subvert the position of the oldest son, then so be it! The stories of the patriarchs illustrate perfectly the conflicts that can arise from the newly established marital practice known as polygyny. Women of varying standing find themselves in bitter competition; they produce sons in order to ensure their status. Of course, they also give birth to daughters or suffer miscarriages, but the Bible doesn't find it necessary to mention these events.

If recent Bible research has emphasized that women primarily took the initiative when it came to determining heirs,[14] then biology can explain why this is so. The patriarchs themselves could afford to take a more easygoing approach, for their genes passed on equally, regardless of which son became heir. For the women, however, Rachel's plea sums up the desperate nature of their position: "Give me children, or else I die."

Competition Among Brothers

Here we can keep things short, for in the chapter on Cain and Abel, we already examined how the violent conflict for inheritance can play out between brothers. If you read the Hebrew Bible carefully, you soon come to the conclusion that it describes not exceptions but the rule: Absalom has Amnon murdered, Solomon orders the execution of Adonijah, and Abimelech kills sixty-nine of his brothers (the seventieth managed to hide). The stories of the patriarchs also demonstrate how brothers, once the best of partners, could become mortal enemies.

Isaac doesn't even need to lay a hand upon Ishmael, for his mother has already gotten him out of the way. The twin brothers Esau and Jacob "wrestle already in their mother's womb."[15] They are in a prenatal struggle to see who will be born first. Jacob comes out holding his brother by the heel, but Esau nevertheless wins the race.

The birth of twins in particular offers proof of just how arbitrary the institution of primogeniture can be. The luck of being the first out of the mother's womb decides who will and who will not come into material wealth. It comes as no surprise that Jacob does everything he can to take away the advantage Esau had achieved by mere seconds. When he finally succeeds, Esau, who has just lost his inheritance, "[says] in his heart, The days of mourning for my father are at hand; then will I slay my brother Jacob." Jacob, warned by Rebecca, makes his escape.

And what about Joseph? His brothers hate him from the bottoms of their hearts because he is the favorite of his father, Jacob: "Behold, this dreamer cometh," they say to each other. "Come now therefore, and let us slay him, and cast him into some pit, and we will say, Some evil beast hath devoured him." But right at that moment a caravan passes, so they sell Joseph for twenty pieces of silver.

Once again, biology plays a decisive role: brothers have no qualms about resorting to extreme measures, for in nearly every case they are only half brothers. According to Hamilton's rule, levels of solidarity are lower because there is less shared genetic material in the game. The fact that Joseph is particularly close to Benjamin is equally logical: Benjamin and Joseph are Rachel's only sons, meaning they are full brothers. The Bible's stories provide excellent illustrative material for biology class.

Competition Between Fathers and Sons

So many wives and so many sons provide an unending source of confusion. Added to all this, the patriarchs' younger wives began to arouse the interest of the older sons. They were often of similar age and, not being relatives, could be sexually attracted to each other. Neither had they grown up together, which would have naturally produced an aversion to sexual relations.[16] A young wife would also be interested in this arrangement, for the son would be considerably younger than the father, after all, and if he was the designated heir, then a relationship with him could ensure a better position for her. For this reason Bilhah, one of Jacob's secondary wives, sleeps with his firstborn son, Reuben. Jacob sees this as an attack on his authority and disinherits Reuben.

The reverse was also possible. Judah, another of Jacob's sons, sleeps with his daughter-in-law Tamar. A childless widow, Tamar dresses herself as a prostitute in order to seduce her father-in-law and give birth to an heir. It works: the result of their tryst, Pharez, is one of King David's ancestors. The story also highlights the reigning double standards: whereas Judah's whoring is not seen as an offense, his judgment is quite drastic when he discovers his daughter-in-law's "whoredom": "Bring her forth, and let her be burnt." Only after she proves that Judah himself was the john, so to speak, does she escape death.

The Losers: Sodom and Gomorrah

So far we have only examined the biblical elites, the "happy few." Clearly polygyny, the practice of powerful individuals marrying more than one woman, makes the problem of primogeniture even more poignant. When one son inherits everything, the rest walk away empty-handed. In wealthy families they at least receive gifts, as do Abraham's later-born sons. The patriarch will need their help if he has to defend his property, after all. Most of the later-born sons, however, have no choice but to seek their fortunes elsewhere. And especially poorer families would rarely have enough to shower later-born sons with gifts.

This practice also has a negative impact on society as a whole. The Bible offers proof of this: embedded in the tales of the patriarchs is the story of Sodom and Gomorrah. Both cities represent the flipside of a society in which individual men monopolize land, property, and women. This produces losers, and the number of propertyless and landless men increases. Ultimately they adopt aggressive and risky strategies in the struggle for women, who are forever in short supply. "This will result in higher rates of murder, theft, rape, social disruption, kidnapping (especially of females), sexual slavery and prostitution," anthropologists have determined. Those left empty-handed represent a threat, for unmarried men tend to turn to violence.[17]

Cities offered a melting pot repository for these have-nots. The cities were not only death traps, due to the prevalence of infectious disease, as discussed in chapter 5, but also hotbeds of civil disturbance. As most of their inhabitants no longer lived within their clans, there was a lack of social control.

Even God himself hears about it: "Because the cry of Sodom and Gomorrah is great, and because their sin is very grievous," he sends down two angels to Sodom to see if there are any righteous people residing in the city. They visit the house of Lot, Abraham's nephew: "But before they lay down, the men of the city, even the men of Sodom, compassed the house round, both old and young, all the people from every quarter: And they called unto Lot, and said unto him, Where are the men which came in to thee this night? Bring them out unto us, that we may know them." Interpreters of the Bible are unanimous on this point: here the Good Book is referring to rape—homosexual rape at that. Sodom is, after all, the origin of the term "sodomy." Did the surplus of men in the cities lead to the same phenomenon—men raping other men—that we witness in today's prisons?

Although Lot's desire to protect his guests is quite honorable, the way he sets about it is downright distressing: "And Lot went out at the door unto them, and shut the door after him, And said, I pray you, brethren, do not so wickedly. Behold now, I have two daughters which have not known man; let me, I pray you, bring them out unto you, and do ye to them as is good in your eyes: only unto these men do nothing; for therefore came they under the shadow of my roof." A patriarch must keep his priorities straight after all. Fortunately, the angels put an end to the unseemly spectacle outside Lot's front door, and the next day "the LORD rained upon Sodom and upon Gomorrah brimstone and fire from the LORD out of heaven; And he overthrew those cities, and all the plain, and all the inhabitants of the cities, and that which grew upon the ground." Only Lot's family is allowed to escape.

The editors who composed the Hebrew Bible from so many individual stories were definitely not trying to write an anthropological textbook, but they could not have found a better context than the patriarchs for the story of Sodom and Gomorrah. For the losers created by the polygynous way of life formed the first urban mobs.

A GLIMPSE INTO THE LABORATORY OF CULTURAL EVOLUTION

The problems created by patriarchal families are massive and, when viewed from the perspective of evolutionary history, clearly the product of new living conditions. The patriarchs' desire to sire as many sons as possible produced a centrifugal dynamic: within each clan a number of genetically unrelated people became engaged in a bitter contest over inheritance. Preventing the alliance from collapsing demanded new solutions. If they worked, dynasties resulted. Here, again, the Bible offers us a glimpse behind the scenes in the laboratory of cultural evolution. We can recognize three strategies.

Strategy 1: More!

Strategy number one is obvious: expansion! If you do everything to enlarge your resources, then there will be enough for all to share. The patriarchs are very successful in this regard: the Bible tells us with pride how their wealth grew and grew. Jacob and Esau are certainly proof of the pacifying effect of wealth. Jacob, who came to his riches abroad, returns home in fear. But Esau, who has also long since grown rich, greets him in friendship—not least because Jacob showers him with gifts. As they no longer have to compete for resources, the brothers can go back to being what they should have been: particularly reliable partners thanks to their blood ties.

And one must be able to depend on these blood ties, for the neighbors will have adopted the same expansion strategy. The patriarchs try their utmost to develop a stable system of allies, and relatives offer the best partners. The close relationship between Abraham and his nephew Lot is representative of this arrangement. The pair often had trouble with neighbors over water and grazing areas. When Lot was kidnapped, Abraham "armed his trained servants, born in his own house, three hundred and eighteen, and pursued them unto Dan. And he divided himself against them, he and his servants, by night, and smote them, and pursued them unto Hobah, which is on the left hand of Damascus. And he brought back all the goods, and also brought again his brother Lot, and his goods, and the women also, and the people."

Patriarchs had to demonstrate strength to deter competitors. Two particularly disturbing episodes in Genesis illustrate this strategy. The first appears under the heading "The Rape of Dinah and Shechem's Bloodbath." Steven Pinker provides a succinct telling of the story:

> Isaac's son Jacob has a daughter, Dinah. Dinah is kidnapped and raped. . . . [T]he rapist's family then offers to purchase her from her own family as a wife for the rapist. Dinah's brothers explain that an

> important moral principle stands in the way of the transaction: the rapist is uncircumcised. So they make a counteroffer: if all the men in the rapist's hometown cut off their foreskins, Dinah will be theirs. While the men are incapacitated with bleeding penises, the brothers invade the city, plunder and destroy it, massacre the men, and carry off the women and children. When Jacob worries that neighboring tribes may attack them in revenge, his sons explain that it was worth the risk: "Should our sister be treated like a whore?"[18]

The message of the story to the neighboring tribes is loud and clear: "Don't mess with us, for we know no pardon! And most importantly: keep your hands off of our women!"

The "Binding of Isaac," the story of Abraham's willingness to sacrifice his son, is an expression of the same strategy. The episode is universally considered proof of Abraham's unconditional loyalty to God. Even so, it also raises a number of questions. Why does God present Abraham with such a moral dilemma? Why does God insist on more evidence of Abraham's loyalty? After all, he doesn't need it, for he knows what he has with Abraham; otherwise he would never have chosen him. In truth, Abraham's action is a signal to others, for it demonstrates how far he is willing to go.

A story from the Old Testament's book of Kings underscores this message. It tells of Israel's war against the Moabites. An Israelite victory is at hand, but then something outrageous happens: "And when the king of Moab saw that the battle was too sore for him, he took with him seven hundred men that drew swords, to break through even unto the king of Edom: but they could not. Then he took his eldest son that should have reigned in his stead, and offered him for a burnt offering upon the wall. And there was great indignation against Israel: and they departed from him, and returned to their own land."[19]

Biologists would call this a "costly signal" strategy or a credibility-enhancing display (CRED).[20] By investing his most precious good—his firstborn son—the king was signaling to his enemy his willingness to go to any lengths. At the same time, he obliged his god to perform a major favor in return, and a god willing to engage in such an exchange must be truly powerful. The enemy had no choice but to flee.

We will elaborate on this point later, for the New Testament presents us with a second and thoroughly familiar story of a father sacrificing his son: God lets Jesus die on the cross. Christianity will draw enormous persuasive power from this act, for it is impossible to doubt the credibility of a God who is willing to sacrifice his own son for our sins. Most believers consider Jesus's crucifixion a proof of love on the part of God that cannot be outdone. He really must care very deeply about humanity if he is willing to give up his only son for us.

This interpretation of Abraham's willingness to sacrifice Isaac as costly signaling shows how nicely this event fits within the context of the stories of the patriarchs. The message to the audience is loud and clear: don't even think about messing with Abraham's clan.

Internally, the strategy of expansion draws the family together, for all members profit from increased property. Externally, it tends to provoke competition with other clans and can unleash a spiral of violence. The losers in the inheritance game become so many willing combatants in the wars that follow. Violence is often their only hope of gaining access to women and property.[21] Welcome to the era of despotism!

Strategy 2: Let's Stick Together

The second strategy in the stories of the patriarchs represents a rather surprising move toward incest. Even theologians nowadays remark on how the patriarchs' aimed to establish "an endogamous marriage within one's own larger family group." On the other hand, "exogamous marriages with 'foreign' women (outside of one's own clan or [one's] own tribe)" were frowned upon, as reflected by the case of Esau.[22] His marriage to two Canaanites brought a great deal of "grief of mind" to his parents. This can best be explained by the fact that the stories were reworked during and after the Babylonian exile, for in the Diaspora endogamy becomes more important since it "ensures that wealth remains within the family and—what is even more important—Jewish identity is maintained."[23]

No doubt this is correct, but a much older mechanism is also at work here. Massive competition had always forced the clans to keep their property together and strengthen internal forces of solidarity, and marriage among relatives was a tried-and-true means of doing so. For example, when a man only had daughters, he could keep property in the male line of the family by marrying his daughters to the sons of his male relatives.[24] The bride price paid in patriarchal societies also plays an important role. A wife from among one's relatives is generally cheaper than one from a foreign clan. Most importantly, the bride price stays within the patriline, ensuring that property is not broken up and the clan is drawn closer together. Seen in this light, marriage among relatives makes sense.

Let us examine the biblical evidence: Abraham was married to his half sister Sarah, with whom he shared a father. Isaac and Jacob both took relatives as their wives. The latter even married two sisters, something not unusual in polygynous societies. As sisters share genetic material, this helped minimize competition among wives. Unfortunately, in the case of Leah and Rebecca, it did not quite work out because Leah gave birth to a number of children, whereas Rebecca was infertile until late in life.

In this context, we should also mention Onan, who achieved infamy. He refused to impregnate his sister-in-law Tamar, whose husband had died without having sired a child by her, even though it was Onan's duty to sleep with her. The institution of the "levirate" bound brothers to marry their brothers' widows in the service of the clan. This ensured continuation of the deceased brother's line and secured the family's property. Onan's refusal to fulfill this family service amounted to breaking the rules of the patriarchal world. Thus God punished Onan, who let his semen fall upon the ground, with death. As described above, the childless widow Tamar must resort to a trick to become impregnated by her father-in-law Judah to have children at all.

Finally, the stories of the patriarchs present us with the closest form of relations among relatives. When Lot flees Sodom together with his family, he loses his wife. She is transformed into a pillar of salt because she does not heed God's warning and looks back after he passes judgment on Sodom and Gomorrah. Left all alone with his two daughters, the latter decide to get him drunk—at least according to the official version of the Bible—so that he will get them pregnant. Each gives birth to a son. The children of their couplings become the progenitors of the hated Ammonites and Moabites. The Bible's message could not be more explicit: the neighboring tribes are despicable bastards!

It now becomes clear: the stories of Genesis attempt to work out how closely a couple can be related and still engage in sexual intercourse. This comes as no surprise. Throughout history, wealthy families in particular have shown a tendency to marry within the family. This is true of the Egyptian pharaohs of the past as well as the European royal households of the modern era. But the results of these familial marriages are equally infamous, because incest carries considerable costs. The incidence of hereditary diseases increases, and inbreeding also leads to impaired intelligence.

Were people really aware of this in biblical times? We think they were. The effect of inbreeding is sometimes blatant. Although no one in those days knew about genetics, people could certainly have figured out that it was bad for children if parents were too closely related. Such easily recognizable disabilities as harelips and susceptibility to disease occurred throughout history. People could make the connection because evidently such marriages were common. Analysis of teeth from over fifty skeletons found in a 9,000-year-old site at Basta, Jordan, shows that incest was not unusual in the settlement. The consequences were literally written on their faces: more than a third were missing their maxillary lateral incisors.[25] The gods evidently did not like incest.

If incestuous marriages were not all that unusual in ancient Israel,[26] even though humans have a natural aversion to them and are aware of the potential health risks, then their advantages must have been very great indeed. Here, too, culture and biology find themselves in a dialectic process. From a cultural perspective, marriage among close relatives offers a means of keeping the clan together, particularly in times of increased competition. The biological health risks counter this benefit. These two conflicting forces give rise to a compromise solution: marriages among close relatives are taboo, but a certain degree of relationship remains allowed, especially when wealth is at stake. To this day, marriages between cousins are considered normal in many parts of the world. The positive social effect seems to outweigh the heightened risk of genetic defects.

Later in the Torah, in the book of Leviticus, we find the rules that forbid sexual contact among close relatives. Interestingly, intercourse between in-laws is forbidden, even though their offspring would suffer no adverse health effects.[27] Here we might be dealing with a case of overgeneralization: because people did not understand the biological reasons for avoiding sexual relations between close relatives, the rule was stretched out a bit just to cover all the possible bases. But such relationships may also have been forbidden because they engendered too much conflict, for they made it too difficult to maintain the sensitive network of family loyalties.

At any rate, Abraham was lucky to have lived in the time before Moses, for the verdict that God handed down to his prophet was unmatched in severity: "And if a man shall take his sister, his father's daughter, or his mother's daughter, and see her nakedness, and she see his nakedness; it is a wicked thing; and they shall be cut off in the sight of their people."[28] As he had married Sarah, his father's daughter, Abraham should have been put to death—and the Abrahamic religions would have never come to be.

Strategy 3: Reconciliation

Expansion and cohesion: these two strategies go hand in hand and ensure the establishment of clear external boundaries. But the story of Joseph presents us with another strategy. Considered great literature, it is most likely the work of a single author. Joseph, the eleventh son of Jacob, is conspicuous for his intelligence. His ability to interpret dreams and his gift for wise counsel lead him out of the dungeon directly to the top of perhaps the most powerful state of the day, Egypt. Joseph has emotional intelligence in spades: he is willing to forgive his brothers—the same ones who once wished to kill him—and support them in their time of need.

The story illustrates a collision between two worlds. Joseph's brothers are bound by the patriarchal doctrine of revenge (remember the vengeance in the story of Dinah), whereas Joseph strives for reconciliation. He represents a new type of player in the Bible: the social genius able to generate solidarity with nothing more than the power of the word. It is no coincidence that Joseph has made a career for himself in the highly developed state of Egypt, which valued competence much more than family ties. Joseph is no patriarch and certainly no despot. He seems to foster a family model like that of most regular folks at that time: he has only one wife. Yet these qualities ensure that Joseph will remain an exception for some time to come.

The Advantage of Monogamy

Once again, the stories of the patriarchs appear at just the right moment in the Bible. They offer impressive examples of the enormous problems resulting from the monopolization of women and property. The patriarchs thus have what it takes to make an anthropological classic. They illustrate the difficulties produced by the new institution of polygyny and help us understand why monogamy established itself in the long run.

But monogamy's hegemony was by no means inevitable. In many of the world's economically advanced states, monogamy may be the standard today, but throughout most of history things have looked rather different.

Here it is important to be specific, for we are talking about what we today understand by monogamy: a lasting, exclusive relationship between two partners. Most of the marriages among the hunter-gatherers were also monogamous, but they rarely lasted a lifetime. Importantly, polygyny was not forbidden—on the contrary, unusually successful hunters had two (or, very rarely, even three) wives. Anthropologist Frank Marlowe has evaluated the available ethnological data: around 85 percent of all societies ever studied by anthropologists have allowed a man to have multiple wives. In contrast, polyandry (one woman has two or more husbands) is found in less than 1 percent of all societies.[29]

Why does this change? The phenomenon is often described as a "puzzle": "Historically, the emergence of monogamous marriage is particularly puzzling since the very men who most benefit from polygynous marriage—wealthy aristocrats—are often those most influential in setting norms and shaping laws."[30] However, if you read the stories of the patriarchs and thus gain an impression of family reality under polygyny—even if only a literary one—you will immediately appreciate the advantages of monogamy. It reduces competition for women among men. It also reduces competition between women and among their sons. Furthermore, there are fewer unmarried men in monogamous societies, providing less fuel for violent conflict.

But there was still a long way to go before monogamy established itself as a common living arrangement among the elites. Cultural evolution, so the Bible tells us, first bet on other horses: the despots and warlords, such as Samson and David, who simply took everything that pleased them—especially women.

END OF GENESIS: THE POWER OF A GOOD YARN

We have arrived at the end of Genesis and thus at the end of the first part of our book. The stories of the patriarchs have demonstrated one of cultural evolution's most powerful tools: the power of a good story. "In the artistry of the biblical narrative, the children of Abraham, Isaac, and Jacob were indeed made into a single family," write Israeli archaeologists Israel Finkelstein and Neil Silberman. The patriarchs were united by the "power of legend" "in a manner far more powerful and timeless than the fleeting adventures of a few historical individuals herding sheep in the highlands of Canaan could ever have done."[31] There is no question that the power of legends is mighty indeed, but only by the power of biology are these stories alive and well today.

This is true of all of Genesis. It stands as a literary memorial to human hardship that continues to fascinate to this day. It chronicles the misery that began when the species *Homo sapiens* began living a life for which it lacked the necessary biological adaptations. Humans are not made for sedentary life and all of its attendant drags, and this is why they are damned to improvise. We have proven surprisingly good at it: human culture is one of the planet's greatest wonders. But there was also a price. Our modern existence often suffers from a lack of natural self-evidence. Mismatch appears to be our destiny.

Now that we have taken stock of the problems people carried around with them, we are ready to examine the strategy that gave rise to the cultural-protection system, which makes up the very heart of the Hebrew Bible, the core of the Old Testament. It strives to define new ways of social cooperation and to curb catastrophes, disease, and violence. It aims at controlling human behavior so as to bring an end to a situation so terrible that people believed a raging God was out to get them "with the sword, with the famine, and with the pestilence," in the words of the prophet Jeremiah.[32]

The Torah is both impressive and seminal, and in Moses and the exodus we again encounter the power of a good yarn. Let us now turn to another masterpiece of biblical storytelling.

PART II

MOSES AND THE EXODUS

Becoming the One and Only

A more formidable story than that of Moses and the Israelites' exodus from Egypt is hard to find. Let us refresh our memory. For 430 years since Joseph brought them there, the Israelites have toiled as slaves at the Pharaoh's great building sites; in the process they have become transformed into a great people. When the Pharaoh orders every newborn Hebrew male put to death, little Moses is cast adrift on the Nile in an basket of bulrushes. Of all people, the Pharaoh's daughter fishes him out of the water. Once an adult, Moses is ordered by Yahweh, who appears in a burning bush, to lead his people out of Egypt, but Pharaoh refuses to let them go until after ten plagues have visited the land. God enables Moses to part the Red Sea so that his people can cross and later presents Moses with the Ten Commandments at Mount Sinai. But because the Israelites lack the necessary discipline, they must wander about the desert for forty years before finally reaching the borders of the Promised Land.

With the 1956 film *The Ten Commandments,* Hollywood dedicated one of the all-time greatest monumental films to the story of exodus, with 14,000 extras, 15,000 animals, and Charlton Heston in the lead role. Movie audiences were overwhelmed with emotion when Moses defied the Pharaoh and told him, "Let my people go!" or smashed the tablets of the Ten Commandments on discovering his people dancing around the golden calf. Since then the story has been filmed again and again.

Moses is, after all, "the most prominent character of the most important part of the canon of the Hebrew Bible."[1] He is considered the founder of monotheism, and his story occupies four of the five books that bear his name: Exodus, Leviticus, Numbers, and Deuteronomy. Even the modern knowledge that he did not author the five books of the Torah (also known by their Greek name, the Pentateuch) has done little to harm his reputation. Moses stood directly before Yahweh and was in constant dialogue with him; no one else in the Old Testament achieved such closeness to God.[2]

Exodus, that great, epic story of freedom, influences Judaism's self-image to this day.[3] Retold each year at Passover, it has tied families to their traditions for thousands of years. Exodus is deeply rooted in Christian tradition as well: peoples as diverse as the Americans, the Armenians, the English, the Dutch, and the Boers, social and religious movements such as Protestantism, Pietism, Puritanism, and the civil rights movement, and even Latin American liberation theology all identify with the tale. But even if this led Egyptologist Jan Assmann to conclude that "the book of Exodus contains what is probably the greatest and most consequential story that humanity has ever told,"[4] we believe that the epic itself is not the most spectacular part of Exodus. In fact, in our view, it is just a vehicle for the actual masterpiece: the law.

Genesis offered an impressive demonstration of the challenges people faced in the millennia following the Neolithic Revolution. The other four books of the Torah lay out the strategies developed to come to terms with all of the diseases, wars, and other catastrophes that accompanied these great changes. Part II of this book argues that the Torah represents a highly refined cultural-protection system in the form of 613 proscriptions and prescriptions. We examine this system from an anthropological perspective and show what a remarkable milestone of cultural evolution it truly is. The Torah pointed the way ahead for the development not only of religion but also, and more remarkably, of science.

The exodus epic has three main protagonists: Moses, Yahweh, and the stubborn Israelites who aren't all that willing to follow the other two and for this reason are constantly "murmuring" throughout the Torah's pages. Each of the three parties merits a chapter of its own. We begin by scrutinizing the great prophet Moses and carefully examine the tremendous corpus of laws that he delivered in God's name. We then turn to Yahweh himself and investigate how he became so uniquely monotheistic. Just think about it: Why did the god of the "fewest of all people" (the Torah's own words) make a career that would change the face of the world? Finally, we consider the murmuring people themselves. Why did they keep resisting the religion revealed to them at Mount Sinai? And what did women have to do with it? In pursuing these questions, we discover a great deal about people's true religiosity. Toward the end of this part we inquire about the portentous consequences of this cultural-protection system set forth in the name of Yahweh. But let us first begin with God's greatest prophet.

7

MOSES

God's Will Becomes Law

WHAT IS THE MOST IMPORTANT SCENE IN THE STORY OF THE exodus? Is it when Moses parts the Red Sea and leads the Israelites across its expanse? Or when he smashes the Ten Commandments on finding them worshipping a false god? No, the decisive scene takes place on Mount Nebo at the very end of their desert wanderings. Moses looks out over the Dead Sea and into the distance, where the heat dances over the hills. There, across the River Jordan, lies the Promised Land—the land of milk and honey. Moses has waited for this moment for forty years. "And the LORD said unto him, This is the land which I sware unto Abraham, unto Isaac, and unto Jacob, saying, I will give it unto thy seed." God pauses a moment before adding, "I have caused thee to see it with thine eyes, but thou shalt not go over thither."

We can only imagine Moses's disappointment. Forgotten are the suffering of slavery in Egypt, the tribulations, the thirst in the desert, the resistance Moses had to endure because his people refused to follow him. He had finally reached his goal! God had led his prophet to Mount Nebo, so he could catch this one glimpse of the Promised Land. And then he had to die. That was God's will.

It was all because of the story with the water of Meribah. When Moses and his brother Aaron conjured water from a rock, they must have done something wrong, for God was furious: "And the LORD spake unto Moses and Aaron, Because ye believed me not, to sanctify me in the eyes of the children of Israel, therefore ye shall not bring this congregation into the land which I have given them."[1] God sentenced Aaron to die atop Mount Hor, and now it was Moses's turn.

God makes no exceptions, not even for his most loyal servant—despite the fact that Moses has remained steadfast for forty years, and the two have spoken "face to face, as a man speaketh unto his friend."[2] No, God shows no mercy. Moses stands atop Mount Nebo, and he knows it is all over.

The Bible doesn't make much of a fuss about it. "So Moses the servant of the LORD died there in the land of Moab, according to the word of the LORD. And He buried him in a valley in the land of Moab, over against Bethpeor: but no man knoweth of his sepulchre unto this day." Joshua ultimately leads the Israelites into the Promised Land. At the head of the group walk the priests, who carry the Ark of the Covenant containing the written laws that God had given Moses. "Only be thou strong and very courageous, that thou mayest observe to do according to all the law, which Moses my servant commanded thee," and all will be well, God promises Joshua.[3]

Dry as the Bible's telling of the tale may be, it represents nothing less than a turning point in world history. "And there arose not a prophet since in Israel like unto Moses, whom the LORD knew face to face, In all the signs and the wonders, which the LORD sent him to do in the land of Egypt to Pharaoh, and to all his servants, and to all his land."[4] The law replaced Moses. God no longer needed a human being to herald his will: he had written everything down to the very last detail, as we shall see. Abstract letters took the place of men of flesh and blood. From this moment on, the world would never be the same again.

WHO WAS MOSES?

For centuries Bible scholars have speculated about the truth behind the story of the Israelites' exodus from Egypt. The wonders have garnered the most attention. There have been numerous attempts at scientific interpretation: the transformation of the Nile into blood (first plague) must have been an algal bloom, which caused the frogs to leave the river (second plague), and because all of the fish had also perished, the mosquitoes reproduced unchecked (third plague). Some have surmised that strong winds or submarine earthquakes enabled the Israelites to pass through the Red Sea or that they actually merely crossed a marshy sea of reeds. And the legendary manna, which nourished over 600,000 people in the desert for forty years: might this not have been the sweet secretion of scale insects living on the tamarisk tree?[5]

Such speculation is entertaining yet futile. Researchers unanimously agree that "exodus—as described in the Bible—is not historical."[6] The list of arguments is long and includes inconsistent details concerning the chronology and the route taken and, most importantly, the complete lack of Egyptian sources describing such a major event. In their classic *The Bible Unearthed*, archaeologists Israel Finkelstein and Neil Silberman note that not even the tiniest shred of archaeological evidence has ever been found to support the tale. They ironically title their chapter on the exodus "Phantom Wanderers?"[7]

At Least the Woman Is Real

And Moses? For his famed sculpture for the tomb of Pope Julius II in the Roman church San Pietro in Vincoli, Michelangelo lent Moses a physical stature that makes even Charlton Heston seem a mere schoolboy. The founder of psychoanalysis, Sigmund Freud (1856–1939), was so inspired by the Carrara-marble figure's "contemptuous, angry gaze"[8] that he dedicated an essay to the "man Moses." According to Freud, Moses was an Egyptian who exported Akhenaten's monotheism to Israel. Here Freud actually drew upon an older idea: as early as the seventeenth and eighteenth centuries, scholars had floated the idea that "biblical monotheism has its roots in Egypt."[9] Akhenaten, the father of Tutankhamen, emptied the Egyptian pantheon and demanded that Egyptians worship only the sun god Aten. But this was only a brief episode in Egyptian history, for after Akhenaten's death in 1336 or 1334 BCE the priests restored the old gods back to the temples.

Freud's thesis has been debunked (we explore the true origins of biblical monotheism in the next chapter). In fact we have hardly any indication that Moses was even a real person. There is no reference to him in any other contemporary source outside the Bible. Furthermore, the narratives of the exodus, the Ten Commandments on Mount Sinai, and the seizure of Canaan originally developed as independent stories. As is so typical of the Old Testament—and cultural evolution in general—we are dealing here with a process of bricolage in which different stories were combined and reworked. The figure of Moses probably appears in just one of these original stories and was retrospectively woven into the other narratives when, during the Babylonian captivity, individual tales were blended to create the epic story we know today. Moses serves as the narrative device that holds everything together.[10]

His name is often viewed as evidence of his historicity. It seems that he might indeed have come from Egypt: "The name Moses (Hebrew: *mošæh*) is derived from the Egyptian verb *mś/mśj*, meaning 'to bear.' It is the short form of an Egyptian name variant, such as Thutmose, 'born of the god Thut.'"[11] This is far from conclusive proof of Moses's origins, however, for Palestine was often under Egypt's influence, and his was probably not an uncommon name in the region.[12]

Moses's family history is a different story and, indeed, might very well be authentic. According to the book of Exodus, Moses married Zipporah of the Midianites. No biblical author would have ever dreamed up such a detail, for marriages to foreign women during the Babylonian captivity were considered objectionable. In fact, Moses himself was at war with the Midianites and commanded his officers to "kill every male among the little ones, and kill every woman that hath known man by lying with him."[13] It's hardly conceivable that anyone would have invented a Midianite wife for Moses.

Another piece of evidence supports this conjecture. The realm of the Bedouin Midianites stretched from the Dead Sea to the Gulf of Aqaba—the presumable homeland of our second protagonist in this monumental epic: Yahweh. Moses biographer Eckart Otto explains that in the Midianite region one "can also search for the origins of the worship of Yahweh as a god of war and weather." This could mean that the historical Moses represents a figure "who was connected to the Midianites' worship of Yahweh and who played a role in transporting this veneration of Yahweh to Palestine north of the Midianite sphere of influence."[14]

While Egyptian sources contain no references to the Israelites' presence in Egypt, they do report on nomadic migrations into the country on the Nile. These sources mention the Shasu, tribes of Semitic small-livestock herders, and the Apiru, bands of outlaws. Egyptian texts tell of "Asiatic and/or Semitic populations living as prisoners of war or as economic refugees in the Nile delta."[15] It is indeed possible that a small group of these Semites fled from Egypt to the land of the Midianites, where they came in contact with the Yahweh religion by means of a man named Moses before continuing on to the north. "If this is true," Otto explains, "this would mean that the biblical image of Moses as a religious founder is accurate, for he played a role in establishing the desert god Yahweh in the cultural land of Palestine."[16] Other theologians suspect that Moses might have been the leader of one of these groups of refugees.[17]

Egypt Is Not on the Nile

But what if the Egypt described in Exodus is not the same as the land of the pyramids? Some Bible scholars read "Egypt" as a code name for the power that brought fear and terror to the Near East in the eighth and seventh centuries BCE, the Assyrians, who expanded westward with brutal ferocity during this time. In 722/21 BCE they conquered the northern kingdom of Israel. The southern kingdom of Judah escaped this fate but was made a tribute-paying vassal state. Judaean kings such as Manasseh (ca. 708–642 BCE) were forced to swear allegiance to the Assyrian rulers in the name of the Assyrian gods.

Open rebellion was not an option in Jerusalem, but a form of literary resistance had apparently been brewing among the intellectual circles of priests and scholars.[18] The exodus narrative, that epic tale of liberation, which originated in the northern kingdom, was reworked into a veritable pièce de résistance that directed its criticism of despots at the Assyrian potentates. The authors were not above drawing inspiration from the enemy, however: the story of Moses's childhood probably derived from legend surrounding King Sargon I of Akkad, who ruled toward the end of the third millennium BCE. The legend tells of how he was set out on a river in a reed basket. Sargon II, who ruled the Assyrian Empire from 722 to 705 BCE, used this story to help cement his own legitimacy. And so the story became woven into Moses's biography. As we have seen before, the best Bible stories are often plagiarized.[19]

After this futile search for the historical Moses, it is tempting to agree with Bible scholar Konrad Schmid that "the Moses-Exodus narrative is not an historical account; it is a collective narrative of origins." We agree with Schmid "that so many later elements have been concentrated in the biblical figure of Moses that there are limits to what can be gained by insisting that he was a historical figure."[20]

GOD'S WILL

The search for the historical Moses has not done the prophet any favors: "He becomes a barely recognizable dwarf very different from the giant whose name is on the Pentateuch and all events, from the oppression in Egypt through the beginning of the conquest of the land."[21] A similar fate befell the glorious tale of the exodus. The Israelites, numbering in the hundreds of thousands, dwindled down to a small band of refugees.

But a dwarf-sized Moses also has its advantages. With him out of the picture, it is easier to glimpse what really defines the Torah: a profusion of laws, rules, and instructions spread over four books. The Decalogue, better known as the Ten Commandments, is just the tip of the iceberg. Rabbis have counted a total of 613 *mitzvoth*—365 prohibitions and 248 commandments, to be precise. These include classics such as "I am the LORD thy God, which have brought thee out of the land of Egypt, out of the house of bondage. Thou shalt have no other gods before me." Or the *ius talionis*, the right of equal retribution, usually referred to as "an eye for an eye, a tooth for a tooth." They also include instructions such as "Thou shalt not seethe [boil] a kid in his mother's milk." And then there are the many heavenly directives regarding homosexuality, bestiality, and extramarital sex, not to mention all the detailed instructions dealing with temples, priestly garb, and sacrificial rites.

Readers of the Bible have a hard time with God's bureaucratic tendency to make rules for everything. For some it's the final straw. German poet Johann Wolfgang von Goethe, himself certainly a master when it came to spinning a good yarn, feared that the reader, utterly lost in the forest of regulations, "completely loses sight of the main purpose along with the bewildered people."[22] Goethe's lament may be easy to understand, but we have to disagree with him. The "main purpose" is not to recount Moses's adventures; it is to present the laws. Their true significance has hardly been appreciated—and here we mean their significance not to religion but to cultural evolution. The laws raise the latter to a new level, for without them world history would have taken an utterly different course.

The New Rules of the House

In light of the complexity and sheer number of laws, it makes sense to present and clarify our theory of what the Torah is all about right at the outset. This will make the subsequent detailed analysis easier to follow. The uniqueness of the Mosaic Law comes to light if we compare it with other ancient Near Eastern legal systems. If we hold it up against the Babylonian Code of Hammurabi (eighteenth century BCE), for example, we notice two key differences. First, the regulations set forth in the Torah reach into aspects of life that otherwise had remained untouched by legal stipulations. The ancient Near Eastern legal texts are "strictly secular." They presuppose a "complete separation of legal, religious, and moral-ethical norms" and so contain no "provisions on things like altar construction, sacrifice, cultic taxes and priestly regulations"[23]—unlike the Torah. The latter's obsession with regulation even extends to the spheres of sexuality, personal hygiene, and diet.

Second, the Torah represents God's own laws. After all, he personally put his principles to paper or, more accurately, carved them onto stone tablets, when he wasn't dictating them to Moses. We can thus see all of the rules as reflecting God's very own will—even when it comes to regulations governing the burying of human waste, for example. In the Near East, on the other hand, the gods, though viewed as the "guardians of the law," were never seen as its source.[24] Making laws was the purview of kings. "The world had not previously known a law-giving god," as Egyptologist Jan Assmann explains.[25] That is an invention of the Bible: a God who provides his new creation with a complete set of the house rules.

One has to imagine that the laws form the foundation of the covenant concluded between Yahweh and the people at Mount Sinai. God makes it quite clear: if the laws are obeyed, there will be peace; if they are not, calamity will ensue. The Torah even presents the relevant contract provisions relating to blessing and curse twice.[26] They are much too long to cite here in their entirety; an excerpt will have to do—but it is powerful enough. Whereas God states that he is willing to reward obedience with rain, rich harvests, peace, and liberty, he also tells the people what will happen if they are disobedient: "I will even appoint over you terror, consumption, and the burning ague, that shall consume the eyes, and cause sorrow of heart: and ye shall sow your seed in vain, for your enemies shall eat it. And I will set my face against you, and ye shall be slain before your enemies: they that hate you shall reign over you; and ye shall flee when none pursueth you."[27]

In this context theologians tend to refer to the "deed-consequence nexus," a rather unwieldy expression.[28] Max Weber put it more succinctly when he wrote of a "god of just retribution."[29] Here he refers to the logic of the covenant: if the rules are adhered to, God will reward the contracting parties; if not, he will punish them for breach of contract. Bible scholars have pointed out that the form of the agreement shows a surprising resemblance to the contract the Assyrian kings concluded with their vassals.[30] Here, too, the fate of Judah depended on whether it obeyed the terms of the contract. If it did not, the Assyrian king would unleash his powerful military machine. The parallel between the mighty God and the forceful king is too obvious to ignore, and we shall encounter it again later in our analysis.

We will take God at his word and posit the following thesis: the laws are the contract's fine print guaranteeing the Israelites a world without catastrophe, disease, and violence. The Torah thus represents an exceptionally elaborate cultural-protection system, yet one based on surprisingly simple logic. We have already described the basic mechanisms in our chapter on the flood. In times when people had no knowledge of the real causes of epidemics, droughts, and earthquakes, they resorted to the diagnoses offered by our first nature: they understood calamities as punishments meted out by supernatural actors. This interpretive pattern, still widespread today, was ubiquitous throughout antiquity: all forms of misfortune stemmed from the wrath of the gods.[31] This meant that one of the greatest challenges civilization faced was finding a means of forestalling the gods' anger—the alternative was doom. Our third nature, *Homo sapiens*'s cognitive genius, worked flat out to develop a system of prevention.

The system offered by the Torah relies on a simple assumption: if every catastrophe, every illness represents a divine punishment, then each such instance can be traced back to a human transgression. Thus, everything must be done to identify these sins in order to avoid repeating them in the future. This analysis can take two paths. The first is the path of empiricism, whereby one takes note of the areas where God's wrath is easily aroused and then sanctions them as required. The other is the path of empathy, whereby one tries to get a feel for God's wishes in order to understand the things that make him angry. Pascal Boyer showed that people cannot help but try to understand the behavior of supernatural beings along the lines of human models. With few exceptions, they believe, gods tend to react in the same manner as people. It thus appears safe to assume that God punishes the same behavior that his people also believe to be wrong. Once the actions that aroused God's anger have been identified, it's simply a matter of drawing up laws prohibiting them. If everyone follows these rules, then all will be spared similar misfortune in the future. Without human transgression there will be no divine punishment.

In this manner the priests reconstructed God's will by a process of reverse engineering. According to our cultural-protection thesis, the experienced reality of misfortune shaped the commandments and prohibitions listed in the Torah. This is why God always seems so terribly wrathful. The Torah is therefore the product of the crises people faced after being forced to radically modify their lifeways in the aftermath of the invention of agriculture. And the Torah is a powerful cultural tool for coming to terms with such crises. The goal is to make God—that is diseases, catastrophes, and wars—manageable.

Indeed, this crisis-management system is to be found at just the right place in the Bible. Genesis, which in a way serves as a prologue to the Mosaic Law, introduces us to the dangerous range of problems people faced: violence, disease, and natural catastrophes. Business as usual had become impossible; every reader of the first book of Moses must have understood this. A solution was needed, and this is exactly what the four remaining books of the Pentateuch, with their 613 laws, have to offer.

Why Only Here?

At least two obvious objections can be leveled against our thesis. First, if such fundamental psychological mechanisms—part of the basic configuration of all of us—are at play here, why do we not find these text-based catastrophe-prevention systems everywhere? And, second, haven't the theologians shown that the Torah is not the result of a one-time legislative act and that its contents, especially when it comes to the earliest laws, do not necessarily reflect divine law?[32]

We examine these themes in detail below, so for now a succinct answer will have to suffice. The period in which the Torah was developed does indeed fall between the demise of the northern kingdom of Israel in the second half of the eighth century BCE and the return from Babylonian captivity in the sixth and fifth centuries BCE. That is why it consists of numerous legal documents of varying scope, such as the Covenant Code, the Deuteronomic Code, and the Code of Holiness, as well as shorter texts such as the Ten Commandments, which were presented twice (but did not play as prominent a role as one would expect given this emphatic repetition).[33] All laws were reworked and redacted a number of times, and the authors represented various factions of priests and scholars.

The oldest set of laws in the Torah, the Covenant Code, does indeed limit itself to rather profane areas of law and thus is similar in form to other examples of ancient Near Eastern law. In fact, David Wright, a professor of Bible and the ancient Near East at Brandeis University, suspects that the ancient Babylonian Code of Hammurabi served as a direct template.[34] Only later would biblical laws increasingly encroach on areas of personal life, making them subject to God's will. Among Bible scholars, Max Weber's notion of the "theologization of the law"[35] has established itself as a way to characterize this process. But much more is actually at stake. The process is about nothing less than the theologization *of daily life*—about ferreting out the practices in nearly every aspect of life that might arouse God's wrath. We can now refine the question: How did this come to be, and why did it only happen here in ancient Israel?

First of all, an excessive number of catastrophes visited Israel and Judah, mainly in the form of enemy armies, which would ultimately lead to the breakup of both states. This greatly increased the pressure to uncover, through amazing intellectual funambulism, the root cause of all this misfortune—and maybe thereby to avoid complete destruction. Second, with Yahweh there was only the one god. The catastrophes made him even more mighty—thanks to proportionality bias. More importantly, however, if there is only one actor behind all misfortune, then it is much easier to reconstruct his motivations and create a more consistent system than if one has to sort out the conflicting desires of an entire army of gods. It would have been impossible for the Greeks to reverse engineer the preferences and dislikes of all the Olympic gods with reasonable certainty, for example. Greek and other cultures had to develop other means of catastrophe prevention. We will encounter these as well.

The situation is entirely different when dealing with a single god. As we have just noted, human expectations dictate that he behave consistently. We cannot imagine otherwise. The Torah reflects this, and we shall encounter a number of laws that rest on a simple logic: if Yahweh punishes this behavior, then he won't like this or that either; better if we just forbid those from the get-go. This is compulsive coherence seeking.

Before we begin to test our hypothesis by examining the laws themselves, we would like to forestall a likely misunderstanding. The Torah is only one part of the Hebrew Bible, and the amazingly predictable God we encounter here will undergo tremendous changes over the course of the Good Book. Seen from the perspective of biblical religion's cultural evolution, the Torah represents an intermediate stage—but one of fundamental importance. To repeat our claim: the Torah is most likely the most ambitious attempt ever undertaken to come to terms with the massive problems stemming from humanity's greatest mistake: sedentarization.

In the following we examine the laws of the Torah. Of course, we cannot discuss every single rule, so we will limit ourselves to an examination of their basic categories in order to uncover how the Torah works. Regardless of what we discover, the laws themselves are an exciting area of research, for they lead us straight into the morass of human behavior. After all, laws are only needed when the actions perceived as transgressions actually occur in real life. Viewed this way, the Ten Commandments simply provide a catalogue of the most common crimes: people kill, steal, commit adultery, and worship other gods. But this should no longer come as a surprise, for we already encountered the same litany in Genesis.

The First Set of Laws: Protection Against Violence

We only need think of Cain and Abel or all the chaos, lies, and deception we came across in the families of the patriarchs. Violence, the destroyer of social structure and human lives, is one of the most dangerous visitations of the biblical world. Genesis makes this abundantly clear. With the predictability of a law of nature, violence arises from the injustice of a property-based world in which social control has been thrown out the window. The descendants of Cain make careers as despots—and they nearly bring about the downfall of society.

It therefore comes as no surprise that the majority of the Torah's laws regulate violence. Of course, the same holds for the laws of all other societies that haven't been struck from the annals of history. Nevertheless, the Torah contains a distinctly social impetus and reflects the fear that God could rain down punishment should chaos prevail. That is the message of the stories of the flood and of Sodom and Gomorrah. God simply can't stand it when society descends into violence. This, too, is an ancient experience etched into the human psyche: societies in which inequality mushrooms and violence runs riot are doomed to collapse. Let us begin with the most obvious of the Torah's laws—those that seek to regulate social and family life in order to please Yahweh. Their wide scope offers proof of just how massive and manifold the misery must have been.

The Torah contains laws against typical crimes such as murder, bodily harm, robbery, and theft but has a surprising tendency to veer toward extremes. On one page we find the crimes punishable by death—"He that smiteth a man, so that he die, shall be surely put to death"[36]—yet on another we encounter astonishing examples of leniency. In the case of bodily harm, the perpetrator is not to be punished; he must only pay for the loss of earnings and doctor fees.[37] On the one hand, the massive threat of punishment seeks to deter crime; on the other hand the Torah appears to seek to prevent revenge as a reaction to a crime. Either way, the main goal is to prevent violence from spiraling out of control. Woe to those who ever let that come to pass. God has no reservations whatsoever when it comes to collective punishment should things get out of hand.

Many of the laws deal with the fate of slaves. Slavery was a widespread practice in ancient Israel, and the Torah did not infringe on slaveholders' rights. On the contrary, bodily harm of slaves was punishable only if the victim died on the spot. "Notwithstanding, if he continue a day or two, he [the owner] shall not be punished: for he [the slave] is his money."[38] Even so, we see at least an attempt to provide slaves with a minimum of rights. The Covenant Code orders the freeing of Hebrew debt slaves after seven years. Later laws take this a step further and even order that freed slaves be given animals and seed. Where does this humanitarian impulse come from? Given that many people slid into slavery due to excessive indebtedness, was the law an expression of compassion, or did it perhaps seek to prevent poor freed slaves from becoming a source of turmoil?

We should point out that such compassion extended only to men: "And if a man sell his daughter to be a maidservant, she shall not go out as the menservants do."[39] Indeed, the situation was particularly dire for females, for daughters were the first to be sold into slavery if a family got into financial trouble. They faced not only hard labor but also sexual exploitation.[40] The different treatment of male and female slaves strongly suggests that the biblical authors saw only male slaves as a source of the sort of trouble that could tempt God to interfere. Remember Sodom and Gomorrah?

The existence of slavery points to one of the main problems of ancient societies: poverty. The accumulation of property led to the impoverishment of large portions of the population. In order to keep people from sinking into misery, the Torah prohibits charging interest on loans. Collateral can also only be held for a limited time: "If thou at all take thy neighbour's raiment to pledge, thou shalt deliver it unto him by that the sun goeth down: For that is his covering only, it is his raiment for his skin: wherein shall he sleep?"[41] God expressly places the poor under his protection: "And it shall come to pass, when he crieth unto me, that I will hear; for I am gracious."[42] The same is true of foreigners: "Thou shalt neither vex a stranger, nor oppress him: for ye were strangers in the land of Egypt."[43] Assyrian conquests had forced many refugees into the lands of Judah, mainly from the neighboring kingdom of Israel.[44]

All of these examples offer evidence of just how great the social upheaval was and how much all the injustice upset people—and how they fully believed that God felt the same way about it. That's why he was prepared to punish them by sending chaos and banditry. But our observations would be incomplete were we to ignore the role of the family, whose problems we have already discussed at some length. The stories of the patriarchs showed us the way: "Honour thy father and thy mother," as the Decalogue's fourth commandment demands. This was by no means a given, for even Bible scholars speak of vicious within-family conflicts over inheritance.[45] The importance of family cohesion is underscored by the punishment for such offenses: "And he that smiteth his father, or his mother, shall be surely put to death."[46] It almost seems as if Genesis had hesitated to tell us the whole truth.

Apart from that, the Torah does indeed contain precisely the laws required to rein in the problems appearing at the very beginning of the Bible. It explicitly protects the institution of primogeniture to avoid inheritance struggles. The right of the eldest son, even if born of an unloved woman, could not be infringed. A number of desperately needed directives also define the type of sexual contact considered legitimate. These might have stemmed from experiences with biological damage resulting from high instances of incest. Abraham's risky marriage to his half sister Sarah would have been prohibited by the Torah—and punished with death! Did the people of ancient Israel actually realize that such close relationships within the family could lead to congenital illness and defects, which they interpreted as severe punishment meted out by God? The laws forbidding incest certainly would suggest as much.

We can also make out the social intentions behind these laws. The Torah even proscribes relations between nonblood relatives, which might not endanger the health of any resulting children but certainly could pose a threat to family harmony: "Thou shalt not uncover the nakedness of thy daughter in law: she is thy son's wife. . . . Thou shalt not uncover the nakedness of thy brother's wife: it is thy brother's nakedness."[47] There surely would have been a lot less turmoil within the families of the patriarchs had they lived by these laws as well.

The Humanity of an Eye for an Eye

The Old Testament has something of a bad reputation. Nowadays "an eye for an eye, a tooth for a tooth" is code for an unbridled thirst for revenge. Wrongfully so, however, for this simple logic actually represents one of civilization's greatest achievements. One must understand that before this concept took hold, justice among agriculturalists was not the purview of God or the state. Legal decisions, "including the death penalty and expulsion from the family," were up to the *pater familias*,[48] the patriarch. Conflicts were solved "only through direct negotiation between the participating groups."[49] Justice was thus the right of the strongest. Blood feuds were ever-present.

Courts of elders inside the city gates were intended to put a stop to this state of affairs, drawing upon laws like those we find in the Torah. From this point onward, a man himself was no longer allowed to punish an adulterer caught in flagrante delicto. He had to seek recourse through the courts. The Torah established the need for plaintiffs to produce two witnesses to prevent defamation.

In this sense, the notion of an eye for an eye is not at all archaic. The *ius talionis*, the concept of retributive justice already known in Mesopotamia—"Eye for eye, tooth for tooth, hand for hand, foot for foot, burning for burning, wound for wound, stripe for stripe"[50]—aimed to prevent retribution and ensure just recompense among clans.[51] This clearly rebuffed Cain's descendants such as Lamech, who boasted, "I have slain a man to my wounding, and a young man to my hurt." The end of the spiraling violence of vendetta had arrived.

And the death penalty? Critics of religion such as Richard Dawkins like to point out just how many crimes were punishable by death: "cursing your parents; committing adultery; making love to your stepmother or your daughter-in-law; homosexuality; marrying a woman and her daughter; bestiality. . . . You also get executed, of course, for working on the Sabbath."[52] We should point out that the Old Testament is hardly unique in its severity. The criminal law that began to establish itself in other cultures was just as "exceptionally strict and brutal" and similarly emphasized mutilation and the death penalty as punishments.[53] But in all its strictness, the Torah strove for prevention rather than retaliation. "The severity of the punishment is intended to prevent the breach of norms," explains theologian Eckart Otto.[54]

The emphasis on prevention actually follows directly from the system's own logic: everything had to be done to ensure that infringements never occurred in the first place; otherwise God would punish the entire people for the crimes of an individual. He was, after all, known to go in for collective punishment. Catastrophes and epidemics never only affected the guilty.

Love Thy Neighbor

Let us take stock: the Torah's laws aimed at preventing violence comprise the ethical component of a cultural catastrophe-prevention system. Although the product of our third nature, they were clearly inspired by our first nature, which tends to rise up against injustice. And because the Hebrew people believed that God had a psychology similar to their own, they assumed he felt the same way. The simple empirical observation that murder and mayhem proliferate when the egoists are not reined in supported this notion.

In this context we encounter the famed sentence "Thou shalt love thy neighbour as thyself,"[55] believed to represent the "center of the Torah."[56] Here, in a nutshell, we find the concept of reciprocity describing how we wish to be treated by others: as equals. This fundamental rule of fairness is a core component of our basic psychology and, as such, not an invention of the Bible. In fact, it is the main source of all human altruistic behavior. We demand justice and equality and sympathize with the weak against the alphas in their quest to monopolize power—at least within our own communities. Variants of this golden rule surface in moral systems—secular and religious—the world over.[57] What had been self-evident in hunter-gatherer societies now had to be spelled out explicitly. Once again we see how, because we humans dislike a lack of solidarity, God dislikes it too.

The Torah's egalitarian impulse did not even stop at society's leaders. Whereas in other places kings maintained the right to make laws, the Torah's priestly authors made sure the king himself had to follow its precepts. Indeed, he should beware "that his heart be not lifted up above his brethren, and that he turn not aside from the commandment, to the right hand, or to the left: to the end that he may prolong his days in his kingdom." Furthermore, he should not "multiply wives to himself, that his heart turn not away: neither shall he greatly multiply to himself silver and gold."[58] Such laws could only have been formulated during a time of exile when there were no kings, of course. Nevertheless, the degree to which our first nature managed to break through here is astounding. As in the times of the hunter-gatherer groups, no one should be allowed to rise too high above the others.

While honoring the Torah's civilizing effects, we must not forget that the law also represented a form of compromise. Property in general still enjoyed massive protections. "If the sun be risen upon [a thief], there shall be blood shed for him; for he should make full restitution; if he have nothing, then he shall be sold for his theft."[59] So, when poverty forced someone into thievery, he would end up as a slave.[60] And there were double standards when it came to men and women. Take adultery, for example: "If a man be found lying with a woman married to an husband, then they shall both of them die, both the man that lay with the woman, and the woman."[61] This implies, however, that whereas a married woman who committed adultery was doomed to die, a husband could get away with sleeping with another woman—as long as she wasn't married.[62]

Still, it is safe to state that all this social legislation must have had an effect: if the laws were followed, violence and social inequality would be reduced and cooperation increased. For this very reason we encounter similar rules in other cultures too, with the critical difference that in other places, no one claimed these rules expressed God's will. In the Torah, however, they always had a twofold aim: prevention of violence because it harms people and, most importantly, to keep God from getting involved.

The Second Set of Laws: Protection from Disease

Preventing violence, protecting property, promoting equality—we expect laws to do these things. But if the Torah really is the catastrophe-prevention system we claim it to be, then it also must intervene against a threat that is normally not subject to legislation: disease. In times when people had absolutely no knowledge of microbial pathogens, diseases ranked among the most sinister and deadliest of all threats. Indeed, the Torah devotes a great number of pages to this danger, a fact that we believe has not received the attention it deserves. When scholars discuss the biblical laws pertaining to disease, they usually focus on the extent to which these strictures—with the benefit of hindsight—make medical sense. But these laws actually tell a much more momentous story.

We touched upon this subject in our chapter on the flood: "The ancient world had no knowledge of bacteria," writes microbiologist David Clark. Instead, people believed that "epidemics were one of the main ways the gods expressed their displeasure."[63] American anthropologist George Murdock has described this phenomenon as "spirit aggression," which he defines as "the attribution of illness to the direct hostile, arbitrary, or punitive action of some malevolent or affronted supernatural being." Among 139 historical and contemporary societies, Murdock found only two in which this belief could not be confirmed. Diseases were among the gods' most effective propagandists—and for the greatest part of its history, medicine, as Murdock explains, was nothing more than "applied religion."[64]

The Old Testament, too, finds it entirely natural to interpret every disease as God's punishment for any misdeed by a person or people that qualified as a sin. Illnesses are "God's pedagogical instruments."[65] God himself made absolutely no secret of the fact that he himself was the ultimate pathogenic agent. Here are just three examples:

- "If thou wilt diligently hearken to the voice of the LORD thy God, and wilt do that which is right in his sight, and wilt give ear to his commandments, and keep all his statutes, I will put none of these diseases upon thee, which I have brought upon the Egyptians: for I am the LORD that healeth thee."[66]
- "The LORD will smite thee with the botch of Egypt, and with the emerods, and with the scab, and with the itch, whereof thou canst not be healed. The LORD shall smite thee with madness, and blindness, and astonishment of heart."[67]
- "If thou wilt not observe to do all the words of this law that are written in this book, that thou mayest fear this glorious and fearful name, THE LORD THY GOD; Then the LORD will make thy plagues wonderful, and the plagues of thy seed, even great plagues, and of long continuance, and sore sicknesses, and of long continuance. Moreover he will bring upon thee all the diseases of Egypt, which thou wast afraid of; and they shall cleave unto thee. Also every sickness, and every plague, which is not written in the book of this law, them will the LORD bring upon thee, until thou be destroyed."[68]

And God didn't just leave it at words. When Miriam questions her brother Moses's authority, God punishes her with a snow-white skin rash. God punishes coarser crimes by unleashing a plague onto the Israelites' camp that claims thousands of lives. Where Zeus reaches for bolts of lightning, Yahweh sends leprosy and pestilence. The solution was obvious: "The most important prophylaxis against illness is maintaining an intact relationship with God!"[69]

Clean and Unclean

In the last few years a great deal of research has shown that humans possess what is known as a "behavioral immune system." According to this concept, behavior can prevent infection by helping us avoid contact with pathogens in the first place. This research tradition mentions religion, however, only to highlight the rules limiting contact with people from different faiths—strangers, in other words. This helps prevent the outsiders' germs from spreading while strengthening the social networks that can help minimize an epidemic's impact.[70]

Yet a reading of the Torah suggests that such benefits are actually just a religious side effect. There is no doubt that the Torah serves to protect against disease—that is our main thesis, after all—but the system it presents is much more encompassing than other systems that simply protect the in-group by isolating it from others. Indeed, we can turn Murdock's remarks around: for the most part, the form of religion we find in the Torah is actually "applied medicine."

It is hard to overlook the fact that one of the most fundamental dualisms among religions the world over—that of "clean" versus "unclean"—possesses an overriding hygienic dimension. In the Torah, cleanliness is considered a "basic prerequisite for communicating with the divine world"[71]—just as it is in other religions too.[72] Only clean worshippers may approach God; those who are not may not and will be excluded from the religious community. Theological encyclopedias take a similar view: "Generally speaking, purity is associated with health, life, and (cosmic as well as social) order, whereas impurity is associated with illness, death, and (cosmic as well as social) chaos."[73]

The link is obvious: since disease is God's punishment, diseased persons are therefore unclean and outside God's grace. God's displeasure is contagious. If a person who has been in close contact with the diseased also falls ill, then God has punished him or her for associating with someone who has fallen from grace. As logical as this line of reasoning may be, people have difficulty adhering to it, for it is directly opposed to the empathetic impulses of our first nature. We're supposed to take care of the sick, after all! This, too, is a mismatch problem: infectious diseases were the exception among hunter-gatherers. We shall return to this point below.

Most of what the Torah deems "unclean" (bodily fluids, excrement, carrion) is potentially infectious and produces aversive reactions in people that serve to prevent the spread of disease.[74] Accordingly, lack of cleanliness necessitates elaborate hygienic measures, such as washing and quarantine. If these are not enough to eradicate the disease, then other, further-reaching steps are needed. Theologians put it in a more roundabout way: "If recovery fails to materialize (as is the case with leprosy), then purification rites can also have an eliminatory function in that they ensure that unclean elements are permanently excluded from the community."[75] God, however, does not mince words. After listing all the unclean bodily fluids, he warns, "Thus shall ye separate the children of Israel from their uncleanness; that they die not in their uncleanness, when they defile my tabernacle that is among them."[76]

Below we examine the spectrum of strategies developed to prevent the spread of disease. In doing so, we by no means seek to pursue the "medical materialism" criticized by anthropologist Mary Douglas, which posits that "even the most exotic of ancient rites have a sound hygienic basis" and that this principle can explain everything.[77] We are after something much greater. Let us get down to details.

Blood, Sperm, and Other Bodily Fluids

Readers of the Bible as holy scripture might have trouble understanding God's fascination with bodily fluids. For biblical anthropologists like us, the answer is obvious: infectious diseases, especially sexually transmitted diseases (STDs), are a dangerous scourge, and contact with the bodily fluids of infected persons represents one of the most reliable avenues of transmission. Better, then, to err on the side of caution to minimize the risk of infection.[78] Let us eavesdrop on what must be one of the most unusual bits of guy talk in the history of literature. God gives Moses and Aaron something of a birds-and-bees lecture. We cannot resist quoting it in full:

> When any man hath a running issue out of his flesh, because of his issue he is unclean. And this shall be his uncleanness in his issue: whether his

> flesh run with his issue, or his flesh be stopped from his issue, it is his uncleanness. Every bed, whereon he lieth that hath the issue, is unclean: and every thing, whereon he sitteth, shall be unclean. And whosoever toucheth his bed shall wash his clothes, and bathe himself in water, and be unclean until the even. And he that sitteth on any thing whereon he sat that hath the issue shall wash his clothes, and bathe himself in water, and be unclean until the even. And he that toucheth the flesh of him that hath the issue shall wash his clothes, and bathe himself in water, and be unclean until the even. And if he that hath the issue spit upon him that is clean; then he shall wash his clothes, and bathe himself in water, and be unclean until the even. And what saddle soever he rideth upon that hath the issue shall be unclean. And whosoever toucheth any thing that was under him shall be unclean until the even: and he that beareth any of those things shall wash his clothes, and bathe himself in water, and be unclean until the even. And whomsoever he toucheth that hath the issue, and hath not rinsed his hands in water, he shall wash his clothes, and bathe himself in water, and be unclean until the even. And the vessel of earth, that he toucheth which hath the issue, shall be broken: and every vessel of wood shall be rinsed in water.
>
> And when he that hath an issue is cleansed of his issue; then he shall number to himself seven days for his cleansing, and wash his clothes, and bathe his flesh in running water, and shall be clean. And on the eighth day he shall take to him two turtledoves, or two young pigeons, and come before the LORD unto the door of the tabernacle of the congregation, and give them unto the priest: And the priest shall offer them, the one for a sin offering, and the other for a burnt offering; and the priest shall make an atonement for him before the LORD for his issue.
>
> And if any man's seed of copulation go out from him, then he shall wash all his flesh in water, and be unclean until the even. And every garment, and every skin, whereon is the seed of copulation, shall be washed with water, and be unclean until the even. The woman also with whom man shall lie with seed of copulation, they shall both bathe themselves in water, and be unclean until the even.[79]

Let us recapitulate: first infectious discharge (probably a case of gonorrhea[80]), then nocturnal emission, and finally sexual intercourse. From a modern point of view, the details of the diagnosis and therapeutic steps correspond to the medical severity of the symptoms described. The last two cases are generally unproblematic, but sperm is a potentially infectious bodily fluid that can contain viruses even weeks after an infection. Such overgeneralizations make good sense. Better safe than sorry! Since one cannot be sure how long the risk of infection lasts, it's better to be really cautious.

The Torah's authors were aware that bodily fluids are dangerous, even if they had no idea what made them so. Whenever they are involved, God seems exceptionally prone to meting out punishments. So you had better beware! By the same logic, all women's bodily fluids were under general suspicion. Even the birth of a child made a woman unclean—and the birth of a daughter made her unclean for twice as long as that of a son (sixty-six versus thirty-three days). Mere menstruation made her unclean for seven days (and anyone touching a menstruating woman remained unclean until the evening). Most importantly, any unusual instances of bleeding required additional reparations—an indication that the health risk posed was deemed greater than that of regular menstruation.

Whatever the case, disinfection was required. People suspected that the "unclean" or disease-causing effect was somehow connected to the material itself. The stipulations for dealing with a menstruating woman are as follows: "And whosoever toucheth her bed shall wash his clothes, and bathe himself in water, and be unclean until the even. And whosoever toucheth any thing that she sat upon shall wash his clothes, and bathe himself in water, and be unclean until the even. And if it be on her bed, or on any thing whereon she sitteth, when he toucheth it, he shall be unclean until the even. And if any man lie with her at all, and her flowers be upon him, he shall be unclean seven days; and all the bed whereon he lieth shall be unclean."[81] A modern hospital's hygienic instruction manual could hardly be more detailed.

It is important to stress that additional conscious or unconscious motivations may also have inspired these rules. An ethnographic comparison suggests that menstruating women might have been declared unclean for entirely different reasons. Among the Dogon of Mali in West Africa, for example, the "ideology of menstrual pollution as the supernatural enforcement mechanism to coerce women" is instrumentalized to make women "disclose their menses by going to the menstrual hut." The result is an increase in "coital frequency [with the spouse] around the time of ovulation," which in turn helps ensure the husband's paternity.[82] The rules described in the Torah almost certainly helped to minimize the number of children a woman fathered by men other than her husband—an altogether unsurprising effect, given its patriarchal intentions.

Hygiene as a form of disease prevention is a comprehensive theme of the Old Testament. There are even hygienic instructions pertaining to warfare:

> When the host goeth forth against thine enemies, then keep thee from every wicked thing. If there be among you any man, that is not clean by reason of uncleanness that chanceth him by night, then shall he go abroad out of the camp, he shall not come within the camp: But it shall be, when evening cometh on, he shall wash himself with water: and when the sun is down, he shall come into the camp again. Thou shalt have a place also without the camp, whither thou shalt go forth abroad: And thou shalt have a paddle upon thy weapon; and it shall be, when thou wilt ease thyself abroad, thou shalt dig therewith, and shalt turn back and cover that which cometh from thee: For the LORD thy God walketh in the midst of thy camp, to deliver thee, and to give up thine enemies before thee; therefore shall thy camp be holy: that he see no unclean thing in thee, and turn away from thee.[83]

Biblical scholars have had such a difficult time interpreting this passage that some have suggested God himself drew up a "principle according to which everyone is responsible for getting rid of their own waste."[84] God as the first environmental activist?

We believe the following alternative is more plausible. The army camps of antiquity were epidemics waiting to happen: "Armies are crowds of men who eat, sleep and work in close contact. Such conditions are ideal for the spread of disease."[85] The Bible offers an exemplary case in the second book of Kings, which describes an incident when the army of King Sennacherib besieged Jerusalem in 701 BCE. "And it came to pass that night, that the angel of the LORD went out, and smote in the camp of the Assyrians an hundred fourscore and five thousand: and when they arose early in the morning, behold, they were all dead corpses."[86] It is a historical fact that the Assyrians were forced to give up the siege of Jerusalem, but the Jewish historian Titus Flavius Josephus (ca. 37/38–100 CE) claims plague, not an angel, was responsible for the deaths (of course, for the Bible's authors, they were one and the same).[87] Even if today's scholars believe that King Hezekiah's tribute payments really saved Jerusalem,[88] the story still illustrates the threat armies faced in those times. Up until the nineteenth century, more soldiers died of illnesses and epidemics in times of war than from the effects of combat.[89] Or, in the words of David Clark, "Only since the twentieth century have improved hygiene and . . . enhanced firepower allowed humans to kill more people than microbes have."[90] This should certainly help readers understand why it was utterly sensible to enact protective measures on the slightest suspicion.

The Torah's instructions regarding the burial of human excrement outside the camp similarly aim to regulate hygiene. Jared Diamond has pointed out that "hunter-gatherers frequently shift camp and leave behind their own piles of feces with accumulated microbes and worm larvae. But farmers are sedentary and live amid their own sewage, thus providing microbes with a short path from one person's body into another's drinking water."[91] Naturally our inborn aversion to excrement ensured a certain degree of sensibility, but the historically recent phenomenon of military camps made it necessary to formulate explicit rules. And because diseases were God's punishment, we encounter him in the Torah as obsessed with digging latrines—which actually greatly improves the protection of the soldiers. The diagnosis doesn't have to be accurate—the effectiveness of the treatment is what matters.

Homosexuality and Bestiality

When it comes to disease prevention, Leviticus finds two other commandments so important that it mentions them twice.[92] These concern homosexuality and bestiality, which are listed one after the other. Here we cite the second passage: "If a man also lie with mankind, as he lieth with a woman, both of them have committed an abomination: they shall surely be put to death; their blood shall be upon them." And "if a man lie with a beast, he shall surely be put to death: and ye shall slay the beast. And if a woman approach unto any beast, and lie down thereto, thou shalt kill the woman, and the beast: they shall surely be put to death; their blood shall be upon them."

We have already seen the important role played by measures to help prevent the spread of sexually transmitted disease and the great extent to which bodily fluids—first and foremost blood and semen—were believed to pose a significant threat. We have also seen that even a simple ejaculation could arouse God's anger. It comes as no surprise, then, that these two sexual practices drew the protomedical attention of the Bible's authors because both increase the risk of contracting and spreading STDs.

Bestiality offers a path for viruses and other microorganisms to make the jump from animal to human hosts (so-called zoonosis). The Torah even forbids bestiality a total of three times—a clear sign that this was not uncommon behavior in biblical times.[93] Why else bother with all the prohibitions? And as research into the spread of AIDS has shown, promiscuous homosexual men rank among the "superspreaders" of STDs, even if we do not know today how promiscuous individual homosexuals were in biblical times.[94]

People must have realized quite early on that promiscuous sexual practices and those likely to produce minor physical damage were associated with a greater risk of infection. According to the logic of the time, this was no uncertain proof that God had a problem with such behavior and thus punished it with illnesses. Interestingly, however, our first nature seems to have no difficulty accepting either bestiality or homosexuality; otherwise there would have been no need for such laws, and they certainly wouldn't have needed repeating.

This makes sense. Bestiality requires the presence of domesticated animals, and homosexuality only poses a danger when it helps spread STDs. As a result, we have no innate aversion to either practice. Ironically, those vilifying these practices renounce them as "unnatural." In fact, religion has etched these aversions so deeply into second nature that fundamentalists regard them as serious violations of divine order and tend to become highly emotional about them. This vehemence shows how our age-old psychological circuitry, which helps avoid disease by eliciting disgust reactions, can be diverted into new uses. This means that homophobia is largely a cultural reaction. Below we discuss another reason the Torah may have had for forbidding bestiality and homosexuality.

Leprosy

Leprous skin can also ooze bodily fluids, and this made it another of the Torah's favorite themes. Leprosy, *Mycobacterium leprae*, itself was probably first introduced into the Near East from Asia during the Hellenic era, so the word in those days most likely stood for a variety of diseases that "resulted in changes to the skin as well as hair loss (eczema, scabies, mange, psoriasis, necrosis, macula)."[95] We hardly need mention that leprosy was considered unclean. When Miriam revolted against her brother Moses, God disciplined her, and she was made "leprous, white as snow."

If you read the Bible's pages of observations regarding leprosy, you might think you were holding a medical textbook in your hands:

> And the LORD spake unto Moses and Aaron, saying, When a man shall have in the skin of his flesh a rising, a scab, or bright spot, and it be in the skin of his flesh like the plague of leprosy; then he shall be brought unto Aaron the priest, or unto one of his sons the priests: And the priest shall look on the plague in the skin of the flesh: and when the hair in the plague is turned white, and the plague in sight be deeper than the skin of his flesh, it is a plague of leprosy: and the priest shall look on him, and pronounce him unclean.
>
> If the bright spot be white in the skin of his flesh, and in sight be not deeper than the skin, and the hair thereof be not turned white; then the priest shall shut up him that hath the plague seven days: And the priest shall look on him the seventh day: and, behold, if the plague in his sight be at a stay, and the plague spread not in the skin; then the priest shall shut him up seven days more: And the priest shall look on him again the seventh day: and, behold, if the plague be somewhat dark, and the plague spread not in the skin, the priest shall pronounce him clean: it is but a scab: and he shall wash his clothes, and be clean. But if the scab spread much abroad in the skin, after that he hath been seen of the priest for his cleansing, he shall be seen of the priest again: And if the priest see that, behold, the scab spreadeth in the skin, then the priest shall pronounce him unclean: it is a leprosy.[96]

The Bible offers the diagnosis, and the priest serves as doctor. The whole procedure also produced striking cases of overgeneralization: even crust-forming structural changes to textiles, leather, or walls of houses were interpreted as leprosy and required the services of a priest. But substantial costs did not accrue from such false positives, which is why the system tolerated them. And who knows—perhaps they also helped keep mildew at bay.

Deformities

The Old Testament does not differentiate between illnesses and physical deformities, which are "ubiquitous in texts of the Hebrew Bible."[97] There was simply a lack of relevant knowledge. Any deviation from normality was seen as God's punishment. This was a particularly sensitive issue when it came to priests:

> For whatsoever man he be that hath a blemish, he shall not approach: a blind man, or a lame, or he that hath a flat nose, or any thing superfluous, Or a man that is brokenfooted, or brokenhanded, Or crookbackt, or a dwarf, or that hath a blemish in his eye, or be scurvy, or scabbed, or hath his stones broken; No man that hath a blemish of the seed of Aaron the priest shall come nigh to offer the offerings of the LORD made by fire: he hath a blemish; he shall not come nigh to offer the bread of his God.[98]

Following the same line of reasoning, only flawless animals were acceptable for sacrifice.

In order to prevent a priest from becoming ill (or, in other words, from bringing down the wrath of God on himself and his people), he was only allowed to marry a virgin,[99] for it could be safely assumed that virgins were free from STDs. This exemplifies the amount of intellectual effort put into creating this system.

Corpses

We have discussed the dangerous nature of corpses in an earlier chapter. It therefore comes as no surprise that biblical health inspectors had a great deal to say about them as well: "He that toucheth the dead body of any man shall be unclean seven days." That person must then cleanse himself with "water of separation" brewed according to an extremely complicated recipe, which begins with the commandment, "Bring thee a red heifer without spot." If the delinquent was unable to meet these demands, his soul was to be "cut off from Israel."[100] For similar reasons, carrion was also taboo.

Strangers

Researchers like psychologist Mark Schaller believe xenophobia is a part of our "behavioral immune system"[101] aimed at keeping foreign germs at bay. When viewed in this light, it would seem God only has his people's best interest at heart when he forbids them from having contact with foreign peoples. The Bible offers a dramatic example of this: "And Israel abode in Shittim, and the people began to commit whoredom with the daughters of Moab. And they called the people unto the sacrifices of their gods:

and the people did eat, and bowed down to their gods. And Israel joined himself unto Baalpeor: and the anger of the LORD was kindled against Israel." A plague would claim 24,000 lives.[102] The reality behind these events may have been much more banal. Variations in bacterial flora and immune systems mean that a meal or sexual intercourse with strangers can more quickly lead to infection. These were seen as punishments. But why? Because God reacted the same way we would expect of our fellow humans: he was jealous! He even admitted as much in the Ten Commandments: "For I the LORD thy God am a jealous God." He simply hated it when his people got mixed up with foreigners and their gods, so he punished them with diseases.

Food

Kashruth, the Jewish dietary laws, obviously forms part of our hygiene discourse as well. The simple act of eating poses a constant health risk. Accordingly, the Torah lists detailed provisions concerning food preparation, offering the final component of its disease-prevention system. We deal with them in detail later, because their explanation requires that we also address an additional aspect.

A Kind of Protomedicine

In light of all this evidence, anyone still doubting the Torah's protomedical function faces the onus of offering a convincing alternative explanation for why God took such an intense interest in things such as leprosy, feces, bodily fluids, sex, dietary habits, and interpersonal relations.

The term "function" is somewhat misleading, for here we are dealing not with a system devised to meet a specific purpose but rather with a side effect arising from our own innate psychology's tendency to seek out agency. When it comes to illness there is always a dependable suspect: some apparently disgruntled demon, spirit, or god. But as soon as we are dealing with the one and only God, we can be sure sickness is punishment for a specific transgression. This allows us to pinpoint the cause. The laws help us develop the appropriate rules of thumb after we have observed the situation surrounding the illness.

For effective prevention, it really doesn't matter whether the cause is God or gonorrhea. It is far more important to identify and regulate potentially infectious situations. Biblical proscriptions were definitely effective, even if they involved such simple steps as quarantine, washing, and avoiding or removing sources of infection. These measures all helped to lower the number of infections. We are therefore dealing here with a form of adaptive protomedicine that introduced survival advantages into a highly sensitive domain.

We must be careful, however, not to project our modern concepts onto historical phenomena. Today we view illness as a phenomenon unto itself, but this was not the case in the past. In Hebrew there is no standalone term for illness, which was always described in other terms that were meant to imply the weakness and therefore often included action words and phrases such as "touch," "strike," "strike down," and "suffer from."[103] To us this clearly indicates that illnesses were originally viewed as the actions of a supernatural actor. They could not have been conceived in any other way—just like there is no blow without someone doing the striking. An additional finding supports this idea. The "word רבד *dævær* 'plague/pestilence/epidemic'" is often used to describe illnesses.[104] Interestingly, *dævær* was once the name of the Canaanite god of pestilence who was later merged with the god Yahweh.[105] In biblical times, diseases were nothing other than the pains or wounds resulting from God's punitive actions. In this sense, the biblical studies of God's agency and of healing—theology and medicine—are one and the same.

The Third Set of Laws: Securing the System

We now turn our attention to a series of laws that today's Bible readers have most difficulty digesting. American author A. J. Jacobs, who once attempted to live by all of the Old Testament's commandments for a year, was utterly floored by all of these proscriptions and asked, "How can these ethically advanced rules and these bizarre decrees be found in the same book? And not just the same book. Sometimes the same page. The prohibition against mixing wool and linen comes right after the command to love thy neighbor." And it really is there in the Bible: "Neither shall a garment mingled of linen and woollen come upon thee," it declares.[106] Why does God find blended fabrics so objectionable?

The Torah presents a whole series of such rules paying homage to a highly obsessive sense of order: "Thou shalt not sow thy vineyard with divers seeds" and "Thou shalt not plow with an ox and an ass together."[107] Or how about the following decree, which sounds particularly disconcerting to modern ears: "A bastard shall not enter into the congregation of the

LORD; even to his tenth generation shall he not enter into the congregation of the LORD."[108]

Until now we have only examined the laws created to respond to existing problems and those that sought to rein in the mismatch problems of violence and disease. Even there we noticed a difference. In the first set of laws, the problem was easily recognizable (pauperization), as was the cause (usury). This made it possible to adopt effective measures to target the problem (a ban on usury). The causal relationship is clear to everyone. God is only needed to add legitimacy to the laws and make sure they are followed.

The situation was different for the second set of laws. Medical symptoms (such as discharge from the penis) were identified as a threat, but the real cause of the illness remained a mystery. No one knew anything about the existence of *Neisseria gonorrhoeae* or any other bacteria for that matter. Nevertheless, these laws had the desired effect (decreased infections) because they were based on the identification of dangerous behavior (sexual intercourse). In this instance, God served as a kind of heuristic. He was the variable that bridged the gap in knowledge and led to the adoption of hygienic measures.

Things are totally different with the third set of laws. Here we face the curious fact that in reality these laws do not address any problems. Still the Torah needs these "laws without a cause." They save the divine logic from inconsistencies and so maintain the protection system's stability.

The Relentless Logic of the System

Perhaps today God moves in mysterious ways, but he didn't in the Torah. Quite the opposite, in fact. The foundation of this belief system is a completely unmysterious God. "His motives were not concealed from human comprehension." He is a "god of just retribution." He doesn't punish people on a whim but has his reasons for doing so.[109] The logic of his laws rests on this supposition. Even God himself must follow it: he must punish transgressions. Remember that he couldn't even make an exception for Moses. The reverse is true as well: wherever there is punishment, whatever the form of the calamity, there must have been a breach of divine law. Anything else would call the system into question.

In those days catastrophes were certainly not in short supply. Epidemics, floods, social injustices, and wars were commonplace. And people perceived these as God's punishments. His arsenal of disciplinary measures appeared inexhaustible. Here is a list of punishments for disobedience threatened in a single, not too lengthy passage from Leviticus: terror, consumption, fever, blindness, withering, loss and theft of harvests, military defeat, foreign rule, paranoia, iron skies, brass earth, infertile land, fruitless trees, children and livestock consumed by wild animals, desolate highways, the sword of vengeance over the people, pestilence, delivery into the hands of the enemy, ruined bread, hunger, cannibalism of sons and daughters, destruction of the high places, death, cities laid to waste, desolation of the sanctuaries, desolation of the land, scattering among the heathen, fleeing from the sword, panicked fear for the survivors, the killing of one another, death or disappearance to everyone.[110] Most of these terrors were a normal part of historical experience in ancient Israel. At any rate, according to the system's relentless logic, human transgression must have been behind them, for God didn't punish without reason. People believed they were all under general suspicion and must have kept desperate watch for potential misdeeds in order to atone for them and forbid them in the future.

This systematic obsessive-compulsiveness had two consequences. On the one hand it led to the expansion of the existing body of laws. Perhaps the provisions weren't precise or comprehensive enough? So they had to be improved upon. We find a whole series of laws that can be viewed as extensions of existing laws based on this internal momentum. The Sabbath, for instance, should not only apply to people but also to animals. Not only the Israelites must be circumcised, but so must their slaves. We have already encountered this phenomenon: it's called overgeneralization.

This compulsive coherence seeking also led to a new type of law. If calamity struck, in whatever form, the priests were put on the spot to present the reasons for God's displeasure to the people. In earlier times, priests would have resorted to practices such as evoking the spirits or consulting oracles, but these only came into play after calamity had struck. The Torah's strategy was prevention. It sought at all costs to avoid God's anger in the first place. In order to avert catastrophe, the priests had to reconstruct his will. They had to map out every last corner of the divine order to preclude violations.

Don't Mix Things Up

This is where the antimixing laws come into play. Not only is it forbidden to wear garments consisting of two types of thread, but God also commands, "Thou shalt not let thy cattle gender with a diverse kind: thou shalt not sow thy field with mingled seed."[111] We can describe this idea as "keeping separate things apart" or "not mixing different things."[112] God made the world the way he did for a reason, so mankind shouldn't mess with the order of creation. That can only lead to catastrophe.

We can comprehend some of the laws we have already examined from this perspective, such as the prohibitions against homosexuality and bestiality. Disease prevention certainly played a fundamental role in their creation. But we can also interpret these acts as violations of divine order. Accordingly, the Torah also forbids men from wearing women's clothing and women from wearing men's.[113]

Here we see how culture arises from a system of moral concepts that humanity imposes on the world, in the words of ethnologist Clifford Geertz, in "webs of significance that he himself has spun."[114] The starting point consists of the real experiences of illness and violence, drought and flood. Our first nature interprets these as God's punishments, whereas our third nature develops the appropriate rules of behavior. Some of these rules have an effect; others don't. They are constantly revised and reworked until a system of order imposed upon reality[115] emerges that divides all things into the categories of pleasing and displeasing to God, clean and unclean. This system offers guidance, provides security, and stabilizes society. From here on there are no more neutral activities. Passed down from generation to generation, this body of laws becomes incorporated into our second nature. Any violation of them inspires fear, for God might punish it.

It's Not Just About Pork

We have now assembled the concepts needed to tackle the Torah's laws governing food. These laws play a prominent role in the reception of the Bible. Nevertheless, the reasons for their existence have never been fully explicated. This is mainly because they are often analyzed individually in search of monocausal explanations.

The desire to keep things clearly separated also influences the division of animals into the categories of kosher and nonkosher, clean and unclean: "These are the beasts which ye shall eat among all the beasts that are on the earth. Whatsoever parteth the hoof, and is clovenfooted, and cheweth the cud, among the beasts, that shall ye eat."[116] Camels, badgers, hares, and pigs are thus deemed unclean, because they do not fit all criteria. The division then extends to all sorts of birds and reptiles. Even in the ancient world, Jewish communities "sought reasons for the incomprehensible taxonomy of the dietary laws," not least because these were major reasons the Greeks and Romans viewed them as "barbarous."[117] Why for the love of God were eels, storks, and swallows categorized as unclean? Let us look briefly at some of the myriad reasons offered in the literature.

In the first century CE, the Jewish scholar Philo of Alexandria proposed that these laws were intended to prevent "gluttony," "an evil dangerous to both soul and body, for gluttony begets indigestion, which is the source of all illness and infirmities."[118] Moses Maimonides (1135–1204), a Jewish philosopher who was also a medical doctor, believed there was a physiological basis to the dietary laws: "I maintain that the food which is forbidden by the Law is unwholesome. There is nothing among the forbidden kinds of food whose injurious character is doubted, except pork, and fat. But also in these cases the doubt is not justified. For pork contains more moisture than necessary (for human food), and too much of superfluous matter. The principal reason why the Law forbids swine's flesh is to be found in the circumstance that its habit and its food are very dirty and loathsome."[119]

Such arguments became increasingly popular after a medical connection was established between undercooked pork and trichinellosis caused by roundworms in 1859. From then on, this was the "most popular explanation of the Jewish and Islamic pork taboo," explains anthropologist Marvin Harris. Popular or not, it is still incorrect, he argues: "There is absolutely nothing exceptional about pork as a source of human disease. All domestic animals are potentially hazardous to human health." Harris believes ecological changes played a much greater role in the ban, because pigs were a source of meat only. When environmental changes made pig farming more difficult, "the creature became not only useless, but worse than useless—harmful, a curse to touch or merely to see—a pariah animal."[120]

Excavations in the mountains west of the Jordan have never uncovered pig bones, whereas evidence suggests the neighboring Philistines, Ammonites, and Moabites did indeed eat pork. Israel Finkelstein and Neil Silberman have therefore put forth an alternative interpretation: "Perhaps the proto-Israelites stopped eating pork merely because the surrounding peoples—their adversaries—did eat it, and they had begun to see themselves as different. Distinctive culinary practices and dietary customs are two of the ways in which ethnic boundaries are formed."[121]

Other scholars have also suggested that dietary laws were actually nothing more than disciplinary measures. They believe the separation was "irrational and purely arbitrary, for precisely this arbitrariness can function as an obedience test for believers."[122] Evolutionary biologists would call these "costly signals": those willing to forgo tasty pork demonstrated that they were serious about their religion.

Finally, in *Purity and Danger* ethnologist Mary Douglas presented the thesis that the separation between clean and unclean animals reflected an attempt to maintain order. Pigs and camels were considered unclean because they did not fit with "the defining characters of livestock"—namely, being cloven-hoofed ruminants. The division of the remaining animals follows the order of creation we encounter in Genesis. "Here a three-fold classification unfolds, divided between the earth, the waters and the firmament. Leviticus takes up this scheme and allots to each element its proper kind of animal life. In the firmament two-legged fowls fly with wings. In the water scaly fish swim with fins. On the earth four-legged animals hop, jump or walk. Any class of creatures which is not equipped for the right kind of locomotion in its element is contrary to holiness." Thus, everything in the water that has no fins and no scales is considered unclean. The same is true of animals that can fly and have more than two legs (such as insects, with the exception of some locusts).[123]

The question remains, however, as to why God employed differing categories. Pigs and camels move over the earth in the correct manner, after all. The same is true of some species of birds that are considered unclean even though they fly through the air and don't have too many legs. And come to think of it, why did God create animals that fail to fit into his own system of order in the first place?

Douglas's argument is nevertheless important. She rightfully stresses the desire for order inherent in this zoological taxonomy. The Torah's authors clearly took issue with anything that presented itself as an "anomaly," that did not fit into existing groups or categories. Ambiguity was a serious problem. For Douglas, this is why anything "out of place" is by definition unclean.[124] Without a doubt, anomalies call the existing order into question; they appear as violations of God's creation and are therefore potentially dangerous. There must be a clear separation between clean and unclean—and by extension right and wrong or good and evil. Ambiguities pose a threat. When actions cannot be unequivocally determined as one or the other, God, too, could misinterpret the situation and in his haste prematurely reach for the lash.

When Slaughtering Was Made Permissible

Let us broaden the discussion. There is no doubt that medical prevention plays a role in dietary laws as a whole. Food has always been a gateway for disease—even our primate ancestors knew exactly what was and was not edible. As *Homo sapiens* was much more flexible when it came to nutrition and no longer had to rely exclusively on innate preferences, our ancestors were able to colonize even the most unusual environments. For this reason it was of the utmost importance that we be able to organize and disseminate empirical knowledge concerning what we could and could not eat. It was a fine line to tread: every newly discovered source of nutrition could have offered an evolutionary advantage, but every unfamiliar bite could equally likely have been the very last. Even today, people die when they mistakenly mix poisonous mushrooms with edible ones or prepare their homemade pesto with lily of the valley instead of wild garlic. To this day there is hardly an aspect of life that people are more sensitive to than food. Little wonder, then, that we encounter all of these stipulations in the Torah. It is the old line of reasoning: those areas of life involving disease and death appear to be under God's particular supervision.

But a full understanding of the origin of biblical dietary laws requires that we also take the historical circumstances into account. In ancient Israel, grains and pulses were the main staple foods. Then came fruit and vegetables. "Meat played practically no role whatsoever in daily life, but was only eaten on specific occasions and under the observance of strict ritual prescriptions." Animals could only be slaughtered upon an "altar built according to religious specifications, that is to say, within a temple area supervised by priests and according to well-defined rules."[125] The priests took exceptional care to ensure that only flawless animals were sacrificed and eaten.

Beginning in the seventh century BCE, the sacrificial sites were centralized, and from then on sacrifices to Yahweh could officially only be made in the temple at Jerusalem. Local temples were to be abandoned, and offerings could no longer be made at them. But this also meant that animals could no longer be slaughtered. As it would have been a step too far to ask people to go without meat altogether, religious authorities soon allowed animals to be slaughtered outside the temple. "It is not accidental," states Frank Crüsemann in his analysis of the Torah, "that we find the first list of animals that should not be eaten precisely at the point where the killing of animals is separated from the cult."[126]

The profanation of an activity that was once the exclusive right of priests is an unusual undertaking. The authorities had to determine precisely which animals could be slaughtered, so the priests drew up lists of the animals that could be eaten without angering God. They began by taking note of common practice, behavior that experience had shown God had no issue with. This explains why the dietary laws reflected customs that had evolved over a long period. "Today zooarchaeology can perfectly show how animals domesticated in the Stone Age—goats, sheep, and cattle—were the animals that were most commonly eaten in the Near East and that wild ruminants with cloven hooves such as gazelles and antelope were among the most important hunted game," explains Swiss theologian Thomas Staubli. Camels, on the other hand, were uninteresting sources of meat due to their low reproductive rates. And pigs were presumably quite scarce in the rather barren hills of the region and also were not kept because they did not produce any secondary products such as milk.[127]

But the priests did not stop at the tried-and-true. In order to avoid any eventualities, they expanded the list of animals in an almost "scientific manner."[128] Exactly which animals made it onto the list of acceptable foods and why is beside the point. It makes just as little sense to seek out monocausal explanations. This is the case with the blood taboo, in which a number of different motifs may have been combined: "Ye shall eat the blood of no manner of flesh: for the life of all flesh is the blood thereof: whosoever eateth it shall be cut off."[129] We have already examined how human blood was problematic from a protomedical point of view. On a symbolic level, blood was the essence of life. During cultic sacrifices, priests ritually butchered animals to cleanly separate life from death.[130] Whereas the meat was consumed communally, the blood was reserved for God. It would have been seen as sacrilege if everyone suddenly could have it.

When it comes to the dietary laws, it is not important to identify a single line of reasoning. What's important is the will to systemization. It is about reconstructing God's logic and distinguishing between those things that please him and those that anger him, then doing the former and prohibiting the latter.

Ethnic Marking

Constructing order implies drawing boundaries. If I am allowed to eat only "clean" animals, then I may not share a table with those who eat "unclean" animals. This reinforces my own identity and is a consequence of the Torah. The Torah had not yet been completed at the time of the Babylonian captivity, but its authors were working under pressure to finish it as quickly as possible. Because the Israelites had lost their own state and temple cult, they now needed "portable" markers of identity. Regulations concerning food and cleanliness and observing the Sabbath were of great importance in exile.[131] They held the community together while it was abroad and signaled its unyielding loyalty to Yahweh. Additionally, these rules functioned as costly signals. They showed who was willing to remain true to his religion, even at great cost, and identified him as a dependable collaboration partner. Accordingly, some of the Torah's laws aimed at recognizability by means of "ethnic marking": "Ye shall not round the corners of your heads, neither shalt thou mar the corners of thy beard"[132] and "Thou shalt make thee fringes upon the four quarters of thy vesture, wherewith thou coverest thyself."[133] Such measures helped prevent assimilation and, from a religious point of view, were of the utmost necessity. The foreign gods of Babylon "had a great deal of appeal," and some of the exiles even named their children after them and placed their businesses under their protection.[134] But Yahweh was a jealous god.

Another aspect of this ethnic marking was the practice of male circumcision. Even to this day its medical value remains controversial. In the conditions prevailing in the ancient Near East, the operation certainly posed a significant risk, which may have supported its use as a costly signal. At any rate, circumcision was an "ancient rite" practiced in Palestine before people began to worship Yahweh.[135] It was also practiced among some of the neighboring cultures, although the Israelites differed by performing the circumcision upon infants: "And in the eighth day the flesh of his foreskin shall be circumcised."[136] Incidentally, the Assyrians, Babylonians, and Philistines did not perform circumcision at all. Thus, during the Babylonian captivity the practice took on an "increased theological and practical significance," as it signaled a demarcation line.[137] Similarly, circumcision was not practiced in the Jewish military colony on the Nile's Elephantine Island.[138] In Egypt, where all males were circumcised, it evidently could not serve as an identity marker.

The Invisible Order

Let us summarize: In this third category of laws we find rules derived from the assumption that God's order permeates the entire world. The antimixing laws and the classification of animals represent attempts at reconstructing this system, for by adhering to this order, the Israelites expected to avoid God's punishments. In addition, the priests could use these laws to extract possible transgressions that could explain why Israel was once again visited by calamity. In doing so they preserve the divine order and make it irrefutable. A marked desire to systematically decipher and categorize reality thus already existed well before the birth of science. Here, too, we see that religion functioned as a form of protoscience tasked with protecting society.

The Fourth Set of Laws: Protection from the Unknown

So far we have observed the great effort the mostly priestly lawmakers invested in defining the rules to promote behavior pleasing to God. But even when these laws were applied, misfortune did not disappear completely. What had gone wrong? The Torah contains a series of laws designed to address precisely this problem. They plug the holes in the catastrophe-prevention system. What, for instance, should they do when unable to uncover a crime or identify its culprit? This was a grave problem, for the sins of the individual affected the entire community. Everything had to be done to bring the perpetrator to justice and expiate the guilt. This was the only way to avoid God's judgment, for there was little use in pleading, "O God, the God of the spirits of all flesh, shall one man sin, and wilt thou be wroth with all the congregation?"[139] Experience had shown that the answer was yes. God was unforgiving when it came to transgression.

Thus no sin could go unrecognized. The priests turned to strategies they might otherwise have avoided. They applied magic, for instance, in cases where a man was overcome by the "spirit of jealousy" and suspected his wife of infidelity, although she vehemently denied it. This situation they could not ignore. If the woman was guilty, that meant she was unclean—an affront to God. So the priests invoked the institution of the "ordeal," the judgment of God. They brought forth holy water and mixed it with dust from the floor of the tabernacle to make a "bitter water that causeth the curse." Then they made the suspect drink the brew. "And when he hath made her to drink the water, then it shall come to pass, that, if she be defiled, and have done trespass against her husband, that the water that causeth the curse shall enter into her, and become bitter, and her belly shall swell, and her thigh shall rot: and the woman shall be a curse among her people. And if the woman be not defiled, but be clean; then she shall be free, and shall conceive seed."[140] Incidentally, the husband whose jealousy was enough to subject his wife to such a life-threatening procedure went unpunished, even if the woman was found innocent.

Another similarly precarious issue arose when a murder was committed but the perpetrator could not be found. The procedure in this case was as follows: The judges determined which town was nearest to the scene of the crime. A heifer was taken from that town and led into an uncultivated valley. There, the elders must strike off its head. "And all the elders of that city, that are next unto the slain man, shall wash their hands over the heifer that is beheaded in the valley: And they shall answer and say, Our hands have not shed this blood, neither have our eyes seen it. Be merciful, O LORD, unto thy people Israel, whom thou hast redeemed, and lay not innocent blood unto thy people of Israel's charge. And the blood shall be forgiven them."[141] This was enough to ward off calamity—or so they hoped.

And what if a commandment had been broken by accident? The Torah shows that all of the continual reflection on questions of guilt brought with it judicial progress. Its authors believed that God only flew into a rage when his laws were broken intentionally. When the infraction was inadvertent, a sacrifice sufficed to avert his wrath.[142]

Finally, as priests were constantly searching for the causes of God's judgment, one thought weighed heavily on their minds: Could it be that God also punished undetected crimes? If so, how could they avoid castigation? A scapegoat was needed! Indeed, the Torah initiated the institution known as the Day of Atonement (the origins of the Jewish holiday of Yom Kippur). On this day, two goats were brought forth. One was sacrificed to God, and the other was sent to Azazel (most likely a demon) in the desert. But before this was done, "Aaron shall lay both his hands upon the head of the live goat, and confess over him all the iniquities of the children of Israel, and all their transgressions in all their sins, putting them upon the head of the goat, and shall send him away by the hand of a fit man into the wilderness: And the goat shall bear upon him all their iniquities unto a land not inhabited: and he shall let go the goat in the wilderness." This ritual freed the Israelites of their sins, even those that remained undetected. When the Torah states, "And this shall be an everlasting statute unto you, to make an atonement for the children of Israel for all their sins once a year,"[143] it is actually calling for "an annual, complete cleansing of the entire nation."[144]

The Last Set of Laws: The Proliferation of Sacrifices and Cult Practice

Our examination of the Torah would not be complete if we ignored the enormous abundance of cultic regulations it describes. Rules for priests and sacrifices run for pages. We will try to be brief and concentrate on the aspects of these regulations that are relevant to our thesis. Sacrifices represent an ancient means of interacting with supernatural actors. Edward Tylor remarked on their gift-like character ("sacrifice is a gift made to a deity as if he were a man").[145] In *The Gift*, Marcel Mauss showed that among people every gift is given in expectation of a gift in return, and the same logic of reciprocity extends to humanity's interactions with the gods.[146] The goal is to keep the relationship balanced.

The catastrophe-protection system known as the Torah ensures this sacrificial balance. In general terms, sacrifice soothes an enraged God. His anger always presupposes human guilt, which means that a deficit must first be neutralized if one is to reestablish a normal relationship to the supernatural. The sacrifice enables people to pay their debt and restore equilibrium. Here, too, reciprocity is the watchword: the greater the guilt, the greater the sacrifice must be. The same principle can also be reversed. People can offer sacrifices in expectation that God will owe them something in return—the principle of *do ut des*: I give so that you will give.[147]

Viewed in this light, the practice of sacrifice is an essential complement to the protection system. It can put right a damaged relationship to God brought about by human transgression and can even be applied preventively. But the arrival of the Torah was the beginning of the end for sacrifice, for obligation to the law was intended to prevent all transgressions, making sacrifice superfluous.

The sheer variety of sacrificial practices is quite remarkable: the Torah differentiates between burnt offerings, food offerings, thanks offerings, sin offerings, and guilt offerings. But what is truly astounding is the complexity of the rules surrounding the cult. The priests' areas of responsibility are precisely defined, as is their clothing ("And he made the ephod of gold, blue, and purple, and scarlet, and fine twined linen"[148]). There are clear guidelines regarding altar style and the form of the tabernacle. Each and every ceremony is regulated down to the very last detail. For example, the blood of the ram must first be applied to the tip of the right ear and on the thumb of the right hand and then on the great toe of the right foot.

On the one hand, this obsession with detail was the product of the same intrinsic momentum that pushes toward ever-greater differentiation. This comes into play when the actions performed do not produce the desired results and thus have to be improved upon. On the other hand, the complexity of many rituals is also an aspect of expert culture, for it allows the priests to prove their expertise. Communicating with God is a truly risky affair, not something to be left to just anyone! Woe to anyone who fails to carry out the ceremonies according to the rules. Even Aaron's sons were killed when they offered God the wrong type of fire. Additionally, all of this complexity offered the priests an excuse if something should go wrong and force God to punish the land despite all of their attempts at atonement. Maybe there was simply a lack of discipline. Was the blood applied to the wrong toe? It had to be the big toe! The one on the right foot!

Finally, Pascal Boyer introduces another factor into the mix and finds within all of these diverse ritual activities a parallel in the behavior of people suffering from obsessive-compulsive disorders who find themselves driven to wash their hands a hundred times a day or to carry out an entire series of senseless activities before they can leave the house, for example. This leads him to hypothesize that our mental contagion system is involved in ritual practices of "cleaning, cleansing, purifying, making a particular space safer, avoiding any contact between what is in that space and the outside." This system is specialized in dealing with "the management of precautions against undetectable hazards" and can elicit feelings of fear and disgust directed toward everything a culture has labeled dangerous. It then instills these feelings in its youngest members. The involvement of these emotions compels people to "perform rituals *in the right way*" for fear of dangers that they cannot see.[149] After all, people have really no idea where a real danger might lurk, which is why they turn to the hyperactive use of ritual, for it is better to perform a ritual one too many than one too few. For us, Boyer's considerations offer one more piece of evidence in support of the idea that health protection plays a key role in this type of institutional religion.

YAHWEH'S ADVANTAGE: PROTOSCIENTIFIC CATASTROPHE PREVENTION

We have now examined all of the Torah's fundamental categories of laws and found no commandments clearly at odds with our hypothesis that the Torah represents a first-class cultural-protection system. As its name indicates, the Torah provides the "instructions" for how to avoid provoking God. The Bible's authors were convinced they had obtained the house rules for living in God's creation. If everyone would only follow them, the world would become a paradise. Indeed, the Talmud later states, "If Israel would keep a single Sabbath in the proper way, the Messiah would come, for keeping to the Sabbath is the same as keeping to all the commandments."[150]

From the perspective of cultural evolution, this leads us to an impressive bottom line: people were confronted with the most dangerous of problems, and although they had no insight into the real causes of these threats, they still developed a functioning system of prevention. God served as a heuristic tool that helped people to find good solutions despite this lack of knowledge. The mechanism is astonishing: a patently false premise (epidemics are not sent by a god) nonetheless initiates the intense observation and analysis of reality that led to procedures whose implementation actually did have a real effect: hygienic measures lowered the risk of contagion, laws governing marriage lowered the incidence of hereditary disease, and social appeals strengthened cooperation. Even if the God who issued these laws might not exist, he still helped a great deal.

The Bible's authors succeeded in large part due to the fact that they were convinced there was only the one God, whose behavior could be reconstructed—at least in principle—just like that of any other person. This would not have been possible in a polytheistic worldview. It is therefore high time that we take a closer look at this monotheistic God responsible for the intellectual masterpiece known as the Torah—if only because some of our readers may be asking themselves if what we are dealing with here can truly be described as a religion. Upon closer examination, it is remarkable how few of the Torah's laws have anything to do with what we today would describe as faith. Theologians have by now concluded that the Torah examined every aspect of reality and all reaches of human life and human experience in the light of Yahweh, but this is simply not the case. Key aspects of human life are missing. When the Torah addresses issues such as fertility, birth, marriage, and pastoring to the sick and dying, these topics are never at the center of the text. But aren't these the classic themes of religion?

In reality, we find ourselves at a crossroads, and here we return to the beginning of our chapter on Moses, when he stood on Mount Nebo. There a new type of religion was born, one that focused not on the spiritual needs of the individual but on protecting society. The particular case was no longer of importance; only the general rule counted. The text mattered, not individuals. This is why the symbolic moment on Mount Nebo was of such importance for world history: the prophet had to die, and in his place came the law. Now let us turn to the figure of Yahweh and investigate his unique career.

8

YAHWEH

The God with Two Lives

WE ARE DEALING WITH A NEW GOD—OR AT LEAST ONE WHO HAS undergone a thorough metamorphosis. The Bible readily admits this. On their first encounter at the foot of Mount Horeb, God speaks to Moses from a burning bush: "I am the God of thy father, the God of Abraham, the God of Isaac, and the God of Jacob." Moses still has to ask him his name, however, whereupon the Lord replies, "I AM THAT I AM."

Theologians have also emphasized this transformation. Whereas in Genesis we encounter a "family- and clan-oriented tutelary deity," we later find ourselves confronted with a "warlike and violent God 'of the people' who demands the rejection—and even the destruction—of other peoples and their gods and who is punitive and violent towards his own people."[1] This sounds a bit like Robert Louis Stevenson's *The Strange Case of Dr. Jekyll and Mr. Hyde,* in which a kindhearted doctor transforms into a violent brute. At any rate, as Harvard biblical scholar James Kugel notes, when Moses approached the mountain, "a new God walked into Israel's life, one who ultimately changed the world's thinking about divinity."[2]

But how did this historic transformation come about? How can a god claim to be the one and only? This moment is often considered the birth of monotheism. That this term is contentious need not bother us here, but we should remember that it is a product of the early English

Enlightenment and thus nearly 2,000 years younger than the phenomenon it claims to describe.[3] We also should not attach too much significance to the Platonically inspired supposition of God's omnipotence, omniscience, and eternal immutability, for this is not the divine being we encounter in the Old Testament. Perhaps this confusion has something to do with the fact that the rise of biblical monotheism has yet to be fully explained.[4]

We now wish to present our own conjecture. It follows from the assumption that catastrophes—everything from epidemics to wars—played a fundamental role in the cultural evolution of religion and transformed the spirits into gods. We believe an overdose of misfortune led to the creation of a magnificent god the likes of which the world had never seen. In order to explain this transformation, we must put the Bible aside for a moment and delve into the environment in which Yahweh first had to prove himself as a cultural product. The historical reality into which he was born was a truly disastrous one.

WHY YAHWEH?

Perhaps the most surprising thing about the God of the Old Testament is that no one seems surprised by his astounding career. How did the "fewest of all people" manage to create a supergod and then, as a people, survive the greatest of calamities through his help? Powerful gods such as Marduk, Amun, and Zeus disappeared in the sands of time, while today more than half the entire world worships the former microstate god known as Yahweh. No one seems to ask just how he managed to make it that far.

It may seem obvious to the Judeo-Christian West—it certainly does from a religious perspective. Plain and simple: God revealed his glory to his prophet Moses. That's all you really need to know. And anyone who doesn't believe this must at least admit that monotheism represents the logical end stage in the development of religion—its purest form—and that alone means its worldwide success was predestined. What could there possibly be left to question?

Today it is increasingly clear to archaeologists and religious scholars that the Old Testament "does not directly reflect Israel's religious history," and the Hebrew Bible does not describe historical reality. Instead it interprets this history, written down a few hundred years later from the perspective of the Judaism of the Persian and Hellenistic eras.[5] The Bible is not a book of history; it is a book of stories. And when God began his career, the circumstances in which he arose were not nearly as unique as the Bible would have us believe.

During the last few decades, archaeologists have uncovered numerous clues suggesting that the religion practiced in the kingdoms of Israel and Judah was hardly unique. Indeed, it scarcely differed from the religions of neighboring states. In the land of Yahweh, too, there was more than one god. People venerated images and sculptures of these gods—and not only in the temple at Jerusalem. Some contemporary theologians go as far as to describe the religion of ancient Israel as a "local variation of north-west Semitic religions."[6] Does this trivialize God? Certainly not—it makes his stellar career even more astonishing. If Yahweh began as an average ancient Near Eastern god, one among many, how did he manage to rise so high? Why not Chemosh or Hadad, the gods of the Moabites and Aramaeans next door? Or what about Assur, Amun, and Marduk, the gods reigning over the empires of Assyria, Egypt, and Babylon? Who in their right minds would have bet that, with such powerful competition, Yahweh owned the future?

GOD WANTED

To understand Yahweh's career, we first have to look around the Canaan of the first millennium BCE, for this is the world in which the Bible was born. Most of these events have been lost to history, but we know that from the tenth and ninth centuries BCE, one encountered here "a colorful mix of urban and regional gods, supplemented with diverse forms of familiar devoutness." When chiefdoms in the hill country west of the Jordan came together to form the small kingdoms of Israel and Judah (only the Bible mentions the great kingdom of David and Solomon), this also had an effect on the pantheon of the region. The two new kingdoms needed an official state god who could offer the religious basis for both king and nation.[7]

They could pick from a long list of candidates. "It is an often observed fact that the name of the people of Israel does not contain the name Yahweh, but the name El as a theophoric element: 'may El (God) reign/prove himself ruler,'" notes Old Testament scholar Otto Kaiser.[8] In the Near East, El was long considered the "god of gods," chairman of the pantheon and wise creator.[9] The fact that the Bible has quite a lot to say about Baal, a weather god from northwest Syria responsible for fertility, makes clear that he was one of Yahweh's main competitors.[10] An Egyptian-inspired sun god was also likely worshipped in Jerusalem.[11] And then, of course, there was Yahweh himself, who originally hailed from the south. In addition, there were several female gods: the remaining goddess cult of Canaan, as well as Asherah, Astarte, and a queen of the heavens.[12] In this patriarchal world, however, it was out of the question that any of these goddesses could contend for the top job.

Let us look at the job description. Palestine found itself between the hammer and anvil of larger empires. For hundreds of years, people lived in "intense anxiety" of merciless conquerors. A constant "pressure of fear" burdened the people and led to "war psychosis."[13] Recall the most significant historical events. First the Aramaeans invaded the kingdom of Israel. Then the Assyrians brought an unprecedented level of destruction to the northern kingdom. In 722 BCE Israel was destroyed. The south, Judah, maneuvered between Assyria and Egypt and was brought to the verge of ruin. The Assyrian king Sennacherib marched on Jerusalem in 701. He laid waste to all of Judah but called off his siege of Jerusalem. From then on Judah was an Assyrian vassal state. When Assyria weakened, King Josiah sought to take advantage of the situation—but with little success. The second book of Kings curtly reports on the events of 609: "In his days Pharaohnechoh king of Egypt went up against the king of Assyria to the river Euphrates: and king Josiah went against him; and he slew him at Megiddo, when he had seen him." Finally, in 587 BCE Nebuchadnezzar of the Neo-Babylonian Empire dissolved Judah and destroyed Jerusalem. The upper and middle classes were marched off into Babylonian captivity. By the time the first of these exiles returned after nearly sixty years, their homeland had become a province of the Persians, who had by then defeated Babylon. And these were merely the military catastrophes of the age. We haven't even mentioned the epidemics, droughts, and suffering refugees.

Only a god of immense stature could withstand this flood of catastrophes, a god whom our first nature could credit with all this misfortune. The endless war and ruin visited upon Israel as an accidental outcome of its geographical location meant that it needed a god of catastrophic temper. The caring God we know today would have been hopelessly out of place in ancient Israel.

Yahweh was the perfect fit for this job description. As a weather and war god, he had the best qualifications among the candidates: rumble, thunder, and brutality he had in abundance. Othmar Keel has pointed out that Yahweh displayed "volcanic attributes" found in no other ancient Near

Eastern god. The mountain of Yahweh smoked like a furnace. In the shape of a pillar of fire, he showed the Israelites the way at night. Mountains melted in his presence.[14] Verily, Yahweh's eruptive personality was capable of anything. And so he became the "god of the nation, the people, and the dynasty" in Israel and Judah.[15]

As a catastrophically inclined god, Yahweh was therefore properly prepared for this particular ecological niche wedged between expansive empires. He perfectly personified the terrors of reality in Israel and Judah. In fact, he thrived on them. As we have seen before, proportionality bias allowed him to burgeon. Just as spirits were transformed into gods by the epidemics that savaged people after the arrival of agriculture, so was Yahweh elevated to a terrible new level by the wars that visited the two kingdoms: "In no other religion in the world do we find a universal deity possessing the unparalleled desire for vengeance manifested by Yahweh," wrote Max Weber.[16]

This not only makes it easier to understand the tremendous rage that defines the God of the Old Testament but also explains his particular tendency to make common cause with the Hebrews' enemies in order to chastise Israel. Yahweh's aggrandizement also arises from compulsive coherence seeking: if defeat is God's punishment, it stands to reason that the great king of Assyria only invaded Israel as an agent of God, for God wished to punish his people. When the Babylonians set Jerusalem ablaze, it was because God used them as his instrument of torture. Even the greatest of kings were mere puppets on Yahweh's string. The Bible clearly states, "O Assyrian, the rod of mine anger, and the staff in their hand is mine indignation. I will send him against an hypocritical nation, and against the people of my wrath will I give him a charge, to take the spoil, and to take the prey, and to tread them down like the mire of the streets."[17] In this world of violence, Yahweh truly was in his element.

TWO LIVES

As logical as this all might be, it also poses something of a paradox. The people who had suffered the most horrendous of defeats imagined the greatest of all gods for themselves. Why don't we find similar processes in other countries facing such a situation? Was it because in those places neither the god nor the people had survived such great horror? But why did they endure it here? The answer, it turns out, is that Yahweh had two lives.

"*One* thing about this Yahweh was unique," explains professor of theology Uwe Becker. "His elevation to the kingdom's official and highest god happened in both Israel *and* Judah. Chemosh in Moab, Hadad in Aram, Milcom in Ammon, Qaus in Edom, but Yahweh in Israel *and* Judah."[18] Yahweh in Israel *and* Judah. He appears to have been leading a double life, a state of affairs that has received little attention to date but whose significance cannot be overestimated.

The fact that Israel and Judah were rivals made the issue particularly delicate. It was not uncommon for the Yahweh of Samaria to ride into war against the Yahweh of Jerusalem.[19] Another delicate issue was that the two kingdoms worshipped Yahweh in different ways. The northern kingdom worshipped him in the form of a young bull in the shrines of Dan and Beit-El (against which the golden calf in the story of Moses polemicized). In the southern kingdom the solar elements of the Yahweh cult were more prominent.[20] In some places the inscriptions indicate that Yahweh also had a paredra, a sacred consort, known as Asherah.[21] This rivalry also helps explain why the proper way of worshipping Yahweh was such an important issue in the Old Testament.

When the Assyrians destroyed the kingdom of Israel in 722 BCE, Judah served as safe haven for many refugees but also for Yahweh. This is a key point, for it meant that Yahweh got a second chance in the south. This was the advantage he had over the neighboring gods of Chemosh, Hadad, and Milcom. Without a doubt, if Yahweh had only been worshipped in the north, the defeat at the hands of the Assyrians would have meant his final end. He would have experienced the same fate as his followers who were marched off to Assyria. These vanished without a trace and now enjoy legendary status as the lost tribes of Israel.

THE FIRST INFERNO: THE ASSYRIAN CATASTROPHE

The Assyrian catastrophe radically transformed Yahweh. The Assyrians, the most powerful military machine of the era, had conquered Israel, and twenty years later they besieged Jerusalem. In this world-changing moment, everything hung in the balance. If King Sennacherib had stormed the city, the world would look quite different today, and there would be no Jews, no Christians, and no Muslims. Most likely only a handful of archaeologists would recognize the name Yahweh. Nevertheless, as already noted, the Assyrians retreated. Whereas historians believe that a last-minute tribute payment saved the day, according to the Bible God sent an angel who killed 185,000 Assyrians in a single night. He must be a pretty impressive god who had such powerful angels at his disposal! We are familiar with the principle: proportionality bias. The enemy's greatness also makes God great.

But this miraculous rescue wasn't the whole story. The Assyrians' influence was critical to Yahweh's aggrandizement. To understand how this could have worked, we must examine the process of state formation. Anthropological research has shown that the transition from village communities and chiefdoms to centralized societies represents an important process of development. In cases where this process occurs for the first time, in the absence of nearby states to serve as role models, anthropologists describe this phenomenon as primary state formation. This occurred in Mesopotamia and Egypt. State formation that happens in the vicinity of existing states is known as secondary formation. The kingdoms of Israel and Judah arose sometime after the tenth century BCE alongside neighboring kingdoms that had existed for millennia.[22] As newcomers, they were secondary states from an anthropological point of view.

The process of state formation also determines the configuration of the godly cosmos: "As on earth, so it is in heaven." There is a "tendency of the divine to track the political."[23] During primary state formation, the various tribal and tutelary deities commonly became merged into a pantheon, but there existed no divine role models upon which to base their new state functions. One can see how they emanated from a world of local potentates and warlords. Just like people on earth, the gods of heaven engaged in cliques, rivalries, and intrigues. The variegated pantheons of the ancient and classical worlds, from Mesopotamia and Egypt to Greece and Rome, reflect this principle.

The situation was completely different in secondary states, which oriented their pantheons along the lines of existing models. When Israel and Judah became states, they faced empires headed by absolute monarchs and lording over huge populations. Although perhaps lacking in might, the two new states turned to these existing states and copied their models when erecting their own institutions—all the way down to their state god.

At the time of Israel's and Judah's creation, the Neo-Assyrian Empire completely dominated the Near East,[24] making it a constant theme for the Bible's authors. The name Assur appears again and again. "The Bible cannot do enough to describe the power of this mighty enemy": its interest in

Assyria fluctuates between fear and fascination.[25] The Assyrian influence had the same effect as when pressurized oxygen is pumped into a blast furnace to produce steel. The Bible's authors took the absolute monarch, the Assyrian king, as a role model, fashioned their own state god, Yahweh, in the manner of an "ancient Near East despot,"[26] and bequeathed him "imperial" power.[27] The stories of the exodus and the conquest of the Promised Land portray Israel as a "devouring superpower"[28] conquering one people after another—exactly as the Assyrians had done in reality.

It is downright uncanny how much the Assyrian king served as a model for Yahweh. The British Museum in London holds Assyrian reliefs that depict the capture of the Judahite city of Lachish by Sennacherib's troops. They show how the city was pillaged and many of its inhabitants taken prisoner; other men were brutally impaled. An inscription on the Rassam Cylinder extols the Assyrian king: "I encircled, conquered, and plundered Altaqu and Tamna. I approached Amqarunna and killed the governor and princes who had allowed this misconduct and hung their corpses on towers circling the city."[29] And what did Yahweh do when his people got themselves mixed up with the Moabites? "And Israel joined himself unto Baalpeor: and the anger of the LORD was kindled against Israel. And the LORD said unto Moses, Take all the heads of the people, and hang them up before the LORD against the sun, that the fierce anger of the LORD may be turned away from Israel."[30] Some Bible translations even mention impalements.

But that's not all. Let us look at the contract the Assyrians made with their vassals. As Othmar Keel explains, the Judaeans were contractually bound to follow "no king other than the Assyrian great king." They had to promise to be "loyal to him alone, 'to love him,' and to denounce and mercilessly punish any attempt to back out of this exclusive arrangement and turn to the Egyptian pharaoh, for example, even if this was done by one's own wife or children."[31] Now let us turn to the fifth book of Moses, Deuteronomy, in which Yahweh speaks to his people:

> If thy brother, the son of thy mother, or thy son, or thy daughter, or the wife of thy bosom, or thy friend, which is as thine own soul, entice thee secretly, saying, Let us go and serve other gods, which thou hast not known, thou, nor thy fathers; Namely, of the gods of the people which are round about you, nigh unto thee, or far off from thee, from the one end of the earth even unto the other end of the earth; Thou shalt not

> consent unto him, nor hearken unto him; neither shall thine eye pity him, neither shalt thou spare, neither shalt thou conceal him: But thou shalt surely kill him; thine hand shall be first upon him to put him to death, and afterwards the hand of all the people. And thou shalt stone him with stones, that he die; because he hath sought to thrust thee away from the LORD thy God, which brought thee out of the land of Egypt, from the house of bondage. And all Israel shall hear, and fear, and shall do no more any such wickedness as this is among you.[32]

It is hard to miss the cue Yahweh had taken from the Assyrian king.

Eckart Frahm, professor of Assyriology at Yale University, considers this a theologically delicate issue. He believes that the "image of an apparently all-powerful Assyrian regent, this ruler of the world . . . , had a direct influence on the biblical concept of God." The result was "the religious use of the idea of absolute rule as was manifested by the Assyrian monarch," which is why the "monocratic character of the Assyrian monarchy" is almost certainly at the "roots of biblical monotheism."[33] Let us put it in more dramatic terms: the Assyrian royal ideology is like the pressurized oxygen in the blast furnace image we used above: it annealed and tempered Yahweh. From then on he was seen as a ruler prepared to bring ruin to his people for their disloyalty.

It is entirely fair to call this idea theologically explosive. Here is the first sentence of the famed Shema Yisrael (which we examine in more detail later), recited mornings and evenings by devout Jews: "Hear, O Israel: The LORD our God is one LORD: And thou shalt love the LORD thy God with all thine heart, and with all thy soul, and with all thy might."[34] In Christianity, too, this sentence plays an important role. When asked which commandment is the greatest of all, Jesus quotes precisely this verse (and then adds the commandment about loving thy neighbor).[35]

But where does it come from? Once again, the Assyrians had a hand in it: it is a central motif of the Neo-Assyrian loyalty oath. The demand for "love" refers to the "sentiment of absolute political loyalty and is connected to the formula . . . that one should love the great king as one loves oneself."[36] The Bible's authors thus adopted a key element of Assyrian royal ideology and transferred it to Yahweh in order to instill loyalty to the state god. He demanded absolute fealty—he was no object of love.

It is therefore easy to agree with Eckart Otto's suggestion that the "revolutionizing" moment came when the biblical authors hijacked the

Assyrians' formula and projected it to the God of Israel. His first commandment to love no other but him is somewhat totalitarian, after all. We now know the origins of one of monotheism's main characteristics: the claim that Yahweh was the one and only God, who demanded absolute loyalty.[37] We can thank the Assyrians for God's morbid jealousy. Cultural evolution may take surprising turns.

MISSION TORAH

Yahweh's priests in Jerusalem transformed the shock brought about by the downfall of the kingdom of Israel into a powerful narrative. They presented an explanation for the catastrophe: the north met its demise because it had worshipped Yahweh inappropriately. If Yahweh did not shrink from wiping out the militarily and economically stronger Israel,[38] then correct and fervent worship of Yahweh was now a matter of life and death for the weaker Judah. Such veneration alone, the priests believed, would avert the same terrible fate.

This was the starting signal for the Torah's great mission.[39] Old laws were taken up and reformulated as an expression of God's absolute will. The priestly elite viewed this as not just a pressing theological need but a social one as well. The Assyrian army had depopulated entire swathes of territory. Families had lost their link to land and property, to graves and ancestor cults. This resulted in the destruction of "the traditional foundation for the ethos of solidarity in the genealogies of the extended families and clans," which in turn led to escalating social disparities. The new legislation in Deuteronomy, around which the Torah eventually crystallized, served as a response to these problems.[40] The terrors of war and the refugees' suffering necessitated the creation of a first-class cultural-protection system.

Let us summarize the results of the Assyrian catastrophe. Yahweh won out in the divine struggle for ascendancy because he adopted the monocratic demands of the Assyrian king, proceeded jealously against all other gods, and, just like his role model, demanded absolute loyalty from his followers. The existence of the other gods was not yet called into question (for here we are still dealing with monolatry, or worship of one god among many), but they were taboo for the people of Israel from this moment onward. To ensure Yahweh's loyalty and thus avoid any further episodes of divine rage, his priests began setting every detail of his will to paper.

THE SECOND INFERNO: THE BABYLONIAN CATASTROPHE

Following the demise of Israel, Yahweh was tempered a second time in 587 BCE when the Babylonians eliminated the southern kingdom of Judah. Their armies laid waste to the capital Jerusalem and destroyed its temple. They killed the sons of King Zedekiah before blinding him so that the last thing he saw was the princes' deaths. Finally, they marched Zedekiah and parts of the population off to Babylon.[41]

In ancient times such a defeat would seriously weaken, if not obliterate, the cult of the vanquished god, but in this case the opposite happened: "in the Bible, the power of the God of Israel was seen to be even greater after the fall of Judah and the exile of the Israelites," explain Israel Finkelstein and Neil Silberman. "Far from being humbled by the devastation of his Temple, the God of Israel was seen to be a deity of unsurpassable power."[42]

How did Yahweh pull this off? The standard answer is that he manipulated "the Babylonians to be his unwitting agents to punish the people of Israel for their infidelity."[43] But why were priests in other parts of the world unable to save their own necks by explaining the demise of their own gods in a similar fashion? Because no one would have believed them! It makes much more sense to defect to the victorious gods who had proven their superior power. Why should one remain loyal to a god who had doomed his own people?

How exactly, then, did the priests of Yahweh manage to transform their god's direst defeat into his greatest triumph? The best explanation is, as mentioned above, that Yahweh had two lives. He had already survived the downfall of one kingdom, a belief etched into the minds of the priests and believers. Of course those who were sent into exile were unsure whether Yahweh had simply decided to sit back and watch the destruction of their city or instead lacked the power to protect himself and his people. But the people forced into exile were of the middle and upper classes, and they had always been Yahweh's biggest backers.[44] The historical-theological narrative of a god who was willing to punish his own people, cultivated after the fall of the north, had become a solid part of their second nature. They were no longer capable of seeing the world from any other perspective. Didn't the victory of the Babylonians confirm in the most terrible of ways the belief they had always propagated—that it was God's will that the people of Israel, unwilling to display the necessary loyalty, be delivered into the enemy's hands? Of course this also meant they could conveniently place the blame on the common people, who continued to worship false idols.

As outlandish as such an interpretation might seem, it complies with confirmation bias and was much more comforting than the alternative, which was to admit that for centuries they had worshipped a weakling. It also made Yahweh much more powerful than the gods of Babylon and allowed the exiled losers to "step out of the role of mere sufferer." If the catastrophe was God's punishment, then they had to avoid repeating the same transgressions. This offered the opportunity for a "turn to the future,"[45] which would become the core concept of the Babylonian captivity. The priests set to developing a literary interpretation of the catastrophe, initiating the "most productive phase in the history of Israel." "More than half of all the texts in the Hebrew Bible were either written during this time or revised to produce their current versions."[46] Through all this intellectual activity, the authors strove to explain what had caused the loss of their country and to perfect the cultural-protection system of the Torah so that the people of Israel could live in peace in the future. Priests and scholars rewrote the old stories and inserted God's laws into the great epic about the exodus of the Israelites out of slavery.

CAST OFF THE SCORIA

And God? The tempering of the catastrophe in 587 BCE radically changed him once again. After the destruction of his temple, his city, and his state, he could no longer remain the same God he had once been. He had to relinquish his physical form, which was melted down in this second round of tempering and cast off as scoria.

In order to understand the great intellectual challenge of coming to terms with the destruction of the temple of Jerusalem, one must remember that the temples of the ancient Near East were seen as the gods' residences: the gods actually lived in them in the form of statues or cult symbols. But these figures were not just cold stone monuments; they were living gods (even if they were able to leave their statues when they wished).[47] Old Testament scholars are increasingly calling into question the traditional belief that Yahweh has always been worshipped without the aid of statues or figures. During the Judaean monarchy, a "normal" ancient Near Eastern cult "that was fundamentally the same as contemporary cults in Phoenicia, Syria, or Transjordan" was practiced in the temple of Jerusalem. And such a temple cult presupposed the existence of a statue or cult symbol. Accordingly, there must have been some material representation of Judah's primary god, Yahweh, in the temple of Jerusalem.[48]

When flames consumed the temple, the fire would have also demolished whatever visually represented Yahweh, be it a statue, the Ark of the Covenant, or an empty cherub throne (three often-mooted possibilities). His home had been destroyed, the sacred objects removed, his capital leveled, and his country occupied. Maintaining an intrinsic connection to these material realities would have spelled Yahweh's end. It was therefore impossible for him to be bound by any aspect of physical existence, and so his followers had to "completely rethink his entire nature."[49] From that moment onward, God was no longer of this world. As a transcendent being he was rendered indestructible. Henceforth God was invisible, and his home was heaven. Only his name remained on earth.[50]

His power, too, became unearthly, and the power of other gods dwindled to naught in his presence. The compulsive coherence seeking of an abstract God led to a prohibition of images and foreign gods. When in the Ten Commandments Yahweh demands, "Thou shalt have no other gods before me. Thou shalt not make unto thee any graven image, or any likeness of any thing that is in heaven above, or that is in the earth beneath, or that is in the water under the earth: Thou shalt not bow down thyself to them, nor serve them: for I the LORD thy God am a jealous God," he really had no alternative. No other gods were to be permitted, of course, for they could have played a role in Jerusalem's demise. But jealousy of other gods is no longer the only reason for his insistence on absolute loyalty. If divine power were located in cult figures and symbols, the destruction of his temple would have shaken him badly. So this meant that forbidding the production of graven images was a matter of survival—it was pure self-defense on Yahweh's part. Anything else would have meant his demise.

In this tempering process, monolatry (worship of one god among many) became reworked into monotheism (belief in the existence of only one god). The other gods, accepted until then, were declared false idols and cast aside. Not even the members of Yahweh's own entourage were allowed to carry on. Yahweh returned from exile a "widower"—there would be no more talk of his former consort Asherah.[51] God was now unique. Indeed, he had become abstract, no longer of this earth, and he was well on his way to becoming a universally applicable, purely spiritual principle. Robert

Wright offers an apt summary: "Israel's exilic theologians made the most of their disaster." The defeat at the hands of the Babylonians would turn out to be "the best thing that ever happened to Yahweh."[52]

FRESHLY STEELED

Let us summarize: Predestined by his catastrophic temper to do so, Yahweh won the race to become the state god of war-ravaged Israel and Judah. He survived two infernos, whereas other gods wouldn't have survived one. In the first inferno he was alloyed with the claims to absoluteness and demands for unconditional loyalty of an Assyrian potentate. In the second he shed his material form, casting it off like scoria and simultaneously blasting away the competition. Steeled by the fires, he emerged a transcendent being. All of this could only come about because the work on the Torah, the textualization of his will, had long since set this process of abstraction in motion. God had become word. He was well on his way to becoming an intellectual principle, and the Torah would serve, in the words of German poet Heinrich Heine, as a "portable homeland."[53] From this moment on, God was universal, all purpose.

The monotheistic Yahweh was the intellectual masterpiece of a small elite. He was not an intentional creation but one born of the absolute necessity of taking all possible steps to protect the people from definitive destruction. We believe this process constitutes further evidence in support of our thesis that catastrophes played a critical role in the cultural evolution of religion. They had made gods out of spirits, and now an overdose of misfortune had transformed a local divinity into a supergod who lorded over the entire cosmos. Had it not been for these calamities, Yahweh would undeniably still be sitting in his burning bush at the foot of Mount Horeb.

9

THE MURMURING PEOPLE

Our Clamoring First Nature

ERIK AURELIUS ONCE WROTE THAT IN THE PAGES OF THE OLD Testament, we encounter a people "who throughout the entire story is in open and passionate dispute with its own religion."[1] The professor of theology and retired bishop of the Church of Sweden was probably thinking of the exodus, in which the Israelites could not stop "murmuring" during their wanderings through the desert. In other words, they were protesting against the single God that Moses had introduced them to. Even if he claimed to be the God of the ancestors, he seems to have changed almost beyond recognition.

Depending on how you count them, there are a good dozen "murmuring stories" in the Torah. The children of Israel murmur about a lack of water and long for the fleshpots of Egypt. They dance around the golden calf, get mixed up with foreign gods, and rebel against Moses's privileges.[2] The Bible couldn't make it any clearer that the Israelites are dissatisfied with the new state of affairs. In the end, God decides that he has no choice but to let all of those he led out of Egypt die in the desert. Only the next generation will make it to the Promised Land. The relationship between the people of Israel and Yahweh really was quite precarious.

All this murmuring makes a wonderful subject for biblical anthropologists, for it allows us to observe just how difficult it is to establish a new god or, more precisely, a new religion. Reality and biblical story become congruent. The story of the exodus focuses on the same issue close to the hearts of the Bible's authors: how to impose monotheism upon the people. That the exodus is such a first-class propaganda tool illustrates the resistance Yahweh's priests had to overcome in their mission to create the Torah.

GOD COMES HOME

As we saw in the previous chapter, Yahweh's monotheistic metamorphosis took place during the Babylonian captivity. But the exile did not last forever. The Persians, who conquered Babylon in 539 BCE, allowed the deportees to return home. In Jerusalem, now part of the small Persian province of Yehud, the returnees began reconstructing the temple with the Persians' financial assistance in 520. But not all of the deported Judaeans returned home, for many had made lives for themselves in Babylon. Why should they return to a war-ravaged country they only knew from stories? Those who did decide to risk the journey into the unknown, however, were absolutely convinced that they were returning to the land of their forefathers, which, according to their holy scriptures, Yahweh had promised to them. Wasn't the end of their captivity further proof that God directed everything? This time he had used the Persians to reward his people's obedience by returning them to the Promised Land.[3]

Although the Bible seems to suggest the opposite, Yahweh's loyal followers did not return to an empty country, for a large part of the population had stayed behind. What followed was a collision of two worlds: the educated elites, born and raised in the sophisticated state on the Euphrates and Tigris, and the poor rural population that had never been abroad. The elites were fundamentalists on a mission: they wanted not only to regain their possessions but to ensure that only Yahweh was worshipped in God's own country. They met with resistance, however, for the people of Judah found the demand that they worship an abstract deity who tolerated no other gods utterly unreasonable.

In the modern world, a monolithic understanding of religion holds sway. We are used to official religions and their accepted doctrines, which treat everything else as mere superstition. The Bible also takes this stance: anyone refusing to worship Yahweh in the prescribed manner is a servant of false idols. But this is a biased view. Archaeologist William Dever clearly sums up this dichotomy: "The Hebrew Bible, written by elitists (and propagandists), is an ideal portrait, not of what most people actually believed and practiced, but of what they should have believed and practiced—and would have, had these theologians, these nationalist orthodox parties, been in charge." An accurate overview of the true state of affairs, says Dever, who has directed excavations in Israel for over fifty years, "is the one we can and must now derive from information supplied by modern archaeology."[4]

For many years the Bible's descriptions of ancient Israel were accepted without criticism, which is why the evidence of a rich and colorful religious world uncovered by archaeologists in recent years came as a real surprise. Religion "is not uniform but pluriform," explain Francesca Stavrakopoulou and John Barton in *Religious Diversity in Ancient Israel and Judah.* It "can vary from place to place—whether temple, tomb or home—and within and among different groups of people—from rural households to royal households, from garrison troops to women's local networks." It is a colorful world of multiple gods: "Indeed, it is now understood that the religious worlds of these groups were likely populated by different combinations of various deities and divine beings, such as Yhwh, Asherah, Baal, specialized craft deities and household ancestors."[5]

The Three Spheres of Religion

To appreciate the religious diversity of ancient Israel during the time of the kingdoms—that is to say, in the centuries prior to exile—we must differentiate between the various forms of religion.[6] Everywhere we find a cult of "decentralized personal and family religiousness" directed at the family's ancestors and gods. Then there was the "level of the local cult at the village level." This included the hilltop cult sites reviled repeatedly in the Bible where various gods—but also local manifestations of Yahweh—were worshipped. Finally, there was the "supraregional official cult connected to the city and state" where the state god Yahweh was worshipped at central cult sites.[7]

The boundaries between these spheres were fluid. The people "could worship various gods at various levels . . . depending on their area of responsibility."[8] This comes as no surprise, for "the whole of life was permeated by the presence of the gods,"[9] and these manifested themselves in the social spheres in their own specific manner. Let us examine these spheres more closely.

The lowest of these was the sphere of daily life, dominated by what is often described as folk, family, or household religion.[10] This is not to be disparaged, as biblical scholar Carol Meyers explains, for the household, "as the basic unit of both production and consumption, . . . was the single most important economic and social unit."[11] People lived their lives at this level. Dever agrees: "Their entire life revolved around the family, the village, the clan, the world of nature, and the rhythms of the changing seasons." Most people never made it to Jerusalem, and the household shrine was "the only temple they knew."[12]

"Living in antiquity was *being* 'religious,'" as Dever puts it.[13] Danish biblical scholar Anne Katrine de Hemmer Gudme speaks of a "religion of everyday social exchange" that emphasized the "welfare of the family, clan and friends, the health of children and animals and good crops." This "mode of religion" focused entirely on existential questions: Will I recover from this illness? Will my wife conceive? Will it rain? The people's religious activities were "a modification of everyday practices, such as gift-giving, food-preparation, feasting, honouring and petitioning, and the know-how that is required is the kind of everyday knowledge and experience one needs to get on in the social worlds."[14] And since women played an important role in the household economy, they were the true "ritual experts" when it came to household religion.[15]

Among the remains of ancient Israelite settlements, archaeologists have come across a plethora of figures, amulets, and seals. As it turns out, the notion that the ancient Israelites did not produce idols couldn't be further from the truth.[16] Ancestor figures—the *teraphim*, often translated as "household gods"—stood in bedrooms.[17] Then there are the female terracotta figurines unearthed by archaeologists. When Frederick John Bliss excavated one of these figures in 1882, he noticed, "much to his regret," that the figurine was unclothed.[18] Today they are known as "Judaean pillar figurines." Their most striking characteristic is the manner in which they flaunt their breasts. As there are no existing written records, we can only presume whom they represent. Most likely they are images of "Mrs. God,"[19] as sociologist Robert Bellah has called the goddess Asherah. Inscriptions describe her as Yahweh's companion, and her image was not removed from the temple at Jerusalem until the end of the seventh century BCE. The Bible mentions her around forty times, but never does the Good Book have anything nice to say. Although over 3,000 such female figurines have been discovered and every household probably had one in those days, the Bible does not mention them once. "Here the silence of the biblical texts is deafening," says Dever. For him there is no question as to their function: "They clearly have to do with reproduction: the desire of their users to be able to safely conceive, bear children, and lactate. These are in effect 'prayers in clay': talismans to aid women in having children, nursing them, and rearing them through childhood."[20]

Over time, societies become functionally stratified, and the same holds true for religion—which brings us to the second sphere. Holy sites developed out of the religion of everyday life in places where there were people with a regional reputation for their special connection to spirits and gods, where the clan or tribe assembled for festivals, or where the actual home of a particular god was believed to be. As a result cults developed for fulfilling distinct needs at these special sites. One temple acquired a reputation as a particularly curative site; another might be a house of herbal lore to which women came on pilgrimage when they had trouble conceiving, and so on.

With the rise of larger territorial dominions, we also see the establishment of the third sphere: politically active gods. In their temples one found professional priests. This allowed expert knowledge to accumulate in the form of holy texts once writing had developed. The monarch often served as the state god's chief priest. As state formation often involved centralization, it was in the system's interest to rid itself of intermediate institutions—and this was true in the religious spheres as well. "In short, supernatural pluralism was an enemy of royal power," says Robert Wright. "If every prophet of every god went around broadcasting divine decrees, and every clan in Israel consulted the spirit of its most revered ancestor on policy matters, the king would have trouble staying on message."[21] The abolishment of local and regional temples strengthened the monarch and freed the priestly elite from bothersome competition.[22] However, because the state god was only responsible for the big picture, the lowest of the spheres remained untouched. When it came to everyday life, the religious practices of the king and his people were probably not all that different.[23] In times of existential crisis, even a ruler like King Saul sought the aid of a witch to summon the spirits of the dead.

Minority Report

This was the variegated religious world of ancient Israel, but the Torah, the work of priests and scholars in the service of Yahweh, presents only the monochromatic perspective of the third sphere. "It provides much information about 'national' practices at the top of the socio-political pyramid but little about the practices of the myriad of households comprising the bottom."[24] This conforms to our own reading of the Torah. Its laws do not cover "the main concerns of domestic cults," such as "problems surrounding health and fertility, the quest for children, and important life events such as marriage, childbirth and death."[25] The reason is simple. Yahweh didn't have to worry about these things; that's what household religion was for. For as long as he was prepared to tolerate it, at any rate.

The focus on the third sphere also explains why women play little more than a marginal role in Yahweh's holy scripture. The Hebrew Bible mentions women by name at a ratio of one to twelve relative to men.[26] The Bible's authors simply did not focus on areas in which women played a central role. In the Old Testament one finds no ritual texts or laws pertaining to pregnancy or birth, although these themes were certainly central to people's lives.[27] Despite being an all-time best seller, the Hebrew Bible actually is, in the words of William Dever, a "minority report."[28]

Intuitive Religion Versus Intellectual Religion

The Babylonian captivity was a religious experiment in which the various belief spheres that had once coexisted and complemented each other moved apart over the course of two or three generations. In Judah, the practice of household and local cults continued in the countryside. Yahweh, too, was still worshipped in the ruins of the temple at Jerusalem. Thus, everything remained as "pluriform" as ever. This comes as no surprise, for we are dealing with a form of religion tightly interwoven with our innate psychological dispositions and their belief in a numinous world inhabited by countless spiritual actors. This world is self-evident—belief in it requires no extra cognitive effort. We can describe this phenomenon as *intuitive-individual religion*, which arises directly out of the innate intuitions with which every one of us is endowed. Its psychological foundations are therefore an essential part of our first nature. Its practices had been a part of our second nature for centuries, if not millennia.

The previous chapter explains what happened to Yahweh during his time in exile. He lost his form, evolved into the one and only God, and became thoroughly totalitarian. Although the Assyrian influence had already provided a strong impetus to make Yahweh the sole religious focus, some kings continued to sympathize with the older polytheistic cult, if only because acceptance of foreign gods facilitated international trade and diplomacy.[29] But during exile, the priests began to believe this lack of loyalty to Yahweh had led to the Babylonian catastrophe in the first place. Hence, the deported elite, perched in its Babylonian ivory tower, worked to perfect the codification of divine will. As our analysis of the Torah has shown, this kind of religion is an astonishingly rational cultural-protection system aimed at averting future catastrophes. Both the formulation of its basic principles and the development of its 613 laws required a great deal of strictly logical thinking. It was anything but intuitive: institutionally embedded experts had to explain its precepts. Here we are dealing with a new form of religion—a cultural product of our third nature. We call it *intellectual-institutional religion.*

For centuries, the intuitive and intellectual forms of religion had complemented one another quite well. Put simply, one dealt with the private worries of human individuals, whereas the other focused on the welfare of society as a whole. But now the most fervent representatives of Yahweh's avant-garde had returned from Babylon, convinced that following the Torah was the perfect recipe for avoiding catastrophe. They returned to a homeland that was no longer theirs, but with the aid of the Persians sought to reassert their power. This meant enthroning Yahweh as the one and only sovereign and forbidding anything this newly monotheistic god might find offensive—particularly when it came to the intuitive religion of the family sphere.

One can in fact read about this conflict in the Torah, which was completed around this time. It forbids all other gods, casts out all idols and graven images, and dictates that the cult may only be practiced in Jerusalem. It bans evocation, sorcery, and all forms of divination. The indigenous religious spheres, the family and local cults, were to be eradicated root and branch to implement the new rule-based divine concept. The books of Ezra and Nehemiah, which tell of the return to Jerusalem, even call for the dissolution of interfaith marriages. Is it any wonder that the common folk began to murmur?

Theological Incorrectness

No matter how much effort Yahweh's advocates invested in their undertaking, they were up against a very powerful opponent: human nature. Human nature stubbornly relies on its intuitive conceptions of the design of the supernatural world. Admittedly, when monotheism was able to establish itself and become a part of second nature through socialization and by anchoring itself in daily practices, intuitive religion soon found itself on the defensive. But it never disappeared altogether, for it reflects our innate religious substrate. Justin Barrett, a leading exponent of the cognitive science of religion, has shown how these assumptions remain in the minds of even those believers who know full well what is "theologically correct" in the eyes of the dominant intellectual religion. Even though they *know* that God has no body and exists beyond time and space, they can *feel* God's reassuring presence right next to them at particularly spiritual moments.

Experiments have shown that religious adults actually have two sets of representations when it comes to God: "One set is the fancier theological set about an all-present, all-knowing, and radically different kind of being that comes up in reflective situations, and the other set is the one that looks much more like a human and is easier to use in real-time situations." This theological correctness is very similar to political correctness. "When our intellectual guard is up," we know what types of ideas we can express in public. When our guard is down, we might allow ourselves to laugh at a joke that might not be all that PC.[30]

Be they Christians, Hindus, or Jews, people tend to envision the divine in more human terms than those set forth by their official religion.[31] This "theological incorrectness" is simply the "natural byproduct of the cognitive tools" in our heads.[32] Pascal Boyer describes this as the "tragedy of the theologian": "people, because they have real minds rather than literal memories, will always be theologically incorrect" and will always adjust official religious concepts to fit their mental structures.[33]

The Mosaic Distinction

Cognitive science has a great deal to offer to the discourse of religious history. Egyptologist Jan Assmann caused a great deal of discussion when he introduced the notion of "Mosaic distinction." He believes the tale of the exodus reflects the transition from polytheistic to monotheistic religions, "from cult religions to religions of the book." Assmann believes this particular transition has had a much more "profound impact on the world we live in today" than all political changes. In his view, polytheistic religions are "primary religions," which "evolve historically over hundreds and thousands of years within a single culture, society, and generally also language, with all of which they are inextricably entwined." "Secondary religions," on the other hand, owe their existence to an "act of revelation"; though based on primary religions, they distance themselves from them, disparaging them as "paganism, idolatry and superstition."[34]

Assmann has convincingly worked out the fundamental distinction. Primary religions—basically the same phenomenon that we have described as intuitive religions—have an instinctive, natural appeal. "No one would ever contemplate denying the existence of divine forces. They are there for all to see in the form of sun and moon, air and water, earth and fire, death and life, war and peace. They can be neglected, insufficiently venerated, sinned against in a hundred different ways, for example by breaking one of the taboos associated with them, but one can choose neither to initiate nor terminate a relationship with them. We are all irrevocably born into such relationships, which can therefore never be made the object of an inner decision." On the other hand, secondary religion (our intellectual-institutional religion) has to struggle with the fact that it is based on a revelation "that cannot be seen or experienced, but that must simply be believed in."[35]

Even though we agree with Assmann's notion, we want to qualify one point. The Mosaic distinction suggests that secondary religions completely replaced primary religions, but this is simply not the case. These events are fascinating when viewed from the perspective of cultural evolution because we see how a cultural product (intellectual religion) is superimposed on a phenomenon much closer to our biological foundation (intuitive religion) *without* causing the latter to disappear. Whereas secondary, institutional religion requires belief in its postulates, the continued presence of intuitive religion ensures that people have persistent doubts about the new belief system. Human nature simply cannot imagine abstract beings that exist outside time and space.

The Problems Caused by God

Regardless of how intellectually convincing the rational concept of monotheism might be, it confronts our first nature with a number of problems. Let us list the key aspects of monotheism that grate against our intuitive psychology.

First, there is no convincing evidence for the existence of a monotheistic god. He is abstract and transcendent. We are forbidden to make images of him. Indeed, he is truly inconceivable. Anthropomorphism, the tendency to imagine spirits and gods in a human form, "is one of the best known traits of religion."[36] The gods might be a bit counterintuitive, but not too much. "An abstract, impersonal God may be theologically possible, but psychologically not compelling."[37]

Second, he is the one and only. The people of the ancient world were completely unfamiliar with this form of individualism. Personality was founded collectively, and every person could only be imagined as a member of a group.[38] A god who didn't even have a royal household was implausible—which is why people had to make sure he at least had a few angels at his disposal. Above all, however, such a form of monotheism negates our innate Hyperactive Agency Detection Device (HADD). Based on millions of years of experience, our HADD knows that there are a lot of mischievous beings out there—the complex happenings around us can't all be the work of just one entity.

Third, it's hardly possible to communicate with him directly—this is the job of the priests. The Torah says next to nothing about the subject of individual prayer. But such a "secondhand"[39] religion is dubious, for the priests can say whatever they want. Furthermore, other possibilities for the individual to influence God, most of them magical, are eliminated. What can people do with a god to whom they have been denied direct access?[40]

Fourth, the mono-god creates a vacuum. Yahweh, the rational state god, and his male priesthood could not—and, indeed, did not want to—fill the gap created by the elimination of household religion and its deities. This explains the lack of typical religious themes that we remarked on in our analysis of the Torah's laws. In doing so, God made women—once the traditional "ritual experts"—redundant. He had just as little to offer them as he did regarding the many questions of everyday life. This contempt for women is particularly fatal, for the "study of religious believers is the study of women," according to psychologist Benjamin Beit-Hallahmi.[41]

All of this explains why the process of monotheism's establishment was so long and arduous. The Bible's authors would have to improve on their concept. For many years, religion in ancient Israel remained a pluriform affair, but monotheism would eventually triumph. The story of the exodus played a significant role in this victory, for it truly is a great and highly refined effort to quell the murmuring of the people. This is something worth scrutinizing in more detail.

THE GLORY OF GOD

After their return from the Babylonian captivity, Yahweh's followers shared the same fate as Moses. After years in the Midian desert, where he had encountered Yahweh, Moses returned to Egypt and set out to convince his people to accept a new god they had never before seen. Whereas Moses had the advantage that he could rely on God's miracles, the Bible's authors could only turn to their own inspiration. Given all we have examined so far, it should by now be clear just what a monumental uphill struggle they faced. Yet they passed this test with flying colors. Their brilliant piece of propaganda contains an impressive series of heavenly design features aimed at making Yahweh as believable as possible and so assuage the people's first nature.

Actions Speak Louder Than Words

The new monotheism's number one propaganda instrument was the idea developed by Yahweh's priests and scribes of wrapping the Torah's laws into a story that continues to inspire Hollywood directors some twenty-five centuries later. And as everyone knows, packaging is extremely important when it comes to selling a product. Not only does Exodus present the laws as the sacrosanct, personal will of God, but it is also a liberation epic ("Let my people go!") whose criticism of despotism has inspired humanity ever since. The message is clear: adherence to the law is the royal road to freedom. Who wouldn't want to obey the law?

But to be convincing, a story must be more than just good. What rhetorical device did the biblical authors' employ to accomplish this? They wove their problem (how do we convince the skeptics of this new God?) into the story itself. When Yahweh entrusted Moses with the task of leading the people of Israel out of Egypt, the latter was reluctant: "But, behold, they will not believe me, nor hearken unto my voice: for they will say, The LORD hath not appeared unto thee." So God showed Moses how he could transform a rod into a snake. And for good measure, he also showed him a few more tricks (how to make his hand leprous before healing it again and transform the water of the Nile into blood). And guess what—it worked!

The people needed only three miracles to believe Moses. Here the Bible is reiterating a bit of wisdom that evolution has etched deeply into our psyches: actions speak louder than words.[42]

Even God Needs CREDs

Anthropologist Joseph Henrich has alluded to a serious problem that inevitably accompanied the evolution of language. Before our ancestors had words at their disposal, they transmitted information by acting out. Someone craftily fashioned a hand axe while someone else looked on. This required effort, but it showed that the teacher was serious and simultaneously proved that the solution actually worked. The development of language, however, removed the need to act out everything. This made learning easier, but it also introduced a new danger: manipulation. Talk is cheap! Anyone can say just about anything.

Henrich offers us an example: "Since prestigious individuals can influence the beliefs (and other mental representations) of many learners, a prestigious Machiavellian could dramatically increase his fitness with well-designed culturally transmitted 'mind viruses' that strategically alter others' beliefs and preferences. For example, people in many places believe 'the wishes of our dead ancestors must be obeyed.' A manipulator might transmit the belief—not held by him—that he is 'the mouthpiece for the ancestors, and they will talk through him; their first command is to pay the mouthpiece for his service to the ancestors with one pig from each house.'"

To defend themselves from such attempts at exploitation, people developed a "cultural immune system"[43] that required credibility-enhancing displays (CREDs) before they would accept a new idea. CREDs are actions confirming that things really are as claimed. We want to see with our own eyes that things really do behave as we have been told they do. We want to know that people really believe in the things they claim to represent or really can deliver what they promise. "In short," as Ara Norenzayan puts it, "we want to see that people walk the walk and not just talk the talk."[44]

Sensory Pageantry

We've just seen an example of such a CRED. As long as Moses only *told* his people he had met a mighty god, they did not believe him. Anyone could say such a thing. Moses had to show the doubting people that his god had transformed a rod into a snake, healed his leprous hand, and turned the water of the Nile into blood. Doubting Thomas, who first had to touch the wounds Jesus received on the cross before he could believe in the resurrection, was merely a typical representative of the species *Homo sapiens.*

Suspicion of chicanery has always attached to those claiming to be in the service of a supernatural power.[45] In biblical times—during which there was no lack of prophets—you had to come up with something big if you wanted to be seen as anything but a charlatan. Thus, Exodus is literally bursting with CREDs. Whereas God on the whole was quite stingy with miracles, in Exodus he missed no opportunity to perform them. Every single doubter had to be convinced—and to accomplish this, God needed a whole lot of "sensory pageantry."[46]

This explains why God had to send down a whopping ten plagues to move the Israelites to leave Egypt. Couldn't he simply have ordered the Pharaoh to chase them out? Of course he could, but he didn't want to. Boils, locusts, darkness—the plagues represent nothing more than a series of CREDs. They are proof of God's greatness. And to ensure that no one hatched the suspicion that Yahweh was simply trying to wear down the mighty Pharaoh's will, he repeats the same pattern again and again. First God sends a plague. The Pharaoh wants to let the Israelites go, but then comes the decisive moment: "And the LORD hardened the heart of Pharaoh king of Egypt, and he pursued after the children of Israel." Yahweh needed the Pharaoh as a marionette to demonstrate his sheer power.

The same mechanism lies behind the greatest of all of the exodus's miracles: the parting of the Red Sea. This is traditionally interpreted as a rescue action: didn't God want to prevent the treacherous Egyptians from slaughtering the Israelites? No. God committed cold-blooded mass murder in order to remove any last remaining doubts about his great power. The Bible tells us so: the Pharaoh lets the Israelites go after the last of the ten plagues, but this is not enough for God. He commands Moses, "Speak unto the children of Israel, that they turn and encamp before Pihahiroth, between Migdol and the sea, over against Baalzephon: before it shall ye encamp by the sea. For Pharaoh will say of the children of Israel, They are entangled in the land, the wilderness hath shut them in. And I will harden Pharaoh's heart, that he shall follow after them; and I will be honoured upon Pharaoh, and upon all his host; that the Egyptians may know that I am the LORD. And they did so."[47]

And so it comes to pass. God closes the trap, and once again he commands Moses,

> Speak unto the children of Israel, that they go forward: But lift thou up thy rod, and stretch out thine hand over the sea, and divide it: and the children of Israel shall go on dry ground through the midst of the sea. And I, behold, I will harden the hearts of the Egyptians, and they shall follow them: and I will get me honour upon Pharaoh, and upon all his host, upon his chariots, and upon his horsemen. And the Egyptians shall know that I am the LORD, when I have gotten me honour upon Pharaoh, upon his chariots, and upon his horsemen.[48]

Not a single Egyptian survived this divine credibility-enhancing display.

The principle should be obvious by now. God also applied it against his own people. Their misdeeds were so great in number that they presented God with more opportunity for merciless crackdowns. He first ordered Moses to slaughter 3,000 Israelites for dancing around the golden calf and later sent down a plague that took another 24,000 people for getting themselves mixed up with the daughters of the Moabites and their foreign gods. And when Korah, Dathan, and Abiram and "two hundred and fifty princes of the assembly, famous in the congregation, men of renown" protested against Moses, "the earth opened her mouth, and swallowed them up, and their houses, and all the men that appertained unto Korah, and all their goods."[49] For the record, these punitive acts were not expressions of God's lust for revenge, as so often claimed—they were CREDs. Like the miracles, they served a narrative function aimed at giving Yahweh credibility. Once he had established himself, he no longer needed to hyperactively perform miracles or mete out collective punishment. This is typical in the establishment of a new religion.

Immune Against Refutation

The Bible incorporates yet another type of divine design feature, this one intended to prevent Yahweh's refutation. Following philosopher Karl Popper (1902–1994), we dub this strategy—intended to place God beyond all possible doubt—immunization against refutation (IAR).[50] For example, it aims to prevent a new prophet from coming along and presenting a new set of laws in God's name. That's why the Bible categorically states, "And there arose not a prophet since in Israel like unto Moses, whom the LORD knew face to face, In all the signs and the wonders, which the LORD sent him to do in the land of Egypt to Pharaoh."[51] Of course the priesthood was protecting itself with this IAR; no one should be allowed to undermine its power, which descended from Moses himself.

The priests certainly had to be prepared for a lot of competition. What happened if, for example, "there arise among you a prophet, or a dreamer of dreams, and giveth thee a sign or a wonder, And the sign or the wonder come to pass, whereof he spake unto thee, saying, Let us go after other gods, which thou hast not known, and let us serve them"? In other words, what are you to do when a new prophet tries to prove his power by means of CREDs? The answer: "Thou shalt not hearken unto the words of that prophet, or that dreamer of dreams: for the LORD your God proveth you, to know whether ye love the LORD your God with all your heart and with all your soul."[52]

But how can we be sure that Moses himself wasn't a snake oil salesman? Why didn't God show himself to the Israelites? That certainly would have done away with any remaining doubts. What if Moses merely climbed to the top of one of the smoking volcanoes that did indeed exist in biblical Midian? Once again the Bible's authors had an IAR at the ready: It was the people's will! God had wanted to show himself to everyone atop Mount Sinai, but his magnificence was just too much for the people to bear. "And all the people saw the thunderings, and the lightnings, and the noise of the trumpet, and the mountain smoking: and when the people saw it, they removed, and stood afar off. And they said unto Moses, Speak thou with us, and we will hear: but let not God speak with us, lest we die."[53]

We already encountered the IAR effect of the impossibility of looking on God in the story of Lot's wife, who was transformed into a pillar of salt for turning around to witness God's judgment on Sodom and Gomorrah despite his warnings not to do so. The Torah tells us again and again that God is a risky matter. Even Aaron's sons had to die for offering the Lord "strange fire." The people would have been at a complete loss if not for the priests. Wasn't it better to err on the side of caution and give them complete trust? The great sophistication of the cult rules also offered immunity against criticism. If a priest's actions had no effect, there was always an explanation: one of the ritual details must have been performed incorrectly! And if misfortune did indeed strike, more discipline and greater attention to detail were needed at the next attempt. This kept people on the defensive and ensured they always had a guilty conscience. With 613 rules to keep in mind, there's always a good chance that you may have forgotten to follow one of them.

We do not intend to claim these IAR strategies were actually cases of intentional manipulation. Many simply resulted from compulsive coherence seeking: if something doesn't work, one looks for a plausible supplementary explanation instead of throwing out the entire system. But more importantly, as evolutionary biologist Robert Trivers has shown, it is easiest to convince others of something if one is himself completely convinced. Self-deception is the best foundation for effective proselytizing.[54]

Making It Second Nature

The goal of all propaganda is to ensure that someday it is no longer needed. When this day comes, credibility-enhancing displays become superfluous, for all doubts have been cleared away. Immunization against refutation will likewise not be needed anymore, for no one will even think about questioning the status quo. The Bible aimed to make Moses's God a part of our second nature. Anthropologist Harvey Whitehouse developed the concept of the "doctrinal mode of religiosity," which holds that constant repetition works to anchor nonintuitive doctrines in people's minds.[55] We also find this in the Torah. Exodus draws on the ancient family festival known as Pesach, connected to animal husbandry and pastoral life,[56] which was reinvented as an annual remembrance of the departure from Egypt. From then onward, during Pesach the community reexperienced the exodus.[57]

But a yearly festival alone could not anchor a new tradition. For it to be internalized, more intense repetition was needed. In addition to its integration in the earliest stages of a child's upbringing, daily autosuggestion offered the best approach, exactly as prescribed by the Shema Yisrael:

> Hear, O Israel: The LORD our God is one LORD: And thou shalt love the LORD thy God with all thine heart, and with all thy soul, and with all thy might. And these words, which I command thee this day, shall be in thine heart: And thou shalt teach them diligently unto thy children, and shalt talk of them when thou sittest in thine house, and when thou walkest by the way, and when thou liest down, and when thou risest up. And thou shalt bind them for a sign upon thine hand, and they shall be as

> frontlets between thine eyes. And thou shalt write them upon the posts of thy house, and on thy gates.[58]

To this day pious Jews repeat the Shema Yisrael every morning and evening. Internalization functions as a form of embodiment; it seeks to anchor the social institution of religion inside the individual's consciousness.[59] Intuitive religion, on the other hand, has no need of this process, for its basic principles are a part of our nature from birth. The fascinating thing about the Shema Yisrael is that the message is not only internalized by means of recitation but incorporated by physical means as well. The phylacteries, also known as *tefillin*, bound to the head and arm, contain texts from the Torah. The law is engrained into the individual by both psychological as well as physical means. It becomes his second skin, or second nature, in the truest sense of the word.

In the end Yahweh was well aware that the need to implant belief culturally was not fitting for a god of his stature. First, it was a risky, error-prone process; second, it led to the question of why God didn't just resolve the issue when he created humanity in the first place. Thus he told the prophet Jeremiah of his plan to do everything differently next time. "I will put my law in their inward parts, and write it in their hearts; and will be their God, and they shall be my people."[60] Then his divine plan would at last be a part of our first nature. The law would be a genetically anchored institution in the psyche of *Homo sapiens*. The discrepancy between what we should do and what we want to do would cease to exist—and his people would no longer murmur all the time. And no one would have to learn all 613 laws.

10

THE TORAH'S LEGACY

The Lesson of Exodus

GIVEN ALL WE HAVE LEARNED ABOUT THE PERSISTENT RESISTANCE that greeted Yahweh and his new form of religion, the eventual success of monotheism is little short of amazing. What began as a minority report later became the most successful book in human history. Was this due to the story in which it was packaged? That's not such a bad guess, for Exodus is a truly fascinating narrative. The same cannot be said for the Torah's contents, however. Of the 613 commandments and prohibitions, more than a few have completely lost their significance. Although others were useful helping reduce violence or the spread of disease, no one could claim that the success of monotheism rests on its hygienic efficacy.

Obviously, the Torah is not the entire Bible. In the coming chapters we will marvel at the enhancements—but also the corrections—Yahweh undergoes. Nevertheless, the Torah's content represents a true milestone of cultural evolution. Based on the assumption that there is only one true God, its makers created a highly dynamic, rule-based cultural-protection system. Here we are not referring to the concrete effects of one or the other law but rather to the specific legacy of the Torah as a whole, which was seminal for Western culture. Its consequences in the realms of cooperation, rationalization, and violence deserve closer examination.

FIRST LEGACY: INCREASED COOPERATION

It is often said that the Judaism formed around the Torah was a "religion of self-exclusion" and that its laws functioned as a "high wall around its people."[1] Particularly in the foreign culture of the Babylonian captivity, the Torah's laws undoubtedly made assimilation more difficult, and this remained the case under Persian, Hellenistic, and Roman rule. The main goal was to remain loyal to God and his laws and demonstrate this loyalty in the most visible possible manner. The Torah had become a recipe for cultural survival. It ensured identity and solidarity.

But the laws served as more than a barrier against outside influences. Adherence to the laws also had enormous social consequences. Take the commandment "Thou shalt not seethe a kid in his mother's milk." The Torah mentions this rule no less than three times.[2] Regardless of whether this commandment reflects traditional association in the Levant of the nanny goat with the goddess Astarte, a desire to prevent mixing life and death, or pity for the killed creature (three existing explanations),[3] the correct construal does not really concern us here. What matters is that devout Jews rigorously separated "milky" from "meaty" and even had separate pots, pans, dishes, and cutlery for each—rules observed to this day.

This commandment has therefore come to function as a "costly signal," a relatively deception-proof sign that something is truly near and dear. Anyone following this commandment about the seething of kids and really going through the trouble of having two separate sets of cookware in order to keep the milky apart from the meaty, day in, day out, must take his or her religion very seriously. At the same time, the need to follow this rule will strongly discourage free riders, those who try to reap the religion's advantages without themselves making an effort. This in turn strengthens the community and mutual trust. In short, it promotes cooperation. Complicated rules requiring a great deal of effort to follow—and the Torah's 613 *mitzvoth* can hardly be surpassed as costly signals—are a first-class source of social glue.

This is particularly true when the goal is not to bring the community together around formulaic regulations but to instill in it a core of common moral values. The Torah is also a handbook for proper behavior aimed at preventing poverty and violence, at reducing suffering and disease. It seeks to protect the weak—and this even includes animals. While it does have the collective in mind, the actions of each and every individual matter too. The Torah's high ethical standards were greatly admired even in ancient times.[4] And as we have seen, they are largely in tune with the code of conduct of the hunter-gatherer world—in other words, the code of our first nature, which looks askance at inequality and injustice and relies on solidarity: love thy neighbor.

SECOND LEGACY: RATIONALIZATION

Nowadays we often view religion as the embodiment of irrationality. However, the cultural institution of religion, what we call intellectual religion, got its start as a thoroughly rational affair. In protoscientific fashion, religion devised a system of misfortune-reducing actions based on careful observation. In doing so it created "cultural resistance."[5] If you will, the Torah resembles a programming code, which, if executed perfectly, will lead to a peaceful world. The God of the Torah could no longer be tempted to behave in a particular fashion by magic—individual intervention in other words. This God is incorruptible; he is a "god of just retribution"[6] who doesn't make exceptions—not even for his own stalwarts. Indeed, even Moses himself was punished. In a region of despotic states rampant with arbitrariness and nepotism, these laws introduced major civilizing elements.

Adherence to the rules replaced self-aggrandizement. From our modern perspective, it is easy to overlook the revolutionary nature of this shift. The God of the Torah is predictable, and only after people eventually realized that reality refused to follow human notions of justice were God and his ways declared mysterious. But optimism still pervades the Torah that all will be right with the world—in the here and now on earth—if only people follow God's laws. When the Torah was created there was still no concept of a Last Judgment that would ensure God's justice in the afterworld. This came much later.

The notion of a unified principle that permeates the whole world, a principle that can therefore be reverse engineered, had far-reaching consequences. Since the one and only God is responsible for everything and thus liberates the entire world from potentially high-handed gods, spirits, and demons, the world becomes manageable. The notion of a rational world in which causes predictably produce consequences paved the way for further forms of rationalization. This is the fertile ground in which the seeds of science, medicine, and legal and political theory could germinate.

This idea is certainly not new but has hardly penetrated our collective consciousness. It was mentioned by Émile Durkheim ("the fundamental categories of thought, and consequently science, are of religious origin"[7]) and a century later by, among others, Ara Norenzayan: "Secularism based on science and reason is often portrayed as anathema to religion." But "you can see that secular societies are really an outgrowth of prosocial religions."[8] Theologians agree: the "roots of modern societies' rationality are to be found in the rationalizing pragmatism of the Israelite understanding of God and the world."[9] Max Weber famously referred to this rationalization as the "disenchantment of the world." The intellectual religion manifested in the Torah was at the forefront of cultural progress. We should induct the authors of Yahweh's Bible into the science hall of fame.

THIRD LEGACY: VIOLENCE

The Torah's third legacy is something of a touchy subject: violence. In recent years there has been a great deal of discussion in Europe about whether violence is inherent to monotheistic religions.[10] The debate was triggered when Jan Assmann presented his theory of the "Mosaic distinction," which he describes as the differentiation "between the true god and false gods, true doctrine and false doctrine," which is a fundamental aspect of monotheism.[11] This results in a "power of negation," which injects "the principle of *tertium non datur* into a sphere where it had previously been neither found nor even suspected: the sphere of the sacred and the divine, the religious sphere."[12]

Monotheism accepts only one truth—everything else is a lie. Such fanaticism regarding truth is not found in polytheistic religions; there is violence of course, but it concerns struggles for power and not truth.[13] It is only in monotheistic religions that "the truth to be proclaimed comes with an enemy to be fought," explains Assmann. "Only they know of heretics and pagans, false doctrine, sects, superstition, idolatry, magic, ignorance, unbelief, heresy, and whatever other terms have been coined to designate what they denounce, persecute and proscribe as manifestations of untruth."[14] With respect to the exodus story, Assmann suggests, "The fact that monotheism tells the story of its own foundation and consolidation by drawing on all the registers of violence must surely be of some significance." The story of the Israelites' exodus and occupation of the

Promised Land is told as "a history of violence punctuated by a series of massacres."[15]

We do not want to get into the monotheism debate, which is primarily fought out among theologians. But we do want to stress how our evolutionary approach introduces a previously unobserved factor into the mix. The "series of massacres" we observe in the tale of the exodus, which Assmann himself emphasizes were purely fictitious, were actually credibility-enhancing displays (CREDs). They were intended to impress upon the reader what people were capable of if finally prepared to swear absolute fealty to Yahweh. Instead of always being the victim—the historical reality—the Israelites finally thought they had the opportunity to humiliate their enemies. And as we have shown, CREDs were sorely needed, because the Torah's mission was nothing less than exchanging the older system of beliefs for that of the new God.

Here we have actually identified one of monotheism's vulnerabilities, for it will never attain the self-evidence—the implicitness—of intuitive religions. Even if a monotheistic religion has been established, possesses powerful institutions like the church, and has anchored itself in people's second nature, it remains intellectual and therefore will always be plagued by nagging doubts. Intuitive religion simply cannot be completely silenced in believers, for it belongs to their first nature. It continues to whisper theological incorrectness into their ears.

There's a little heretic inside every monotheistic believer. It's completely natural to doubt counterintuitive truths—like the maxim of the one and only God who exists beyond time and space. This may appeal to our rational side, but our first nature sees things entirely differently. In the end, monotheism only offers words and for this reason can never fully dispel the suspicion that it's nothing more than cheap talk. This is why monotheism has such a particular interest in appearing "absolutely serious."[16] An intellectual religion must always prove that it is more than mere words. This it can only do through actions.

"Absolute seriousness is based on the threat of death," classicist Walter Burkert writes.[17] So, when absolute seriousness must be established, there is hardly a way for intellectual religions to avoid the topic. Death is the one absolute certainty, and this makes it the ultimate CRED. Two of the Bible's most important scenes rest on this principle. One is when Abraham displays his willingness to sacrifice his son Isaac after God orders him to do so. The fact that he is already holding the knife when the angel tells him to stand down proves how deadly serious Abraham is about his religion. Who wouldn't follow him in his belief if God was more precious to Abraham than his own flesh and blood?

And this brings us to the second scene: the cross—a symbol of which millions and millions wear around their necks. Most of these people would say that it represents the resurrection of Jesus Christ, the son of God. But more than anything, the cross symbolizes his death. The fact that Jesus was crucified for his beliefs assures Christians that they haven't been taken in by an impostor. It's a terrible thing to have to say, but death is the only costly signal that human nature finds absolutely convincing. The crucifixion is what Burkert calls an "ineradicable seal" certifying Christianity's truth and "absolute seriousness."[18] This is why early Christians were obsessed with the fact that Jesus died as a real human being. Had he been a mere "divine tourist" on earth, the crucifixion would have been an easy trick for him.[19]

The finality of death makes it so attractive as a purveyor of credibility. The same principle holds for martyrs and, unfortunately, holy warriors, who are more willing to sacrifice others' lives than their own. This explains why the history of monotheistic religion is full of holy wars, crusades, and smoldering stakes at which witches and heretics were burned. Earthly motives have certainly played a role in these events too, but each and every instance also shows how people believed that killing was a surefire means of proving the seriousness of one's faith and dependability.

God himself had shown the way. In the Torah he even puts to death those dearest to him. He lets Moses die on Mount Nebo while gazing out over the Promised Land—and not because his prophet had grown too old, as the Torah emphasizes: Moses's "eye was not dim, nor his natural force abated."[20] Let us not beat about the bush: Yahweh punishes Moses with death. He kills him. Why?

If we recall the biblical tale, the Israelites had once again begun to murmur after suffering from thirst in the desert of Zin. Yahweh ordered Moses, "Take the rod, and gather thou the assembly together, thou, and Aaron thy brother, and speak ye unto the rock before their eyes; and it shall give forth his water." But what did the prophet do? "And Moses lifted up his hand, and with his rod he smote the rock twice." The water gushed forth, and the Israelites slaked their thirst—but God was enraged: "And the LORD spake unto Moses and Aaron, Because ye believed me not, to sanctify me in the eyes of the children of Israel, therefore ye shall not bring this congregation into the land which I have given them."[21]

Both had to die because Moses struck the rock with his stick instead of merely speaking the words God had ordered. A punishment that might seem rather excessive to today's reader was simply a matter of consistency. In Yahweh's eyes—in other words, in the eyes of intellectual-institutional religion—the prophet was guilty of sacrilege. He didn't trust in God's word but instead performed a magical act. In doing so he gave free rein to his first nature, which believes only concrete actions, not mere words, can bring about physical changes. But the written law was based on the opposite principle: it's all about the word of God, about which there cannot be the slightest doubt.

This is why God could not tolerate flare-ups of first nature, not even in his greatest prophet. Even more importantly, his letting Moses die also served as unmistakable evidence of just how dead serious he was about his own laws. Moses's death was the ultimate CRED to persuade every last doubter. With this feat, God stamped an ineradicable seal upon his written laws. This, too, is the legacy of the Torah.

PART III

KINGS AND PROPHETS

Morality Made Divine

At the end of the Torah we find the people of Israel standing on the banks of the River Jordan after forty years of wandering in the desert. At last, God allows them to set foot in the Promised Land. The second main part of the Hebrew Bible, the Nevi'im ("prophets"), relates what happens there. It is most famous for a single episode: the battle between David and Goliath. We all know the story. It is a struggle of small versus big; yet against all odds, the little guy wins. Malcolm Gladwell, one of the world's most successful authors, has recently written a book about them. Its subtitle tells us what their story is really about: the art of battling giants.[1] To this day, David and Goliath remain an archetype of Western culture.

The story is almost too good to be true. The shepherd boy David was supposed to take food to his brothers, who were fighting as soldiers in King Saul's army.

Before he knew what had happened, he found himself standing before Goliath, the Philistines' mightiest warrior. No one had dared to accept Goliath's challenge of one-on-one combat. But David, who that morning had still been tending his father's sheep, did not hesitate for a second: "The LORD that delivered me out of the paw of the lion, and out of the paw of the bear, he will deliver me out of the hand of this Philistine."

He had neither sword nor armor, only his sling. Goliath, on the other hand, "had an helmet of brass upon his head, and he was armed with a coat of mail; and the weight of the coat was five thousand shekels of brass. And he had greaves of brass upon his legs, and a target of brass between his shoulders. And the staff of his spear was like a weaver's beam; and his spear's head weighed six hundred shekels of iron." All that gear didn't do him any good, however, because David "put his hand in his bag, and took thence a stone, and slang it, and smote the Philistine in his forehead, that the stone sunk into his forehead; and he fell upon his face to the earth." David proceeded to lop off Goliath's head with the giant's own sword. "And when the Philistines saw their champion was dead, they fled." What a hit!

Just like the Bible itself. The triumph of these holy texts, written by a small group of people from the backwaters of world history, is just as sensational as David's victory over Goliath. Both share a complete lack of respect for despots and other big shots. The secret of their success? Their faith in God, a theme worth examining more closely.

The Nevi'im consists of two parts. The former prophets (the books Joshua, Judges, Samuel, and Kings, which make up the "history books" of the Christian Bible) tell the story of how the Israelites conquered and ruled the Promised Land, then lost it again. Here we meet kings such as Saul, David, and Solomon, as well as prophets such as Samuel and Elijah. The latter prophets include the "classical" prophets, ranging from Isaiah and Jeremiah to Zechariah and Malachi. The Christian Bible places them at the end of the Old Testament in order to pave the way for the coming of Jesus as the fulfillment of their prophesies.

Below we address whether David actually defeated Goliath and whether he existed at all. But neither question is the main issue here, for we are dealing with the same phenomenon we encountered in our examination of Moses: the stories are just the packaging, and no matter how ingenious that packaging may be, their content is key. The Bible's authors were unaware of how the invention of missile weapons such as slings and spears represented a milestone in human evolution.[2] These missiles not only improved the effectiveness of hunting but also enabled the weak to slay the strong for the first time. Once these weapons appeared on the scene, every Stone Age Goliath had to weigh the risks of acting tyrannically. Might no longer made right. Instead, egalitarianism, cooperation, and justice were now far more likely.

The Bible takes up the story here. David's sling may have been the direct source of his victory, but ultimately his faith in God allowed him to stand up to Goliath. This is the Bible's great civilizing achievement. Justice becomes one of God's priorities. He doesn't support the powerful; he stands behind every shepherd boy—provided the boy is prepared to place his fate in Yahweh's hands, of course. The story is especially exciting in that when David himself later came to power, justice was no longer one of his main concerns. We can only admire the Bible's criticism of this development. Based on a careful examination of reality, the Good Book reveals yet again what a formidable problem this newly evolved phenomenon of "despotism" posed for humanity.

We begin this part of the book with the rulers of ancient Israel, then turn our attention to the prophets, God's charismatic earthly advocates. We end by focusing on one of the most controversial issues in the cultural evolution of religion: the gods' moral qualities. We can already tell you this much: Richard Dawkins was wrong when he wrote, "The God of the Old Testament is arguably the most unpleasant character in all fiction."[3] It is just easy to misunderstand him.

11

JUDGES AND KINGS

God Gives You Strength

THEOLOGY ONLY HAS ITSELF TO BLAME IF IT IS NOT AMONG THE MORE popular scholarly disciplines. The Bible offers readers fabulous stories of trumpets that bring down the walls of Jericho, of Delilah cutting Samson's hair, of David peeping at Bathsheba in her bath, and of Solomon rendering even the Queen of Sheba speechless. And what dry, unexciting title did theologians select for this treasure chest of wonderful stories? The Deuteronomistic History.

We need to be aware of the theological idea behind this lackluster title if we are to understand what it's all about. Deuteronomy, the fifth book of Moses, offers the clearest formulation of the logic of the "monotheism of loyalty"[1] that we have seen so far. Yahweh enters into a covenant with the people of Israel. He will reward whoever remains loyal to him with victories, power, and riches. But anyone breaking the covenant will meet his demise at the hands of God. This means that if Israel was responsible for its own misfortune because it did not remain faithful to God, then God must have rewarded it for good behavior in earlier times. How else could there have been kings in Jerusalem and Samaria? This conclusion derives from our compulsive coherence seeking. It is also the historical-theological leitmotif that dominates the Deuteronomistic History.

The Bible's authors followed a double strategy while in exile. On the one hand they put God's laws to paper in the form of the Torah, and on the other they presented evidence in the Nevi'im of how God had kept a close eye on these rules in the past. The latter is pure fiction if only because there was no way the rulers of this bygone era could have complied with the Torah: with the exception of a few laws, the Torah didn't exist yet. The Nevi'im is therefore not a historical treatise in the modern sense of the word but instead represents an attempt to trace back all historical misfortune to failure to follow Yahweh's will. Nevertheless, the history is remarkable for its empirical approach. After all, one must perform a historical analysis to identify the offenses responsible for the demise of the kingdoms of Israel and Judah. Herein lies the anthropological value of the stories behind the rise and fall of these kings.

GOD'S HERO: JOSHUA

The Bible is a terribly weighty tome, which is why we now have to go into time-lapse mode. Let us begin with Joshua. Yahweh selected Moses's former assistant and scout to lead the Israelites to the Promised Land. God promised that all would end well for Joshua as long as he meditated upon the book of laws day and night and "observe[d] to do according to all that is written therein." And with few exceptions everything did indeed go well for him. He led the people through the Jordan with the Ark of the Covenant at the head of the column. The river's waters parted just as the Red Sea once had, opening the way to Canaan. First all of the males were circumcised (producing a "hill of foreskins"), and Pesach was celebrated; then the cities of Jericho, Ai, and Hazor were conquered. The goal was to drive out and exterminate Canaan's population—with God's active support. He let stones rain down from heaven upon the enemy and halted the course of sun and moon "till the nation avenged itself on its enemies." The list of defeated kings runs to an impressive thirty-one names. Joshua built an altar for Yahweh and impressed the Mosaic Law upon his people word for word. After Joshua divided the country among the twelve tribes of Israel, he exhorted them once again to remain faithful: No foreign gods! No miscegenation with foreign people! Then the perfect ruler died, and the Israelites continued to reside in the land that God had promised them.

SAVIORS IN THE NICK OF TIME: JUDGES

The following generations once again forgot about Yahweh and "followed other gods, of the gods of the people that were round about them, and bowed themselves unto them, and provoked the LORD to anger." This sounds familiar: the people murmured and murmured. The old religions were just too attractive. "And the anger of the LORD was hot against Israel, and he delivered them into the hands of spoilers that spoiled them, and he sold them into the hands of their enemies round about." But before their hardships could become too great, God "repented" the situation. He sent a "judge," a hero, who defeated the enemies and steered the people back to the straight and narrow. But never for too long: "And it came to pass, when the judge was dead, that they returned, and corrupted themselves more than their fathers, in following other gods to serve them, and to bow down unto them." God proved surprisingly patient, however, for the pattern of "falling off, salvation, and relapse" repeats more than a dozen times in the book of Judges.[2]

These were not judges in the judicial sense, however. They were military leaders, chieftains, and local rulers.[3] They were powerful, at times obscure heroes, such as Samson, who slew 1,000 Philistines with the jawbone of an ass before Delilah robbed him of his power by cutting off his hair after a night of lovemaking. Then there was the female judge Deborah who led attacks against the Canaanite kings. Judge Ibzan had thirty wives, Abdon forty, and Gideon "threescore and ten sons of his body begotten: for he had many wives." With one of his concubines Gideon also sired Abimelech, who killed sixty-nine of his half brothers in a bid to become king (he couldn't find the seventieth). Things didn't work out the way he planned, however, for Abimelech had to order his own armor bearer to pierce him with his sword in order to avoid the shame of being killed at the hands of a woman who had thrown a millstone at his head. And then there was judge Jephthah, "son of an harlot," who sacrificed his own daughter as a burnt offering because he had sworn a rash oath. Interestingly, the Bible doesn't mention whether Yahweh made an effort to stop this instance of human sacrifice.

THE TEMPTATIONS OF POWER: SAUL, DAVID, AND SOLOMON

Things really couldn't go on like this. The elders of Israel's twelve tribes begged Samuel—priest, prophet, and judge—to bring them a real king like other nations had. Samuel was skeptical, but God commanded him to anoint Saul from the tribe of Benjamin as king, for he was "an impressive young man without equal among the Israelites—a head taller than any of the others." As a military leader he had brought fear to the Philistines. But he was plagued by depression and violent fits of temper and in religious matters didn't exactly toe the party line—he was known to seek counsel from witches, for example.[4]

God felt he had to pick a replacement, and he chose David while Saul was still alive. David was capable of more than simply slinging stones at giants—he could also drive the tormenting spirit out of Saul by playing his harp. While in Saul's service, David met with one success after another, which earned him the king's jealousy. David fled into the wilderness and collected a band of outlaws around him. He proved a talented bandit chief, an unscrupulous warlord who did not shy away from serving Israel's enemy—the Philistines.

This is probably why the Goliath incident is incorrectly attributed to him.[5] The Bible actually admits as much: "And there was again a battle in Gob with the Philistines, where Elhanan the son of Jaareoregim, a Bethlehemite, slew the brother of Goliath the Gittite, the staff of whose spear was like a weaver's beam."[6] No, Elhanan was not David's second name; nor did Elhanan slay Goliath's brother, as has been conjectured, in order to preserve David's reputation.[7] The Goliath incident was foisted onto David in order to whitewash the stain of his collaboration with the Philistines.

King Saul lost the war against the Philistines in the Gilboa Mountains and fell upon his sword. The Philistines cut off his head and fastened his corpse to the walls of Bethshan. David then became king of Judah in Hebron. Later—after the murder of Saul's last son—the tribes of the north also elected him king of Israel. The Bible makes a huge effort to erase the slightest suspicion that David had staged a coup d'état. He conquered Jerusalem and made it his capital, and God promised that David's dynasty would endure for eternity. His kingdom stretched from the border of Egypt to the Euphrates.

The biblical story of David is surprising in just what an "unvarnished image"[8] it presents. David committed murder and adultery, after all. As we shall see, his succession would devolve into a fierce competition from which Solomon eventually emerged victorious. The latter may have had the law of primogeniture against him, but he had a seductive mother on his side. The brilliant statesman Solomon succeeded the brilliant warrior David. During her visit to Jerusalem, even the Queen of Sheba allegedly could not stop marveling at Solomon's power and wealth. His wisdom became proverbial, and Yahweh allowed him to construct the first temple in Jerusalem. But women—his harem included seven hundred wives and three hundred concubines—would be Solomon's undoing: "It came to pass, when Solomon was old, that his wives turned away his heart after other gods: and his heart was not perfect with the LORD his God, as was the heart of David his father." The punishment was soon in coming: the great kingdom collapsed, and from that time on there was a kingdom of Israel in the north and a kingdom of Judah in the south.

THE ROAD TO PERDITION: KINGS IN NORTH AND SOUTH

Before we continue we must ask how historical these stories actually are. Scholars generally agree that the conquering of Canaan never actually happened, for example. According to Israel Finkelstein and Neil Silberman, "The early Israelites were—irony of ironies—themselves originally Canaanites!" After the collapse of Egyptian rule over Canaan in the twelfth century BCE, the region saw the sedentarization of nomadic herdsmen in inhospitable mountainous areas.[9] Regional powers did indeed develop over time, but in light of the archeological evidence, most researchers assume that the biblical assertion of a great united monarchy is a "literary construct." The two kingdoms of Israel and Judah developed independently from humble beginnings.[10] And the real Jerusalem at the time of the biblical David and Solomon was no more than a modest mountain village.

No historical source outside the Bible mentions either Saul or Solomon. David's name has been found once, on a stele found at Tell Dan in 1993. The triumphal inscription (ca. 835 BCE) mentions a House of David defeated by an Aramaean king.[11] If the legendary kings really did exist (Saul as the ruler of the north, David of the benighted south), then they have hardly anything in common with their biblical portraits.[12] Their kingdoms were not nearly as powerful as the Deuteronomistic History would have us believe.

The story is different, however, when it comes to the rulers of Israel and Judah, whose histories we hear in the first and second books of Kings. Whereas the story of the division of the kingdoms, which occurred under the reign of Solomon's son Rehoboam, is almost certainly a fiction, the succession of the later kings generally agrees with the historical record. The Bible's editors most likely resorted to chronicles. Kings first tells the story of the kings of the north, about whom the Bible has very little good to report.

After all, God sent the Assyrians to destroy their kingdom (in 722 BCE)—surely they must have done something wrong. They had fashioned an image of Yahweh in the form of a steer and worshipped other gods. The kings of the south don't look so bad, but because they, too, did not always take their loyalty to Yahweh all that seriously, Judah eventually fell to the Babylonians as well (in 587 BCE). We needn't go into the savage details of all the misfortune described in the books of Kings.[13] The point of the Deuteronomistic History, which the two books of Kings end, is to explain how the Promised Land was won and lost again. The Bible's authors are trying to impress upon us that loyalty to Yahweh was the decisive factor. They may have done so by recreating history, but this makes the tale no less stirring.

THE BENEFITS OF GOD

The colorful stories of the Deuteronomistic History demonstrate how the one and only God had always directed the fate of Israel. It was he, rather than the expansionistic Assyrians or the poverty of the tiny kingdom of Judah, who had been responsible for every single turn of its history. Once again, God appears as the same heuristic principle that so impressed us in the Torah. During times when no one knew the true causes of historical transformations, God functioned as a variable, as a substitute actor, who enabled the Bible's authors to carry out systematic analyses and propose solutions. Here, too, intellectual religion functioned as a form of protoscience—in this instance, as protopolitical science or protosociology—aimed at deescalating social conflicts, avoiding defeats, and even preventing the collapse of the state.

"The Bible, like Homer's *Iliad* and Virgil's *Aeneid*, was an ancient attempt to imagine the origins of the nation," writes the historian David Biale in his conclusion to the voluminous *Cultures of the Jews*. And the Deuteronomistic History certainly ranks among the core texts of this "collective biography of Israel."[14] Even if political scientist Benedict Anderson had more modern phenomena in mind when he developed his famous concept of nations as "imagined communities," it is also useful here. As most of a society's members would never know the majority of its other members, the people needed a powerful concept of the community to keep in mind. And the Deuteronomistic History gave them just that: it produced an imagined community. "We are God's chosen people."

"In fact, all communities larger than primordial villages of face-to-face contact (and perhaps even these) are imagined,"[15] Anderson writes. We are dealing here with the same mechanism we encountered when discussing the genealogies of Genesis. The biblical authors constructed a collective identity by inventing a shared origin that could activate the kinship-fixated psychology of our first nature. In essence, the carefully composed history of God's people as an imagined community created a fictive kinship and thus simulated the feeling of security that hunter-gatherer groups once shared.

All this is extraordinary in that the imagined community took on the aura of existential urgency—thanks to Yahweh. He is not only the community's social glue, he also monitors its adherence to his orders. No other part of the Bible makes this as clear as the book of Joshua, which describes the Israelites' preparation for their legendary attack of the city of Jericho:

> And the LORD said unto Joshua, See, I have given into thine hand Jericho, and the king thereof, and the mighty men of valour. And ye shall compass the city, all ye men of war, and go round about the city once. Thus shalt thou do six days. And seven priests shall bear before the ark seven trumpets of rams' horns: and the seventh day ye shall compass the city seven times, and the priests shall blow with the trumpets. And it shall come to pass, that when they make a long blast with the ram's horn, and when ye hear the sound of the trumpet, all the people shall shout with a great shout; and the wall of the city shall fall down flat, and the people shall ascend up every man straight before him.

And that's exactly what happened. The people followed God's orders, and on the seventh day the mighty wall fell and the Israelites sacked the city. Archaeological excavations have shown that Jericho at the time of the Bible's tale was neither fortified nor even occupied, but that is not the issue here.[16] What matters is the story's message: God alone was responsible for the victory. He rewarded the Israelites for following his instructions down to the very last detail. This was not a war; it was a procession! The conquest of Canaan was in truth a "gift" from Yahweh.[17]

God, not a more powerful enemy, was also responsible for losses in battle. Defeat was God's punishment for disobedience. The Jericho story reflects this fact as well. Later, Yahweh ordered the city "utterly destroyed" and "all that was in the city, both man and woman, young and old, and ox, and sheep, and ass" put to the sword. The rest was to be consigned to flames—only valuable objects were to be saved for the "treasury of the house of the LORD." The Israelites next set out to conquer the city of Ai, but their army met with defeat. Horrified, Joshua tore off his clothes and threw himself to the ground. "Get thee up," God commanded. "Israel hath sinned, and they have also transgressed my covenant which I commanded them." Their crime? Theft, apparently, "for they have even taken of the accursed thing, and have also stolen, and dissembled also, and they have put it even among their own stuff." Joshua immediately set off to interrogate his people. A man called Achan confessed: "Indeed I have sinned against the LORD God of Israel, and thus and thus have I done: When I saw among the spoils a goodly Babylonish garment, and two hundred shekels of silver, and a wedge of gold of fifty shekels weight, then I coveted them, and took them."

The message is unmistakable: woe unto all if even a single individual fails to follow the law! All of Israel would pay for his sins. This was the old pattern of collective punishment. Wars, catastrophes, and epidemics as an expression of divine rage provided proof that the entire community must always pay the price. Moreover, the individual's responsibility for the actions of his fellow tribe and clan members was an age-old principle. Wrongful actions had led to blood feuds and vendettas since the dawn of prehistory.[18] The totalitarian insistence on absolute conformity encountered in the Bible is therefore not a conscious invention aimed at promoting proper social conduct but a reflection of our first nature. Yahweh simply made use of our preexisting tendency to monitor the behavior of others to ensure their errors did not bring us all down.

Achan's confession was not the end of the story, however: "And Joshua, and all Israel with him, took Achan the son of Zerah, and the silver, and the garment, and the wedge of gold, and his sons, and his daughters, and his oxen, and his asses, and his sheep, and his tent, and all that he had: and they brought them unto the valley of Achor. And Joshua said, Why hast thou troubled us? the LORD shall trouble thee this day. And all Israel stoned him with stones, and burned them with fire, after they had stoned them with stones."

Collective stoning is frequently explained as allowing for "deindividuation": if the community as a group punishes an offender, then the victim's relatives can no longer seek revenge.[19] But in Achan's case his entire clan was killed, leaving no one to avenge his death, which suggests a different intention. By executing Achan's group collectively, the people sought to prove how much they detested the perpetrator's crime and demonstrate to God that he no longer had the slightest reason to punish them. The stoning was therefore a credibility-enhancing display[20] aimed at God. Of course, it also sent an unmistakable message to all the other members of the community: no one should even think about following in the footsteps of the thief Achan. The stoning thus communicated something language could never achieve: it demonstrated just how serious the people were and stamped an "ineradicable seal" on the event—the fact of death.[21]

Thus the story of Joshua also deserves a place in anthropology textbooks, for it offers a prime example of the theory that religion is a "powerful cultural institution for the promotion of group cohesion."[22] The collective belief in God functions as social glue, especially when the religion is capable of carrying out such a massive program of "supernatural monitoring." We have already discussed in our chapter on the flood the idea that "watched people are nice people."[23] Belief in a "punishing god" who notices every act of theft brings undeniable advantages.[24]

THE SECRET OF *ASABIYA*

Let us now turn to one effect of religion as a form of protopolitical science that we believe is as revolutionary as it is prescient. Here we enlist the aid of Ibn Khaldun. The great historiographer, who was born in Tunis in 1332 and died in Cairo in 1406, attempted to explain the rise and fall of states in his major work, *The Muqaddimah.* In particular his notion of *asabiya* is now enjoying increasing popularity among sociologists. *Asabiya* describes the group solidarity that holds a people together; without it, a group cannot realize a shared goal. Ibn Khaldun derived the term from the word *asaba,* which refers to male relatives who are closely bound by intense "kin spirit."[25] *Asabiya,* Ibn Khaldun wrote, "leads to affection for one's relations and blood relatives, [the feeling that] no harm ought to befall them nor any destruction come upon them. One feels shame when one's relatives are treated unjustly or attacked, and one wishes to intervene between them and whatever peril or destruction threatens them."[26] Ibn Khaldun knew that other "social glues" beyond biological kinship hold groups together and promote solidarity: "The real thing to bring about the feeling of close contact is social intercourse, friendly association, long familiarity, and the companionship that results from growing up together, having the same wet nurse, and sharing the other circumstances of death and life."[27] Ibn Khaldun reserves the task of instilling *asabiya* in large groups for religion, for it eliminates rivalry and jealousy and "causes concentration upon the truth."[28]

The Arab scholar had learned this by observing the never-ending cycle of wars in the Maghreb.[29] The tribes of the steppe and desert have the greatest *asabiya*: "In the desert, each tribe can rely only on itself for survival against the harsh environment and depredations of other tribes," writes Peter Turchin in *War and Peace and War*. "Ibn Khaldun stressed that 'only tribes held together by *asabiya* can live in the desert.'"[30] The conditions of state formation in Israel and Judah were not all that dissimilar from those in the Maghreb. The political situation in the mountains of Palestine was turbulent and unstable. Historians speak of a "dimorphic" tribal system. Egyptian sources repeatedly mention the peoples from which the proto-Israelites emerged: the Shasu, nomadic steppe and highland herdsmen vilified as plunderers, and the Apiru, "uprooted peasants and herders who sometimes turned bandits, sometimes sold themselves as mercenaries to the highest bidder." The parallels with the Bible are obvious. There is much to suggest that David's bandit past reflects the actions of an Apiru leader who would first "be recognized as popular leader over the sparsely settled southern hills" before taking his Yahweh religion to Jerusalem.[31] We can even speculate that David's career also owes much to the fact that he could, thanks to Yahweh, instill *asabiya* among his followers. Ibn Khaldun wrote, "This is because aggressive and defensive strength is obtained only through a group feeling which means affection and willingness to fight and die for each other."[32] Indeed, in the Bible we find a list of David's closest comrades in arms, a sworn community prepared to lay down their lives for one another.[33]

Let us summarize: The historical framework of the Palestine of the twelfth to tenth centuries BCE and its tribal culture, with its harsh conditions on the edge of the steppe and desert, may have engendered a high degree of *asabiya* among the tribesmen, which etched itself deeply into the culture of the population that would later form the Israelites. By means of a "kind of natural-selection mechanism," every tribe lacking in internal solidarity was eliminated.[34] This is one reason why the Hebrew Bible places so much value on cooperation—and also why the Bible's Israelites were able to survive the worst oppression meted out by the Assyrians, Egyptians, Babylonians, Ptolemeans, Seleucids, and Romans.

THE CURSE OF LUXURY

Ibn Khaldun offers even more inspiration to biblical anthropologists. He noticed a cyclical pattern in the history of the Maghreb. Empires blossomed only to disappear again. He believed the "corrosive effect" of increasing luxury on group solidarity lay behind this phenomenon. He argued, "As the former tribesmen forget the rude ways of the desert, and become accustomed to the new luxurious life, they somehow become 'enervated.'"[35] He recognized that dynasties usually had remarkably similar life spans and rarely held power for more than three generations. Arab scholar Peter Enz sums it up in a nutshell: "The first generation is characterized by its wild, strong nature and remains under the influence of the original *asabiya.* The second generation lives as sovereigns in luxury and makes *umran* [civilization, culture] blossom. The luxury reaches its high point in the third generation, but at the same time *asabiya* is lost and an all-encompassing degeneration sets in."[36]

The reigns of Saul, David, and Solomon mirror this pattern remarkably well. The first generation—represented by Saul and by David during his stint as a mercenary leader—is that of the warlords, the powerful warriors, whose unspoiled *asabiya* inspires the troops to defeat even the most powerful of enemies. David's kingdom represents the second generation. He rules over a powerful realm and makes Jerusalem into his capital, but trouble soon began to brew at the court. The third generation, in which luxury reaches its zenith before leading to downfall, is the reign of Solomon. The Bible presents him as the epitome of a king: full of wealth and wisdom, he builds the magnificent temple for Yahweh. But he indulges too much; he keeps women and horses in excess. In his arrogance, Solomon is disloyal to Yahweh. He serves foreign gods and thus squanders the source of the original *asabiya.* The realm collapses, and the tribes of the north establish a new kingdom. David's dynasty is left with its own tribe of Judah, in which genetic kinship ensures a last dram of *asabiya.*

IT HAPPENS IN THE BEST OF FAMILIES

We do not wish to overstretch Ibn Khaldun's theory of history, given that the Bible presents us with more fiction than fact. But isn't it an unusual coincidence how well the theory also fits the stories of the patriarchs? Here, too, we are dealing with three generations: Abraham, Isaac, and Jacob. And once again the clan goes downhill in the third generation. Only Joseph's fortuitous career in Egypt prevents it from disappearing completely. And in Kings we come across the same problems analyzed in detail in our chapter on the patriarchs. The temptations of amassing property get plutocrats into trouble with downright scientific predictability. Despite the fictional enrichment, the Bible once again presents readers with human experience in condensed form—this time as protopolitical science.

We can see the same familiar patterns noted in Genesis, now magnified two or three times, in the stories surrounding King David's succession. Typically, David's ascendancy was accompanied by—for want of a better word—hoarding of women. David married Michal, Saul's daughter, for political reasons (the bride price was "an hundred foreskins of the Philistines"). He then took the clever Abigail, whose husband died under mysterious circumstances, and then went on to marry Ahinoam of Jezreel. After being made king upon Saul's death, David married four more women. And when he became king of all Israel, a huge surge in reproductive capital accompanied his rise in power: "David took more concubines and wives in Jerusalem, and more sons and daughters were born to him."

The story of Bathsheba illustrates the "corrosive" effect power has on rulers: "In the spring, at the time when kings go off to war, David sent Joab out with the king's men and the whole Israelite army. They destroyed the Ammonites and besieged Rabbah. But David remained in Jerusalem." As if that wasn't bad enough, David took it a step further: "One evening David got up from his bed and walked around on the roof of the palace. From the roof he saw a woman bathing. The woman was very beautiful, and David sent someone to find out about her." He learned that she was Bathsheba, the wife of Uriah, who was at that time fighting for his king and appears in the Bible's list of David's particularly loyal "mighty warriors." Knowing no scruples, David had Bathsheba brought to him and got her pregnant.

Never mind that adultery was a capital offense according to the Torah; never mind that the king had his way with the wife of a soldier who was away fighting. David then had Uriah brought back from the front to sleep with Bathsheba and thereby hide his crime. When Uriah, proud of his military ethos, refused to sleep with his own wife, David had no qualms about having him killed (murder—another capital crime, needless to say). He sent Uriah back to the front with a letter addressed to his commander, Joab, with the following order: "Put Uriah in the front line where the fighting is fiercest. Then withdraw from him so he will be struck down and die." And that done, David then married Bathsheba.

Here we have a king who remained at home while his people took to the field for him. A king who took advantage of the wives of soldiers who were away risking their lives for him. A king who couldn't give a damn about his God's commandments. A king who ordered his own officers to let his best man be killed. What better way to annihilate all *asabiya*? There's no clearer illustration of the problem posed by despotic rulers. They are kleptocrats who sacrifice their best men to slake their own lust—and they bring perdition down upon the states they rule.

Our overview of David's wives would be incomplete without the story of beautiful Abishag the Shunammite, who was put into bed with the senile old David. Rainer Maria Rilke wrote a poem about the young woman that begins, "She lay. And her childlike arms were bound / by servants around the withering king, / on whom she lay throughout the sweet long hours, / a little frightened of his many years."[37] It didn't do any good; David never regained his virility. We have just witnessed the world's oldest form of testosterone therapy, the practice of placing young girls in the beds of decrepit old men, called "shunamitism" for Abishag the Shunammite. This procedure was prescribed up until the Renaissance.[38]

Our chapter on the patriarchs discussed how polygyny leads to competition among the various wives. In the books of Kings, this isn't really a prominent issue. Even so, Bathsheba ensured that her son Solomon was made heir, despite David's having a number of older sons. Thus the potential for conflict is present in this book as well. Conflicts involving brothers, on the other hand, get a lot of attention. Murder, incest, and war between father and son feature prominently. The trouble started when David's firstborn son Amnon raped his half sister Tamar, whereupon her full brother, David's third son, Absalom, had him murdered. The Bible hardly mentions David's second son, Chileab, but he also disappeared. Absalom, however, the most handsome man in all of Israel, instigated a rebellion against his own father. And what did he do when David fled to Jerusalem? He ostentatiously slept with ten of his father's concubines—his stepmothers, in other words (David would later lock them away for the rest of their days for this deed). But Absalom's looks proved his undoing. While he was on the run, his hair got entangled in a tree, and David's commander, Joab, killed him as he dangled helplessly from a branch. "O my son Absalom, my son, my son Absalom!" wailed David, "would God I had died for thee." David had supposedly ordered Joab to spare his son's life, but is it really plausible that the loyal Joab disobeyed the king's orders and killed Absalom on his own?

At any rate, the official heir to the throne, Adonijah, had already had himself crowned king, but Bathsheba thwarted his plans by bewitching the senile old David. In the end her son Solomon was anointed king, and he first planned to spare the life of his older brother. But when Adonijah later asked permission to take David's last wife, Abishag the Shunammite, as his own, Solomon, fearing a coup was imminent, had his half brother murdered.

KINGS IN THE DOCK

The books of Kings make clear that little had changed since the days of Cain and Abel. When rich and powerful alphas take multiple wives, murder and mayhem are the inevitable consequences. Stable states cannot stand on such shaky foundations. Ibn Khaldun was right when he wrote that purely dynastic regimes harbor the seed of their own destruction. The alpha males' unbridled egoism, their kleptocratic accumulation of wealth, and the bloody conflicts among the ruling elite thoroughly destroy loyalty. The kings view the state as their personal treasure chest and allow social inequality to soar. And if one king's rule collapses, the next despot is already waiting in the wings to take over. The result is the cyclical history that Ibn Khaldun described.

Archaeological excavations confirm that the polities of the highlands in the Levant had a will to expand seemingly "written in their genetic code" (remember the first strategy of successful patriarchs in chapter 6: more!). Eventually, these attempts at expansion failed and ended with the elimination of the rulers and their families.[39] Institutions that could rein in despots' excesses and stabilize government were sorely lacking. The books of Kings tell of almost continuous bloodbaths under the successors of David and Solomon; hardly a single ruler managed to die in his sleep. Entire dynasties were wiped out, heads lopped off by the dozens, corpses of kings thrown to the dogs, children sacrificed, victims cannibalized. If we believe the Bible, chaos reigned supreme.

But why were the Bible's authors able to make this terrible history public? Other parts of the world certainly shared a similar past, but their rulers' misdeeds were never put down on paper. Of course we should not necessarily take the Bible at face value (as repeatedly noted, caution is advised when it comes to just about every part of the Good Book). The most horrible tales concern the kingdom of the north, the more powerful Israel eventually wiped out by the Assyrians. We have here a work of propaganda from the south intended to prove that God had no choice but to destroy the kingdom of Israel! This is the origin of the Hebrew Bible's criticisms of the kings, and once again the two lives of Yahweh play a key role. As we have seen, both Judah and Israel worshipped him, but the two states were in competition with one another. A discourse critical of the kings arose in both kingdoms, but remarkably each always directed it at the kings of the neighboring state. The destruction of the northern kingdom of Israel enabled the gradual development of a cogent new and more general narrative: selfish kings are our doom.

Countless Israelites fled to Judah, where priests and scholars soon developed a "pan-Israelite ideology." They combined the northern tradition of Saul with the southern tradition of David, and as everyone in both states was familiar with the negative aspects of the other kingdom's ruler, they had no option but to integrate the two. Hence the shocking candor and the unflattering descriptions of biblical superstars such as David.[40] When Judah fell more than a century later, the reason was staring the religious elites in the face: once again, the kings, with their debauchery, exploitation of the poor, and destruction of the all-important *asabiya*, had brought down God's rage. And because no Israelite king would ever again sit on the throne in Jerusalem—except during the Hasmonean dynasty much later in the first and second centuries BCE—cultivating this discourse was easy. The creative freedom of the exile and postexile periods made it possible to establish a God capable of passing such vehement judgment on the kings.[41]

Despite its somewhat shady origins in the defamation campaign against the opponent's kings, this new political philosophy forms the true core of the books of Kings. It represents a major milestone in the development of civilization. Here we can draw a remarkable parallel with the Torah: we are dealing once again with a catastrophe-prevention system aimed at safeguarding state and society. On the one hand, this system strengthens social cohesion (imagined community) and suppresses transgressions against social norms (supernatural punishment). On the other hand, it impeaches the parties responsible for social misery. The egoistic despots destroyed *asabiya* and drove the state to the verge of complete destruction! They should at least have abided by the laws that the Torah sets forth for kings:

> But he shall not multiply horses to himself. . . . Neither shall he multiply wives to himself, that his heart turn not away: neither shall he greatly multiply to himself silver and gold. And it shall be, when he sitteth upon the throne of his kingdom, that he shall write him a copy of this law in

> a book out of that which is before the priests the Levites: And it shall be with him, and he shall read therein all the days of his life: that he may learn to fear the LORD his God, to keep all the words of this law and these statutes, to do them: That his heart be not lifted up above his brethren, and that he turn not aside from the commandment, to the right hand, or to the left: to the end that he may prolong his days in his kingdom, he, and his children, in the midst of Israel.[42]

Once again our first nature offered the impulse for this type of state criticism. Hunter-gatherers knew no leaders, and only in times of conflict with other groups did they appoint a war chief, who had to relinquish his power again in times of peace. At all other times foragers maintained "leveling coalitions" aimed at preventing dominant individuals from monopolizing power. People simply do not like it when leaders coerce others (unless they themselves make it to the top, of course).

Whereas the Torah offers the theory, the books of Kings in the Nevi'im show us the reality. They operate as a historical and philosophical lecture intended to persuade readers that as long as Israel's leaders stick closely to the letter of the law, as did Joshua, everything will go well for them. In the long run, however, kings are simply incapable of doing so—and the Bible harbors no illusions. Although, like David, they might get off to a promising start, at some point kings fall victim to their thirst for power, wealth, and women. Israel's own leaders are to blame for its downfall—this is the truly revolutionary finding of the Bible's protopolitical science.

HOW COULD GOD LET IT HAPPEN?

Clearly not everyone took the Bible's message to heart. Inevitably, a number of rulers throughout history claimed to be the new Joshua.[43] However, their admiration for Joshua concerned not his pious study of the law but his success as a conqueror. Even greater was the fascination rulers had for David and Solomon. These two "have shaped western images of kingship and served as models of royal piety, messianic expectation, and national destiny."[44] Probably every ruler of the West has dreamed of going down in history for being as bold as David or as wise as Solomon—a beautiful example of the Bible's selective reception. But at best David and Solomon were only loyal to Yahweh during their ascent to power. No sooner had they reached the pinnacle of their rule than their faith began to head downhill.

But we have so far not asked one elementary yet critical question. Why did God even bother placing kings over his people? He must have known the extent of the misfortune this would bring. The Bible's authors had a great deal of practice defusing questions that might undermine God's authority. In this instance they employed the same immunization against refutation strategy as the Torah's writers used to explain why only Moses went to the top of Mount Sinai while his followers stayed behind: it was the will of the people! Once again, God tried everything possible to avoid placing a king above his people. The book of Judges offers proof of this: you can take God at his word. In times of need he sent down a savior á la Superman. Yet the elders had besought the prophet Samuel to provide them with a king "like all the nations." Samuel was not at all pleased, but God told him, "Hearken unto the voice of the people in all that they say unto thee: for they have not rejected thee, but they have rejected me, that I should not reign over them." He then added, "Yet protest solemnly unto them, and shew them the manner of the king that shall reign over them." In other words, he gave his people fair warning about how a king would take their sons and daughters away and impose taxes:

> And he will take your fields, and your vineyards, and your oliveyards, even the best of them, and give them to his servants. And he will take the tenth of your seed, and of your vineyards, and give to his officers, and to his servants. And he will take your menservants, and your maidservants, and your goodliest young men, and your asses, and put them to his work. He will take the tenth of your sheep: and ye shall be his servants. And ye shall cry out in that day because of your king which ye shall have chosen you; and the LORD will not hear you in that day.

There's no denying it: the people should have known what they were getting themselves into!

As already stated, the lesson of the Deuteronomistic History is clear: self-indulgent and lawless rulers bring doom upon the land. But not everyone was prepared to accept this development as inevitable. A new group of men arose who bravely promoted justice and advocated *asabiya.* These ambassadors of our first nature implemented God's program for correcting the kings. They were the prophets, and we now turn our attention to them.

12

THE PROPHETS

God's Word from the Mouths of Men

In the second main part of the Hebrew Bible, the prophets, not the kings, are the true stars. Even its name, Nevi'im ("prophets"), reflects this. As "intermediaries and channels of communication for the divine knowledge,"[1] the prophets not only had a direct line to God but also put the kings in their place. The prophet Samuel anointed Saul as king, yet later also told him that he had fallen out of God's grace: "Because thou hast rejected the word of the LORD, he hath also rejected thee from being king." The prophet Nathan promised King David his dynasty would reign forever but did not shy away from presaging great ignominy as a result of his dalliance with Bathsheba. The prophet Elijah, for whom Jews open the door during the seder on the first evening of Passover, got into a feud with Ahab, the Israelite king. In the end "the dogs licked up his blood"—Ahab's blood, that is. And then there is Ezekiel, who denounced the kings in the tone we have come to associate with the prophets of the Old Testament. The kings are "like a roaring lion ravening the prey; they have devoured souls; they have taken the treasure and precious things; they have made her many widows in the midst thereof." And the princes "are like wolves ravening the prey, to shed blood, and to destroy souls, to get dishonest gain." It's easy to see why theologians consider the prophets the "most striking representatives" of Yahweh's religion.[2]

The prophets' prominence is in fact rather curious. The Hebrew word for prophet (*navi*) means "spokesperson," and the Greek word *prophetes* originally referred to "God's spokesman before the people."[3] The prophets can therefore be seen as the "functional equivalent of pagan magicians, soothsayers, witches, and necromancers."[4] Their chief function is "communicating messages or information from the divine or spirit world to the human world"[5]—meaning they are undeniably related to the "shamans" and "spirit mediums" of the ancient religions.[6] But had not monotheism declared open season on such hocus-pocus?

The end of the Torah is quite clear. The revelation was complete: the prophet Moses died and was replaced by the law. That a future prophet could come along and correct the word of God was unthinkable. Nevertheless, the Torah appears to leave the back door cracked open for the prophets' return when it states, "And there arose not a prophet since in Israel like unto Moses, whom the LORD knew face to face." Although no other prophets came forth who matched Moses in stature, they did not disappear altogether. This truly is strange. What were they needed for? In the context of the intellectual-institutional religion surrounding Yahweh—a rational, rule-based catastrophe-prevention system based on the one and only God well on his way to transcendence—the idea that we needed individual prophets to help us decode God's message was simply preposterous. We have the law! We just have to study it. Why, then, are the stories following in the wake of the exodus and arrival in the Promised Land suddenly teeming with prophets? Isn't God's word enough? Had he not expressed himself with sufficient clarity?

MESHUGA OR REVOLUTIONARY?

The Bible just can't do without them. Establishing monotheism by means of laws alone would simply have been impossible. We have already noted the effort the authors put into embedding the Mosaic Law into the wonder-filled stories of the exodus. God himself was its main salesman; he demonstrated his power by performing miracle after miracle. He got away with this because the flight from Egypt happened during an almost "mythical" prehistory. In the Bible's primeval history, God could also put in personal appearances when he created the world and its inhabitants or sent down the flood. But the stories of Kings took place within historical times—right up to the Babylonian captivity. Daily experience had shown that

Yahweh no longer appeared or performed wonders like he used to, such as when he parted the Red Sea. This meant that God himself could no longer be the main purveyor of credibility.

Who could fill these shoes? When it came to making monotheism attractive and convincing, the kings were out of the question for obvious reasons. Nor were Yahweh's priests and theologians up to the task, since they had profited the most from the religion and were therefore generally suspected of promoting their own interests. It's no coincidence that we never find priests playing a starring role in the Bible. What other choice did the Bible's authors have than to bite the bullet and deploy the prophets, the figures that had always been seen as the real men of God? Ironically, this theological sin turned out to be a true stroke of genius. The authors went on to produce spectacular hybrid characters who played a decisive role in transforming Yahweh into a highly moralistic being the likes of whom the world had never seen.

But we get ahead of ourselves. Let us first have a look at what it meant to be a prophet. The Nevi'im distinguishes between the "former" and "latter" prophets. The former prophets appear in the books of Joshua, Judges, Samuel, and Kings—aka the Deuteronomistic History—with whom we dealt in the previous chapter; they include prophets such as Samuel, Nathan, Elijah, and Elisha, whose writings have not been handed down. The latter prophets appear in the books of the classical "literary" prophets, from Isaiah to Malachi. The prophets drew an unusual amount of attention to themselves, and, indeed, hardly an aspect of them hasn't been the focus of theological study. All this expert attention signals the difficulty of assessing the prophets' proper role in the Bible.

Christians celebrate the prophets as interpreters of the future, as the heralds of Jesus Christ. They therefore placed the books of the literary prophets at the end of the Old Testament canon, so they could serve as a prelude to the New Testament's fulfillment of the prophetical telling of the coming of the Messiah.[7] Jews, on the other hand, saw the prophets first and foremost as "teachers of the law."[8] For this reason the editors of the Hebrew Bible placed the Nevi'im directly after the Torah,[9] despite the fact that the prophets came, for the most part at least, historically before the Torah, making it impossible for them to have taught the Torah's laws.[10]

Since the nineteenth century, a great deal of speculation has surrounded the connection between genius and insanity, and the ecstasies of the prophets have certainly drawn much attention. Some (though not all) of these "religious geniuses" behaved as if possessed. They were believed to be *meshuga* ("crazy").[11] When touched by the spirit of God, they went insane: "Ezekiel smote with his hands, beat his loins, stamped the ground. Jeremiah was 'like a drunken man,' and all his bones shook," Max Weber writes. "When the spirit overcame them, the prophets experienced facial contortions, their breath failed them, and occasionally they fell to the ground, for a time deprived of vision and speech, writhing in cramps."[12] Their actions were downright shocking: Isaiah ran naked through the streets of Jerusalem for three whole years. Ezekiel lost his ability to speak and lay on his left side for 390 days. Jeremiah wore a yoke around his shoulders; others placed iron horns on their heads. Some were known to drool.[13] Psychologists have even begun to read the prophets' texts for indications of post-traumatic stress disorder.[14] After all, the pages of Jeremiah tell us, they were the regular victims of sword, famine, and pestilence.

Last but not least, the prophets were celebrated as social revolutionaries: "Woe to them that devise iniquity, and work evil upon their beds! when the morning is light, they practise it, because it is in the power of their hand," the prophet Micah rages about the kings. "And they covet fields, and take them by violence; and houses, and take them away: so they oppress a man and his house, even a man and his heritage." Theologians, however, have repeatedly questioned whether the prophets were really the "cool-headed social critics" some have claimed. They were under no illusion that they could bring about social change.[15]

A GIFT FROM HEAVEN

There's a problem with all these interpretations: they treat the prophets as real historical figures—something they are not. "According to the current state of research into the prophets, not a single one of the 'books of the Prophets' was written by the prophet who lent it his name."[16] Nor does undisputed evidence support the actual existence of the prophets who did not have writings named after them. The biblical texts are therefore of limited use when it comes to "reconstructing the ancient realities of prophecy." We should therefore distinguish carefully between "biblical prophecy" "as a literary/scribal phenomenon" on the one hand and "ancient Hebrew prophecy" on the other.[17] It bears repeating that the biblical prophets represent literary constructs, colorful hybrids of fact and fiction. But they were constructed to a clear end: to establish monotheism and make this new counterintuitive religion appear plausible.

The historical phenomenon of prophecy surfaces in many other ancient cultures. In general, we can differentiate between four main types of prophets. First there were the group prophets, who put themselves into a trance by means of music and dance. The second group comprises the temple and cult prophets, who were responsible for oracles. Third, there were the royal court prophets, from whom kings sought divine council. Finally, the fourth group consists of "oppositional individual prophets." You could not simply go to the latter and ask about God's actual intentions—as you could with other types of prophets—for they never knew when information would fall from heaven. Their number was the smallest, and during the time of their prophesying, they were held in the lowest esteem. They were definitely not the harbingers of good news.[18]

This is where things get interesting, for nearly all of the Bible's writing prophets belong to this last group. These prophets were in the fringe throughout the Near East, not just among the Israelites. In Mesopotamia, for example, historical evidence only reveals the third type of prophets attached to the royal court. But there we find no evidence of prophets who foretold the end of a royal dynasty or state,[19] even though this is par for the course with the biblical prophets. It was unthinkable that a Babylonian prophet would have spoken like Isaiah: "Therefore is the anger of the LORD kindled against his people, and he hath stretched forth his hand against them, and hath smitten them: and the hills did tremble, and their carcases were torn in the midst of the streets. For all this his anger is not turned away, but his hand is stretched out still." Why then do we come across precisely this type of outsider whose specialty was fright and ruination? The answer is surprisingly obvious.

The Nevi'im was written after the fact. Historically, there existed a number of prophets who were, generally speaking, prophets of salvation seeking to ensure Yahweh's support for the state.[20] Unfortunately for them, their prophesies had one fatal flaw: they were on the wrong side of history. In the Bible we instead encounter prophets of doom. On the roulette wheel of history, these types bet everything on black—on catastrophes, so to speak. And they won: most of the time the ball did indeed land on black. Because their prophesies did come to pass, they were proved true prophets—retrospectively.

This procedure is not as odd as it may seem, because the question of how to determine which words God has spoken receives the following answer in the Torah: "And if thou say in thine heart, How shall we know the word which the LORD hath not spoken? When a prophet speaketh in the name of the LORD, if the thing follow not, nor come to pass, that is the thing which the LORD hath not spoken."[21] Only after the fact can one know if a prophet was real or a charlatan.

We are therefore dealing with a classic case of survivorship bias. With Nassim Taleb, author of *The Black Swan,* we can say that the prophets included in the Bible as God's chosen mouthpieces made it in because the number of available prophets was once so large and the spectrum of foretold prophesies so rich that some inevitably had to be right. The unlucky prophets have been forgotten, lost in the mists of time.[22] Prophets such as Isaiah, Jeremiah, and Ezekiel simply happened to make the cut.

Once again the Bible's authors had an opportunity to do what they did best—put a positive spin on disaster. They recognized what these prophets of doom had to offer. Whether accidentally or not, history had transformed them into "true" prophets whose prophesies had been proven "real." They seemed to have really been in contact with the same divine force that steers earthly history. And what could have been more obvious to the Bible's authors than to enlist these men of God, sanctified by the course of history, and co-opt their prestige for their own mission? The prophets were a gift from heaven.

THE MAKING OF A PROPHET

Let us look over the biblical authors' shoulders as they bring in the prophets to reinforce the persuasive power of monotheism: They began by collecting surviving oracles that history appeared to have borne out. The predictions need not have stemmed from prophets known by name because the authors reworked and expanded upon these tales as works of literature. Most importantly, they embellished these prophesies with the clear message that the misfortune that befell Israel and Judah was actually God's punishment. Different authors continued to rewrite parts of the books of the prophets over the course of generations, and the process finally came to an end around 200 BCE.

The identity of these authors who reworked the biblical texts and used the prophets to put words in God's mouth remains a mystery, but they were certainly theologically educated scholars. Their creation was truly unique: the Old Testament's books of the prophets represent nothing less than a literary and theological innovation—and a very practical one at that. They served as vessels that could be filled with prophesies adapted to fit the new state of affairs. This constant reworking of the texts eventually transformed the biblical prophets into the visionaries of world history that they never were.[23]

The name of an established prophet was like a brand under which other prophesies were traded. In the book of Isaiah we find a second prophet who is considered the first strict monotheist. We know nothing of this person, however.[24] In the words of this unknown prophet, God says, "I the LORD, the first, and with the last; I am he." This prophet's words were integrated into the book of Isaiah, even though Bible scholars have located them a good two hundred years later. This probably reflects an effort to allow the new monotheistic religion to profit from Isaiah's fame. Later theologians would give this anonymous author a name: "Deutero-Isaiah," the second Isaiah. Theologians still debate whether this second Isaiah had been a stand-alone prophet or was simply the invention of a biblical editor. A third Isaiah, who also appears in the book, "Tritero-Isaiah," is almost certainly a later addition.[25]

As Old Testament studies professor Reinhard Kratz notes, "The fact that the words of the prophets that have been handed down do not come directly from the prophets but from a later time and that the authors were not the prophets themselves but anonymous scholarly tradents will presumably confuse some readers."[26] The inescapable conclusion is that the books of the prophets are hybrid products whose prophesies of doom were verified by history and whose prestige served to increase the credibility of the Bible's message.

THE PROBLEM WITH MIRACLES

Let's turn from the books of the prophets to the prophets themselves. They, too, are hybrid creations. The messengers of God's word had to be made especially convincing. A particularly fascinating and sustainable process was at work here. We shouldn't forget that the competition in those days was tremendous. Like shamans, witches, and magicians, prophets had to master credibility-enhancing displays (CREDs). Anyone claiming to have a direct line to the divine always meets with a healthy dose of skepticism.

We have already discussed how the cultural immune system, as Joseph Henrich calls it, generates a demand for CREDs. Talk alone simply is too cheap. Despite the enormous advantages that language has brought us, it has always struggled with a significant disadvantage: "The weakness of the word is the possibility—the likelihood—of lying, of fraud and trickery."[27] Everyone possesses a fine-tuned sensorium keen to expose liars or cheaters[28] and quick to sound the alarm when it hears unverifiable stories intended to provoke concrete action.

We have to remember that the beginnings of language probably reach back some 500,000 years.[29] We pay attention to far more than words alone and have become masters at detecting the slightest behavioral ticks that can signify lies, such as seemingly trivial twitches of the corner of the mouth. We are always on the lookout for actions, for they alone are convincing. Actions speak louder than words: this is a basic truth of human evolution. Accordingly, since the days of the first shamans, anyone claiming to be a medium of the supernatural had to pull out all the stops to convince the community that he or she had been touched by the divine spirit and spoke "God's words in the mouths of men."[30] Thus men (and, to a lesser extent, women) of God were always the masters of all kinds of trickery. To this day we still have a weakness for ecstatics who, possessed by a divine spirit, fall into evident states of rapture.

The Bible's authors found themselves facing a dilemma. How could they convince readers that their subject really had been in contact with the supernatural? By having him perform supernatural acts of course! The classic strategy for lending authority to a prophet was to have him do miracles. Miracles captivate our first nature. But monotheism listed the practice of magic as one of intuitive religion's forbidden acts. It would have been against the rules for any proper biblical prophet to perform wonders—in theory, at least.

The Bible's authors circumvented this problem by resorting to traditions handed down from the times in which such wonders were not yet frowned upon. Indeed, many old stories remained popular because they told of the miraculous works of Yahweh's prophets, and the authors found they simply couldn't do without them. Although not a single text has been attributed to the prophets Elijah or Elisha, the books of Kings nonetheless provide extensive descriptions of their performances as God's magical assistants.[31] These prophets did in fact perform a series of classic miracles: they made sure one woman never ran out of flour in her pot or oil in her jar, they made tainted water potable, and they even brought dead children back to life.

But magicians and wonder-workers were a dime a dozen in the ancient world. To get ahead, a prophet had to prove that he was more powerful than the competition, and engaging in a head-to-head duel with another prophet was the easiest means of doing this. Moses and Aaron had shown the way in their showdown with Egyptian sorcerers, but that was nothing compared to Elijah, who defeated 450 prophets of the god Baal in the service of the Israelite king Ahab. They tried and tried but could not bring Baal to ignite and burn a sacrificial bullock. Yet even though Elijah had poured buckets of water over his bullock and the firewood, the "fire of the LORD" consumed the sacrifice. Elijah had the defeated prophets gathered up and put them to the sword at the brook Kishon. The message was clear: Yahweh's prophet was stronger than nearly half a thousand competitors.

There was even more to the story: Baal's prophets called out the name of their god, danced around the altar, "cut themselves after their manner with knives and lancets, till the blood gushed out upon them," and even entered into ecstatic trances. Yet their entire repertoire of magical acts was of no avail. In contrast, Elijah had only to speak to Yahweh, who heard him right away. Here we can witness monotheism's process of disenchantment in action. The Bible's authors integrated the miracle face-off (it was just too good to pass up) but also impressed on their readers that the word alone was superior to any type of magic.

"THE LION HATH ROARED, WHO WILL NOT FEAR?"

A god who has a problem with hocus-pocus doesn't exactly make things easy for his prophets, for with respect to magic, the Bible's authors couldn't preach water and then let the prophets drink wine. Viewed from this perspective, some of the biblical prophets' typical characteristics start to make sense. Since God was invisible, he could not encourage fantastic visions. This explains why experiences such as those of Ezekiel or Isaiah ("I saw also the Lord sitting upon a throne, high and lifted up, and his train filled the temple. Above it stood the seraphims: each one had six wings") were the exception rather than the rule.[32] Yahweh's prophets had to make do with the voice of the Lord. And because all forms of magic were taboo, they had to rely entirely on their own charisma. Many of their essential features were thus not character traits but CREDs pure and simple. Here are a few of the most striking examples.

A true prophet did not foist himself upon his public. Isaiah was an exception, when, after hearing God ask, "Whom shall I send? And who will go for us?" he answered right away, "Here I am; send me." No, the perfect example of the good prophet was Moses, who balked at the task—not least because of his slow speech and difficulties with pronunciation. Jeremiah, too, was hesitant: "Ah, Lord GOD! behold, I cannot speak: for I am a child." Whereas competing prophets did everything they could to hear the word of God, it simply came upon the prophets of the Bible. Amos summed it up in a nutshell: "The lion hath roared, who will not fear? the Lord GOD hath spoken, who can but prophesy?"[33] When Jonah received his calling, he immediately sought to escape by ship. We all know what happened next: God sent down a storm, and Jonah was tossed into the sea by the crew, swallowed by a great fish, and brought back to land. He then changed his mind and did as commanded. For prophets in Yahweh's service, opting out was not an option. They had to proclaim God's word whether they wanted to or not.

True prophets neither accepted payment nor strove for office. In this they differed from temple prophets in official positions or freelance soothsayers, who offered divine information in return for a fee. "Thus saith the LORD concerning the prophets that make my people err, that bite with their teeth, and cry, Peace," Micah complained of his competitors, "and he that putteth not into their mouths, they even prepare war against him."[34] No one would have even the slightest suspicion that a prophet of Yahweh was actually serving his own interests. He couldn't profit from his prophesying even by gaining in reputation, for a proper prophet was always the subject of ridicule. Amos was reviled as a troublemaker and chased out of town, for example. And prophets were the polar opposites of kings. Whereas the latter accumulated lots of wives, Hosea shared his one wife with other men, for Yahweh had commanded him to marry a harlot. Indeed, God's instructions came with considerable risks. The Israelite royal couple of Ahab and Jezebel had all of Yahweh's prophets killed (only Elijah managed to escape). Jeremiah was threatened and tortured, and people tried to kill him. The prophet Urijah was even executed by King Jehoiakim. Yet nothing could prevent a prophet from carrying out his divine mission.

THE BIRTH OF CHARISMA

Clearly these CREDs were intended to offer proof of the prophets' own selflessness. They show that there is nothing to be gained from proclaiming God's word—quite the opposite, in fact! This signal targets the Bible's readers: these are no tricksters; the prophets are not pursuing their own interests. They care only about the word of God, which is so important to them that they are willing to risk their own lives to proclaim it.

In an astounding twist of cultural evolution, the very process of disenchantment made the prophets into magical figures. As they no longer had any sorcery at their disposal and miracles were taboo, they had to invest everything in their public faces. This is the source of their enormous charisma—their "magical" capital—which we humans find so beguiling. Our first nature is in a constant state of readiness, always on the lookout for manipulators and fraudsters. But it becomes quite enchanted with those who convince it of their total commitment to the cause and lack of self-interest. This signals to us that they serve a cause so momentous that they are prepared to make the highest sacrifice.

This credibility-enhancement strategy makes the prophets the ideal advocates for justice and *asabiya.* They're not bound by outside interests, they are nonpartisan, and they are beholden solely to God's truth. Because they have absolutely no interest in earthly goods, they are incorruptible. They are, quite simply, not in it for the money. Their altruism makes them the enemies of injustice and the opponents of power-hungry kings. As adversaries of the rich, they automatically take the side of the poor and the weak.

In theological terms, prophesying should have been forbidden to them, given that God's revelation was already complete. But precisely because God had become abstract, their presence became necessary. His will made into law might engage our third nature, but we certainly don't find it all that captivating. The Nevi'im of the Bible once again offers evidence of our difficulty in coming to terms with intellectual religion. The prophets show that we still need people of flesh and blood. Thus, the prophets are at their most impressive not when they pronounce judgment after judgment and "grounds for punishment"[35] but rather when they appear as normal people ready to give their all for God's cause. This is why the early Jewish and Christian traditions once considered Elijah the most important prophet alongside Moses,[36] for he behaved like our first nature would expect of a man of God. He performed miracles, fearlessly defied his enemies, and at the end of the story ascended into heaven in a chariot drawn by horses of fire.

So how does one best convince people? With people who dedicate themselves to a cause that speaks to our egalitarian hunter-gatherer ideals—and if they perform miracles to convince us of their mission, then so much the better. This is the foundation of "charismatic authority."[37] Whether this was a trick on the part of the Bible's authors is anyone's guess. At any rate, the Bible's dead prophets helped to defend against the living ones and the ubiquitous charismatic wandering preachers of ancient Israel[38]—and this in turn helped solidify the priesthood's authority. The transformation of the canonized prophets into such extremely charismatic figures raised the bar for contemporary prophetic competitors to dizzying new heights. We are not giving away too much when we say that we will encounter only one other person in the Bible who could stand up to the biblical prophets. His name? Jesus of Nazareth.

GOD'S DEBT TO THE PROPHETS

These heroes of altruism did not leave the being in whose name they prophesized unchanged. The prophets propagated a God "who demanded justice for himself as well as humanity" and, to achieve this, was even "willing to accept the demise of his own people."[39] Such a thing was unheard of in the ancient world. Nowhere else did prophets practice such a degree of social criticism or "take to heart the fate of the *people*."[40] We are dealing here with a major leap forward in civilization. For the first time ever, a God committed himself entirely to the cause of justice.

We have now brought together all of the decisive factors in this process of cultural evolution. First comes the universal biological substrate formed by the egalitarian impulse of our first nature and its deep hatred of despots. Then comes the historical coincidence that Yahweh had two lives, one in Judah and one in Israel, allowing for the unfolding of a critical discourse directed at the kings with an intensity that would have been impossible anywhere else. Next there was the process of monotheization, helped along by the process of secondary state formation in Palestine. It subjected the actions of the one and only God to a unique form of compulsive coherence seeking, since it was impossible that God would operate according to a double standard. Finally, his propagandists, the prophets, those masters of selflessness, transformed him into a pure principle that rose above all other interests. To put it in more provocative terms, this God became so extremely moralistic not least because he, as a counterintuitive monotheistic god, had such a pressing need for credibility.

This is most likely one of the Bible's greatest achievements—an act that was well ahead of its time. Out of the spirit of catastrophe prevention, justice emerged as an abstract principle that generated a universal morality. Rich and poor are equal before God's law, an orientation that warms our old hunter-gatherer hearts. A disposition toward equality and an outrage at tyrants are in our blood, our first nature. And this is why we continue to believe in justice and fairness to this day.

In the Nevi'im we have seen how Yahweh was transformed into the first utterly moralistic God in the history of the world.[41] He shows the tyrants the limits of their power. But here another problem arises: Doesn't this contradict Yahweh's own actions? Why does he so often appear to act like a tyrant himself? To this day many people have a hard time understanding how such a highly moralistic being can display the dark and violent traits they see in the God of the Old Testament. This question deserves a chapter of its own.

13

HOW CAN A GOOD GOD BE SO BAD?

On Heavenly Morality

As stated at the end of the previous chapter, Yahweh was the first thoroughly moral god to appear on the world stage. Where else does one find a deity who stood up for the weak and the poor and subjected kings to his own justice? Who demanded we love our neighbors and whose commandments are still followed to this day? At the same time, this same God made things a great deal more difficult for a lot of people, for we still have the paradox that this paragon of morality also had a penchant for extreme violence. God's wrath ranks among "one of biblical theology's great mysteries."[1] We already grappled with God's fury in our chapter on the flood, but here we wish to take a closer look at the consequences of a good God who does bad. After all, the issue preys on more than a few people's minds.

The pages of the Nevi'im continued what the Torah had begun. The Israelites, under the leadership of the faithful Joshua, captured Hazor: "And they smote all the souls that were therein with the edge of the sword, utterly destroying them: there was not any left to breathe: and he burnt Hazor with fire." Yahweh's direct orders don't leave much to the imagination. The divine judgments foretold by the prophets, those "champions of ethical monotheism,"[2] also represent true acts of barbarism. Hosea threatened that Samaria, the capital of the northern kingdom of Israel, would fall into the hands of the Assyrians and, because it had become unfaithful to Yahweh, its people "shall fall by the sword: their infants shall be dashed in pieces, and their women with child shall be ripped up." Is this the Good Book everyone claims to know?

Steven Pinker primarily drew on the Old Testament as evidence to support his theory that the people of biblical times endured a great deal more violence than we do today. He cites biblical scholar Raymond Schwager: "Aside from the approximately one thousand verses in which Yahweh himself appears as the direct executioner of violent punishments, and the many texts in which the Lord delivers the criminal to the punisher's sword, in over one hundred other passages Yahweh expressly gives the command to kill people." According to Pinker, "The Bible depicts a world that, seen through modern eyes, is staggering in its savagery."[3]

For critics of religion like biologist Richard Dawkins, such verses are grist for the mill. In *The God Delusion*, Dawkins draws a portrait of God that many believers find shocking: "The God of the Old Testament is arguably the most unpleasant character in all fiction: jealous and proud of it; a petty, unjust, unforgiving control-freak; a vindictive, bloodthirsty ethnic cleanser; a misogynistic, homophobic, racist, infanticidal, genocidal, filicidal, pestilential, megalomaniacal, sadomasochistic, capriciously malevolent bully."[4] Quite a few Christians would actually like to get rid of the Old Testament altogether. The Bibles found in hotel nightstands usually only include the book of Psalms and the New Testament, for example.[5] It seems the composers of Gideon's Bible worried that the rest of the Old Testament might induce nightmares.

This concern has a tradition reaching back nearly 2,000 years. One of the very first Christian theologians, Marcion of Sinope (ca. 85–160 CE), wanted to throw out the Hebrew Bible altogether. According to him, the God of laws and prophets was "the initiator of evil, the starter of wars, inconsistent in his decisions, and even contradicted himself."[6] The early church decided differently, however, and declared the Hebrew Bible the Old Testament canon—and Marcion a heretic.

Even though theologians have put a great deal of effort into showing that the God inhabiting the Old and New Testaments is, in the words of pastor Dietrich Bonhoeffer (1906–1945), "one and the same," all the irritation refuses to go away. Theologian Bernd Janowski concludes his book *Ein Gott, der straft und tötet?* (A God Who Punishes and Kills?) with a plea that we face up to the difficult notions of the God we find in the Old Testament, "whose offensiveness continues to challenge us to this day." We would be well advised "to neither suppress nor demonize them, but to approach them with understanding—to the extent that this is possible."[7]

Behind all these efforts is not just the legitimate fear that Yahweh's dark side might corrupt his good one—if not disqualify him altogether. After all, we don't usually praise dictators even though they also might have established functioning social welfare systems or advanced animal protection laws in the shadow of their terror regimes. Could it be that we are dealing here with "tribal" in-group morality, according to which we have to be nice to our own people but antagonistic to all outsiders?[8] The thought is hard to dismiss when reading the Old Testament.

THE DELICATE SUBJECT OF MORALITY

Our lack of understanding of God's cruelty is mainly a modern phenomenon. We saw this in our analysis of the Torah. Yahweh was a "god of just retribution" in Max Weber's characterization. German theologians prefer to refer to the concept of the "deed-consequence nexus" in which every human wrong results in a punishment. In this respect Yahweh was conceived of as having a thoroughly righteous mind. His acts of violence were punishments for humanity's transgressions.

Dominic Johnson, proponent of the supernatural punishment hypothesis, has pointed out, "The term and common conceptualization of 'morality' is a recent and Western phenomenon" that is a feature of WEIRDs (Western, educated, industrialized, rich, democratic people). This concept resulted in a "narrow conception of morality that revolves around caring/not harming, fairness, and justice." For this reason, the question is not whether the existing moral norms fit with today's standards but "whether supernatural punishment affects people's adherence to the norms of their society—whatever those norms may be." Indeed, having a god with a proclivity toward violence often contributed to the success of a society.[9]

Only very few believers will accept this argument and make their peace with the Old Testament God, since this would imply we should view him as a historical figure who underwent a number of transformations. No, God is supposed to be eternal, and his morality should fit with today's standards. Slaughtering enemies is simply unacceptable. At this point we could simply put aside such reasoning to avoid moving the whole discussion into the realm of theology. But the question of the gods' morality is also a hotly debated topic in current research on the cultural evolution of religion, and we simply cannot avoid this important and exciting question.

What, then, is Yahweh's morality really all about? Why do we have so many problems coming to terms with it? The question merits our attention, for it can offer deep insights into both godly and human nature. In the end we will have to conclude that the God of the Old Testament really doesn't deserve his bad reputation.

GOD, TOO, IS ONLY HUMAN

The modern difficulty in accepting that Yahweh's character has both a dark and a light side reveals a lot about how our minds work. Our displeasure at coming to terms with a good but violent God stems from the simple yet unusual fact that we can only imagine God as a person. As Pascal Boyer reminds us, we humans rely on ontological categories—mental templates such as "animal," "human," and "tool"—to understand the world, to formulate our knowledge of it, and to help us frame our expectations. Should we encounter an unknown entity that seems to fit in the category of "animal," for instance, then the template offers us information about its features. We know that it has to eat, reproduce, and die, for example.[10]

Not surprisingly we turn to the mental template "person" when we try to imagine God. It fits with an evolutionary genealogy that traces a path from ancestors and spirits all the way up to gods. Thus people all over the world simply know a lot about the behavior of supernatural beings, even if they have never really encountered one. To put it bluntly, the gods behave like people. We find certain, mildly counterintuitive deviations acceptable (the fact that they are invisible, for example) when these serve to increase their attractiveness. But the gods lose their credibility when these deviations become too great.[11]

If we use the concept of "person" for a god, then it brings along the same set of expectations of consistency that govern our interactions with real people. A person's behavior may not be too irregular, for example. By and large it must be predictable; otherwise the alarm system installed by evolution will go off: "Watch out! Impostor!" Or worse, "Psychopath!" This is why we are so confused by Yahweh's apparent split personality. When he appears quite "jealous, wrathful, and punitive, on the one hand, and compassionate, patient, and forgiving on the other,"[12] it seems as if he has a pathological personality disorder. "Careful!" our first nature whispers, "you don't want to get mixed up with a guy like this."

Accordingly, theologians have put a great deal of effort into ironing out these incongruities. "If one does not wish to take seriously the dark, abysmal, and hidden side of our notion of God, then we lack the prerequisites for properly understanding his bright, loving, and uncovered side."[13] Such theological statements might make sense to our third nature but elicit an alarm response in our first nature. These antagonisms simply don't fit with our template for "person"—let alone with the subcategory of "moral, trustworthy person." As a result, many believers prefer to disregard the fact that their God was once a mass murderer (the flood or the killing of the Egyptians in the Red Sea) and master torturer (his directive to impale apostates). This is why over the past 2,000 years the Bible, as Steven Pinker explains, "has been spin-doctored, allegorized, superseded by less violent texts (the Talmud among the Jews and the New Testament among Christians), or discreetly ignored."[14] If we are to trust in God, then we first have to eliminate his inhuman side.

Here one could argue that we should allow a god to be more than a mere person. Shouldn't a god be allowed to burst out of the template? He cannot be bound by earthly concepts of morality. This is indeed the position taken by the catechism of the Catholic Church: "We must therefore continually purify our language of everything in it that is limited, image-bound or imperfect, if we are not to confuse our image of God—'the inexpressible, the incomprehensible, the invisible, the ungraspable'—with our human representations." God is, and shall forever remain, "a mystery."[15] Which brings us to another point that makes our modern difficulty in tolerating divine behavior so illuminating: our inability to accept God's mysterious nature is wonderful proof of the thesis that humans are equipped with a moral outlook that is completely independent of religion.

Many people believe religion is the source of moral values in human society, but this is simply not the case. The gods did not give morals to us—we gave morals to the gods. Did we not just now demonstrate how God has to satisfy human demands? People cannot but seek out God's motives using their own intuition before judging his behavior. That's just how our psychology works. A long-standing assumption of the evolutionary sciences is that our first nature is equipped with the ability to produce moral judgments. Indeed, moral behavior was of the utmost importance when it came to survival, even in the days of the hunter-gatherers. Our ancestors didn't require the aid of a supernatural authority, for the group was capable of taking care of itself. Richard Dawkins sums it up in a nutshell: "We do not need God in order to be good—or evil."[16] And primatologist Frans de Waal writes book after book showing how we can interpret the actions of apes and monkeys as reflecting altruism, empathy, and even justice. In conclusion, morality is much older than religion.[17]

That people are naturally moral beings is no longer something that only evolutionary scientists assume. Egyptologist Jan Assmann is convinced that "justice was not born of religion, but it permeated religion from without." It had "long since been a part of the world, for otherwise human coexistence would not have been possible."[18] We have no choice but to expect God to conform to our innate human morality.

AND THE IMMORAL GODS?

The delicate question of the gods' morality is also a current topic of discussion among scholars of the cultural evolution of religion. One group believes the appearance of morally concerned gods played a decisive role in the advance of civilization. We've already touched on this concept in a previous chapter. According to this theory, "big gods"—powerful, morally concerned deities believed to monitor human behavior—enabled large, anonymous societies to hold together, as older cohesive forces such as kinship and indirect reciprocity maintained by reputation were no longer sufficient.[19] Even if not the "single magic bullet," these big gods were—according to this theory at least—a major source of the social glue that keeps large societies intact.[20] People tend to behave better if they believe they will be rewarded or punished for their actions.

This position is now subject to criticism, however, because so many successful cultures have never had such powerful, morally concerned deities believed to be monitoring human actions. For example, sociologist of religion Rodney Stark has pointed out that the gods interested in human social cohesion "were largely lacking in supernatural conceptions prevalent in much of Asia and in animism and folk religion generally." He also positioned the hedonistic gods of the Greeks and Romans against these presumed moral big gods. "The Greco-Roman gods were quite morally deficient," he explains. "They were thought to do terrible things to one another and to humans as well—sometimes merely for amusement."[21] How could gods such as these have a positive influence on human behavior?

Nicolas Baumard and Pascal Boyer share a similar view: "Recently in human evolution, and invariably in large-scale societies, there emerged organized religions, with codified ceremonies and doctrines as well as specialized personnel. In these traditions the gods were generally construed as unencumbered with moral conscience and uninterested in human morality. That is the case for the gods of classical antiquity, Sumerian, Akkadian, Egyptian, Greek, and Roman, who did not care whether people followed moral codes as long as they provided sacrifices or showed obedience. The same applies to the Aztec, Maya, and Inca gods and to classical Chinese and Hindu deities." These two scholars even go so far as to question whether moral gods are needed at all. As a matter of fact, "the most successful ancient empires all had strikingly non-moral high gods." Could having moral gods who were constantly sticking their noses into earthly affairs actually have posed a disadvantage? "To simplify somewhat, the Romans, with their non-moralizing gods, built one of history's most successful predatory empires. They then converted to Christianity, a moralizing religion, and were promptly crushed by barbarians with tribal, non-moralizing gods."[22]

ZEUS, THE OLD WOMANIZER

How about the supposedly immoral gods of the ancient Greeks? Take Apollo, for example. The satyr Marsyas once had the gall to challenge Apollo to a musical competition. Marsyas lost, and Apollo skinned him alive. The satyr's blood flowed forth and formed the river that bears his name. A horrible fate to be sure, but is there not a moral message behind it? There is, and a clear one at that: don't even think about trying to mess with the gods! Sisyphus, the king of Corinth, also attempted to challenge the gods and even managed to tie up Thanatos, the god of the dead. Suddenly the people of earth could no longer die. Sisyphus's punishment is the stuff of legend; he had to spend his days rolling a boulder up a hill. Whenever he neared the top, the stone rolled crashing back down. And then there was Tantalus. Because he had served up his own son at the gods' dinner table, he was made to suffer eternally the pain now named for him: tantalization. This is a key moral message of Greek mythology: the gods have no patience for hubris. Coincidently, didn't the God of the Bible also respond with a similar fit of pique when Adam and Eve dared to defy God's command and take their fate into their own hands and when his people were audacious enough to build a tower reaching up into the heavens?

Zeus was certainly a notorious adulterer with no qualms about using trickery to get what he wanted. He took the form of a swan to ravish Spartan queen Leda, he kidnapped the princess Europa in the guise of a bull, and he had his way with Danaë in the form of golden rain. All this philandering enraged his wife, Hera, of course. But Zeus did a great deal of good as well. He is known as the god of hospitality and seekers of protection. Hesiod even celebrates him as the father of justice. The Greek gods were by no means lacking in morality—they were just morally imperfect. A bit like us, actually.

And so is Yahweh. The book of Samuel tells of how Yahweh commanded David to perform a census of his people and then punished him for doing so, condemning 70,000 people to die of the plague. The Bible's authors found this so embarrassing that when they told the story again, in the book of Chronicles, they replaced God with Satan.[23] And God's torment of Job wasn't exactly aboveboard either (more on that story later).

We hold our gods to such high moral standards and expect their behavior to be beyond reproach for various reasons. The first is our old hunter-gatherer soul, which demands absolute moral integrity from those telling others how to behave. Preaching water in public and drinking wine in secret—humans have always found such hypocritical behavior scandalous. This is a part of our cultural immune system aimed at protecting us from frauds. The second reason is the phenomenon of monotheistic bias, a kind of halo effect that leads us to expect that God is perfect in all respects, especially when it comes to morality.[24] This is a recent problem, because the idea of God as the embodiment of absolute good is Greek in origin and only began to influence early Christianity in late antiquity.[25] Finally, we have trouble accepting the collective punishment meted out by the God of the Hebrew Bible, because it may (unfairly in our eyes) also target innocent individuals—after all, the Last Judgment had yet to be invented. We will return to this issue in chapters 15 and 16.

FIRST COMES VIOLENCE, THEN MORALITY

Our evolution-inspired reading of the Bible can help readers to better understand these relationships. According to our theory, catastrophes are a key factor in understanding religions. We humans interpret disasters as the actions of the gods, and the notion that "human transgressions lead to God's wrath" in the form of disasters surfaces in the cultures of the entire Near East and Egypt.[26] And not only there: "In every human society ever studied by anthropologists, uncontrollable tragedies have been seen as caused intentionally by a mindful, supernatural actor," Jesse Bering explains. "For most of us, this 'agent' is God."[27] In our chapter on the flood, we explain why the human psyche functions in this manner. Illnesses have been associated with supernatural actors since time immemorial,[28] and there has never been a war in which the gods weren't believed to have played a role.[29] Every earthquake or other natural disaster has been seen as the work of powerful beings.[30] Well into the modern era, "misfortune has been seen as God's answer to the sins of the world."[31] And even in this scientifically enlightened time, we regularly hear voices claiming AIDS, a tsunami, or Hurricane Katrina—to name a few recent examples—are the acts of a punishing God.

Catastrophes *are* the gods. This is why people experience supernatural beings as wrathful, violent, moody, impulsive, and, most importantly, unpredictable. Of course we also attribute to the gods positive events such as rain that comes just in the nick of time, but negative events are more psychologically powerful, and we feel a more urgent need to make sense of them.[32] So if catastrophes represent the gods' main activities, then this has consequences for the question that titles this chapter: we shouldn't at all be surprised when moral gods behave violently and brutally. Our everyday experiences made it perfectly clear that the gods were fundamentally violent. Therefore, the real question we must answer is how these more or less wantonly violent gods could produce a moral effect. The proper point of departure is not the morality of the gods but their violence. And this brings us to the efforts people made to protect themselves from divine terror. Violence made the gods not only great but moral.

PREVENTION EVERYWHERE

According to our cultural-protection hypothesis, every society that wished to survive had to develop means of protecting itself from catastrophes, wars, and epidemics. Our analysis of the Torah showed in detail how such systems work. Only the societies that developed effective prevention systems and prevented outbreaks of divine rage survived. If our hypothesis is correct, we must find catastrophe-prevention systems in other ancient cultures as well, and, indeed, we do. Cultural evolution produced a variety of solutions. Let us now look at the risk-management systems of Mesopotamia, Egypt, and Greece.

It's Written in the Stars: Mesopotamia

In Babylon as well as Assyria, misfortune was "regularly interpreted as an expression of divine rage at the evil, sinful behavior of the king and the people." As complex as the causes for state failure might have been, the explanations were actually relatively simple. Attention centered on the personal relationship between the king and the gods. The monarch's "good or bad conduct was the fundamental deciding factor in the country's success or failure." As a "good shepherd"—such was the ideal image of ancient Near Eastern rulers—the king could not neglect the welfare of the poor and the weak.[33]

Anyone wanting to maintain the gods' favor had to understand their intentions. For their part, the priests of Mesopotamia perfected the art of divination. By analyzing the organs of sacrificial animals, the course of the stars, or the flight of birds, priests could fathom the will of the gods. Mesopotamian augurs enjoyed a formidable reputation in the ancient world. They even made it into the Bible: we know them as the three wise men from the East who followed a star until they found the baby Jesus in Bethlehem. Divination helped the cultures of Mesopotamia to dominate, both politically and culturally, the entire Near East for nearly 2,000 years. But how can it be, asks Assyriologist Stefan Maul, that such a thoroughly irrational procedure, from a modern perspective at least, "could produce such undeniably enduring successes"? His answer: the augurs' use of divination to determine the fate of a particular campaign or construction project led to a process of rational reflection upon the given situation. When misfortune occurred, information about what had happened was archived. Alternatives were discussed if the stars appeared to be against a particular project. This "task-oriented, considerate decision-making process" was not without success; "as a rule, commonsense decisions" were indeed possible, even if "based on false premises." For this reason, Maul believes Mesopotamian augury functioned as a "political and social early-warning system" whose main purpose was to protect the state from harm.[34]

The Weighing of the Heart: Egypt

The daughter of the Egyptian sun god, Ra, was the goddess Maat, the personification of the order of the world. As an abstract term, *maat* can also mean "truth" or "justice." It was the duty of all Egyptians to practice *maat*, for only this could avert *isfet*, a term encompassing, according to Egyptologist Hermann Schlögl, sin, injustice, lies, violence, war, and death. The Memphite theology, a religious text inscribed on the Shabako Stone in the collection of London's British Museum, reads as follows: "So is *maat* given to him who does what is loved, *isfet* is given to him who does what is hated. Thus life is given to the peaceful and death to the evildoer."

The judgment of the dead was the means of ensuring that the people preserved *maat* and spared the country from violence, war, and death. The heart of the deceased was placed on one side of a scale; a feather of Maat was placed on the opposite side. If the scales remained balanced, the deceased was permitted to begin his life in the afterworld. If the heart tipped the scale, its owner had not lived according to the principle of *maat*. The heart was fed to Ammit, the devourer, and the deceased owner was condemned to permanent death.[35] This, too, was a catastrophe-prevention system, for fear of judgment after death ensured that people adhered to social norms to prevent the state from sinking into chaos and misery.[36]

Tribute to the Gods: Greece

The most important ritual in the ancient Greek (and Roman) religion was the sacrifice, which was seen as a means of promoting a successful working relationship with the gods.[37] The correct execution of this cultic activity was of the utmost importance. These gifts to the gods were "regarded as a tribute exacted by their threatening power, which makes it necessary to 'turn off' their impact."[38] If sacrifices failed to materialize, the gods became angry and sent down disasters. Regular sacrifices were the Greek and Roman solution to preventing catastrophes.

This system is fascinating, however, because the Greeks discovered that they couldn't always depend on the Olympians. As Paul Veyne, professor of Roman history at Paris's Collège de France, explains, "The relationship to the deity is like that of a buyer to a more or less dependable supplier."[39] For example, although King Croesus had given more than any other mortal to the immortal gods, he still lost his kingdom to the Persians. Croesus sent his envoys to Delphi to ask Apollo's oracle if the gods were sometimes in the habit of deceiving their followers. "In his answer," explains historian Werner Dahlheim, "a visibly embarrassed oracle referred to the sins of the complainant's ancestors as an explanation: 'No one can escape fate, not even a god.'"[40]

Can this be put down to the particular unpredictability of the Greek gods, or was it merely that a catastrophe-prevention system founded solely on sacrifices was simply not up to the task? Whatever the answer, the Greeks found themselves forced to take their fate into their own hands. And the surge in scientific, political, and philosophical developments that took place in ancient Greece stemmed from this push to develop alternative—that is to say, more dependable—catastrophe-prevention systems. To give an example, in the sixth century BCE the Ionian natural philosophers Thales, Anaximander, and Anaximenes were the first to deny that the gods were responsible for earthquakes and to develop scientific hypotheses instead. They all came from the city of Miletus in Asia Minor (in modern-day Turkey), a region prone to high levels of seismic activity. Here it was relatively easy to deduce how the traditional practice of placating the gods with sacrifices provided no dependable protection from earthquakes. Rather than blaming Poseidon for these disasters, Thales, Anaximander, and Anaximenes posited novel causes, such as fluctuating sea levels or subterranean eruptions of air.[41]

Here we see the next step of cultural evolution in action. Attempts at establishing natural causes for catastrophes lifted the suspicion from the gods. Interestingly, over the course of the following centuries, Greek philosophers began to view the gods as the foundation of all that is good. For the Stoics—who would have no small impact on Christianity—there was no question that the gods' behavior was irreproachable. "The gods are gifted with all virtues and not the slightest trace of evil," wrote Cleanthes (ca. 331–232 BCE).[42] So science made the gods good? We shall return to this question later.

THE ADVANTAGE OF POLYTHEISM

Whether in Mesopotamia, Egypt, or Greece, everywhere we find sophisticated strategies aimed at protecting societies from disaster. A side effect of these prevention systems was their ability to promote cooperation—a moral effect. They led to discussions concerning collective values and produced a social glue that enabled people to live together in ever-larger societies. Rules and rituals regulated human behavior and strengthened the community. The cult of sacrifice also tied people together by obliging them to share common values. Moreover, since they were costly, they also kept free riders in check. If nothing else, fear of the gods' collective punishments encouraged everyone to pay close attention to their neighbors' behavior.

So how was it that the gods—the causes of all the catastrophes—were transformed into moral beings? Remember that they were already moral: we classify them as "persons," and persons are necessarily moral beings. Thus if a flood is the work of a god, we automatically assume he or she must have had a reason for causing it. What comes next is a kind of "profiling" in which people try to reconstruct a personality for the god that fits with observation. Once again, the "person" template prescribes that the gods, too, must have their likes and dislikes—just like the rest of us. Their psychology is the same as that of people; we are simply incapable of imagining anything else. And just as very few people are truly morally perfect, the same must be true of the gods as well.

By this logic, the ancestors of the hunter-gatherers were already moral beings. In the centuries following sedentarization, however, violence and catastrophes surged. This affected the entire society and forced it to do everything possible to figure out the cause of all the misery. With the concept of collective punishment came the gods' responsibility for watching over society's moral values. And the gods' punishments affected even kings, for they too came under divine oversight.

The moment the gods became responsible for everyone's values, people began to demand that these guardians of morality behave properly themselves. We have stated this before, but unfortunately—and this is the decisive point—the world stubbornly refused to work that way: catastrophes, the gods' collective punishments, continued to occur—seemingly out of the blue—and even the most devote worshippers fell victim to them. In a polytheistic world, such events were easier to explain. The catastrophes didn't always have to reflect punishment, for the gods could serve in other ways to explain the world's unpredictability. For example, some catastrophes could stem from disputes between the gods themselves. What caused the Trojan War? Why did thousands of heroes die in the battle for the powerful city on the Hellespont, and why did so many women and children undergo such needless suffering? Because the goddesses Hera, Athena, and Aphrodite fought among themselves for a golden apple bearing the words "For the fairest." And Aphrodite bribed the Trojan prince Paris, the judge of the competition, by offering him the most beautiful woman in the world, but Helen, the beauty in question, was already married to Menelaus, the king of Sparta. The immortal verses of Homer's *The Iliad* memorialize the resulting misery, the collateral damage of divine vanity and belligerence.

The gods act like human beings: imperfectly. Happily, they fit with a world in which catastrophes were completely unpredictable. The historian Paul Veyne answers the question of the moral qualities of the Greco-Roman gods as follows: "One—singular—named god might have his mood swings, his flaws and weaknesses, but the gods as a whole—in plural—respect morality and desire that evildoers are defeated or punished."[43] The same can be said of people. One person might have his flaws and weaknesses, but the society is on the same page. And the bad guys cannot be allowed to get away with their misdeeds. It's the same today as it was 2,500 or 25,000 years ago.

YAHWEH HAD A HARD TIME

Polytheistic religions can explain suffering as collateral damage. But if there is only one god, there has to be a good reason for suffering, and the misery is automatically transformed into a moral event. The very fact that Yahweh was the only available actor automatically made him into the hypermoralistic being that we saw in our analysis of the books of the kings and prophets. Whereas in polytheism a number of moral players have irons in the fire, in monotheism there is only the one judge, who is expected to apply the same standards in every situation.

This makes it easier to understand why the God we encounter in the Hebrew Bible had such a terrible tendency to erupt in fits of violence. He was responsible for everything! Alternative interpretations were simply not possible. Yahweh was no sadist, his wrathfulness was not a character flaw, and the Bible's authors weren't just living out their own violent fantasies. No, the measure of divine punishments reflected real experiences with catastrophes that were inevitably attributed to the one and only God as the only possible cause. And the people of ancient Israel had to deal with a massive amount of suffering. God's punishments, his immeasurable wrath, reflected Assyrian, Egyptian, and Babylonian brutality as well as the epidemics that raged through the land in a time when effective medicine hardly existed.

Yahweh thus personified the misery that plagued humanity. He took everything upon himself and helped the people to stand up against their own demise. As a heuristic tool, he offered a matrix that enabled people to organize their strategies for escaping the sword, the famine, and the pestilence and lent the lessons they learned and the laws they developed the necessary degree of authority.

This is one of the most important realizations to come of our reading of the Bible. At a time when societies were in danger of collapsing under the violent egos of tyrants and emperors, this God was the best thing that could happen to the people. They had to identify a cause for every catastrophe, meaning they had to find out why God had reacted with such rage. And because God got angry at the same things that angered people, his divine morality was a projection of human morality. He hated exorbitant luxury, despised injustice, and demanded that people treat their peers well. After all, this is exactly what our first nature whispers into our ears. And protoscientific knowledge supplemented this divine moral codex. We have already explained how, because certain diseases related to particular habits tended to appear more frequently, people assumed that God had a serious problem with such practices.

Once morality became Yahweh's responsibility, earthly powers could no longer meddle with it. Morality now applied to everyone and was no longer the plaything of the powerful (in theory at any rate, for in practice things often didn't work out that way). This was God's greatest contribution to civilization. What we call intellectual-institutional religion was a heroic attempt to stand up to the new world of sedentary societies in which people found themselves confronting not only catastrophes on a massive scale but also the tyrannical behavior of their leaders. They were like David when he found himself face-to-face with the giant Goliath. Indeed, this form of religion really was an "art of battling giants."[44]

Even atheists should marvel at God's achievement, for Yahweh and the Israelites' empirical catastrophe-prevention system led to scientific advances in the millennia that followed that finally showed us the real causes of the sword, the famine, and the pestilence: lust for power, poor government, climate change, erosion, plate tectonics, and microbes. Ironically, science would eventually exonerate God, acquitting him of his reputation as an impulsive perpetrator of violence. Science became one of the forces that transformed him into the good God we know today.

The next part of our book focuses on another factor that helped bring God's good side to light. In the following chapters we meet a Yahweh so surprising that we might doubt we are dealing with the same deity.

PART IV

PSALMS AND CO.

The Other God of the Bible

A nineteenth-century painting by Hans Thoma shows a countrywoman from the German Black Forest as she sits in her room on a Sunday morning, "knitted stocking in the hand and an opened Bible on her knees in which she is reading." The scene drew the attention of a German philosopher born of a Jewish family, Ernst Bloch (1885–1977), prompting him to ponder how "the Bible can be read and understood everywhere." How likely would it be that a simple farmer's wife elsewhere would ever "read the *Gilgamesh* epic"? Would she "understand elitist Hinduism"? "Would she be in a position to study the Pythagorean forms of religion?" Hardly. The Bible, however, though "written in a foreign language by a foreign people thousands of years before our time," everyone can understand. You don't have to be a biblical scholar to be touched by its stories.[1]

We like to think that the farmer's wife was reading in the Ketuvim, the "writings" that make up the third main part of the Hebrew Bible after the Torah and Nevi'im. Perhaps she perused a psalm. "O LORD, thou hast searched me, and known me," one reads in Psalm 139. "Thou hast beset me behind and before, and laid thine hand upon me." In the pages of the Ketuvim, the reader encounters for the first time a completely transformed God who appeals to believers as individuals. The Torah and Nevi'im we can't really describe as devotional literature, for the sheer terror we encounter in stories such as the flood or the conquest of Canaan would make the poor countrywoman drop her stocking.

The Ketuvim, however, has a completely different feel to it. Here we find texts upon which people can reflect—words that speak to a sense of what we today would describe as spirituality. Here, too, we find the wisdom literature—similar to the texts cultivated in Egypt and Mesopotamia—that focus on virtuous living. Generally speaking, these texts were written late, during the Persian and Hellenistic periods, sometime between the fifth and second centuries BCE. They make up an extremely colorful collection of works. The books of Chronicles, Ezra, and Nehemiah offer a number of historical narratives. And whereas the book of Ruth could be considered a novella, the book of Esther offers up a novel. Then there are the philosophical tracts, elegies, and prayers.[2] All are remarkable for their high literary—and sometimes even poetic—qualities as well as their heralding of intense cultural exchange. "Here the concept of God is expanded from the savior-and-warrior-God claimed by a small group to a universal deity who shows himself in all of creation," explains theologian Thomas Staubli.[3] A recurrent theme throughout most of these writings is a "trust in God's righteous leadership" in a world that often must have seemed threatening.[4]

We simply don't have the space to address all of these texts. It would be presumptuous to claim that we have something important to say about the erotic poetry of the Song of Songs from the perspective of cultural evolution. We also find we must leave aside Ruth or the two books of Chronicles with their theologically correct histories of the royal courts of Jerusalem and Samaria. Instead we concentrate on three books focusing on three questions that are especially relevant to our focus on the cultural evolution of religion. First we examine the book of Psalms, where we discover a new, empathetic God to whom one can take one's personal problems. It's unimaginable to think of having such a heart-to-heart with the God of Exodus. We then turn our attention to the book of Job, which shows us how the question of human suffering got God into a lot of trouble. Finally we turn our attention to the book of Daniel, which not only gives us the familiar tale of the lions' den but also mentions, for the first time in the Bible, the ideas of resurrection and a last judgment. Why does the Bible so badly neglect the subject of death until this point? We will venture the guess that God himself first introduced fear of death into the world—and along the way we'll even make a flying visit to heaven and hell.

The Ketuvim is remarkable in revealing the creation of a highly attractive hybrid religion, one whose holy texts would continue to engross readers some 2,000 years later and that would speak to people from totally different cultures beyond its original target audience, just as it did to Ernst Bloch's countrywoman. This is mainly thanks to the fact that the Ketuvim finally gave our first nature its well-deserved due—and in a truly grand fashion.

14

THE PSALMS

My God

"And about the ninth hour Jesus cried with a loud voice, saying, Eli, Eli, lama sabachthani? that is to say, My God, my God, why hast thou forsaken me?" These were the last words of Jesus on the cross—at least according to the evangelists Matthew and Mark. In his cries we hear his deep despair, but only a few people know that Jesus's words reflected more than mere distress. Never intending to break with Judaism, facing death, he was quoting the Hebrew Bible. The psalm he referenced continues, "O my God, I cry in the daytime, but thou hearest not; and in the night season, and am not silent."

In this psalm, number 22, we hear a person in great need as he turns to his creator. "Save me from the lion's mouth: for thou hast heard me from the horns of the unicorns," he pleads. God reacts to his cry for help and grants him protection—"thou hast heard me!" In gratitude the saved one promises to proclaim God throughout the world, "for he hath not despised nor abhorred the affliction of the afflicted; neither hath he hid his face from him; but when he cried unto him, he heard."

The fact that the evangelists have the dying Jesus quote one of the psalms could not offer better proof of their important role in the Bible as well as in the lives and sufferings of God's people. Indeed, the Psalter, the collection of 150 psalms, is the high point of the third part of the Hebrew Bible, the

Ketuvim. These Psalms present us with a God to whom anyone in existential distress can turn directly. For the Bible, this was something completely new.

Only one chapter ago, we were still dealing with the proverbial wrath of God, and yet here we find a God of moving solicitude. "The LORD is my shepherd"—so begins probably the most famous of all psalms, and the rest is equally familiar: "I shall not want. He maketh me to lie down in green pastures: he leadeth me beside the still waters. He restoreth my soul: he leadeth me in the paths of righteousness for his name's sake. Yea, though I walk through the valley of the shadow of death, I will fear no evil: for thou art with me; thy rod and thy staff they comfort me." Here we see a version of God that anyone would wish for. He offers safety and security and drives out our fear. Is this the same God that sent down the great flood? The God who punished his own people for any act of disloyalty and delivered them into the hands of their enemies? The God who spread fear instead of dispelling it? We think the answer to all these questions is no.

We believe that here we encounter a new type of god, a deity with whom anyone can speak—not just long dead prophets and patriarchs. We even believe that the form of religion we encounter here differs completely from that dominating the Torah. It is the kind of religion that most people have in mind today when they think about their own personal beliefs. We have already mentioned that many Christians feel the psalms are the only captivating part of the Old Testament. "The LORD is my light and my salvation; whom shall I fear? the LORD is the strength of my life; of whom shall I be afraid?" Verses such as these uplift our weary hearts. The psalms present God as "a living You" to whom "one can pour out his heart."[1] Let us now look into the origins of this God of the psalms, what he did to change the religion, and why this was the best thing that could have happened to the Bible. We begin with a short introduction to the psalms themselves.

DAVID'S TORAH

The psalms rank among the greatest works of world literature.[2] The Greek term *psalmós* translates as songs to be accompanied by a string instrument, and the term used in the Christian tradition, "Psalter," can be translated as "collection of songs." The book of Psalms comprises 150 of these songs—the number varies depending on the Bible version. Judaism prefers the term "book of praise"; it considers the poetic texts a polyphonic eulogy to Yahweh—even if prayers of petition and lament predominate.[3]

It is often thought that many of the psalms date back to King David, who, according to lore, cheered King Saul with songs on his lyre. This is pure fiction, but the underlying idea was programmatic. The psalms were grouped together into five books, and the message was that just as Moses gave the Israelites the five books of the Torah, so David gave them the five books of psalms. The Psalter has therefore also been described as "David's Torah,"[4] although it actually represents a radical departure from Moses's book of laws.

According to the current state of research, the Psalter was compiled relatively late, some time between 200 and 150 BCE, even though many of the psalms were probably written earlier. Some psalms take the people of Israel as their subject; others were composed for individuals or the temple cult. Yet others deal with themes of family piety, which lends the psalms the character of a "book of prayer and living" for ordinary people intended to provide comfort and hope in the face of suffering and fear.[5] Even so, the psalms are not all light fare. One requests, "Break their teeth, O God, in their mouth." Or how about this: "O daughter of Babylon, who art to be destroyed; happy shall he be, that rewardeth thee as thou hast served us. Happy shall he be, that taketh and dasheth thy little ones against the stones."

We are interested in the individual verses that express a yearning for divine assistance. "Turn thee unto me, and have mercy upon me; for I am desolate and afflicted. The troubles of my heart are enlarged: O bring thou me out of my distresses." These lines reveal the core of many people's faith: "O keep my soul, and deliver me: let me not be ashamed; for I put my trust in thee."

"FROM HEART TO HEART, FROM SOUL TO SOUL"

Bible scholar Rainer Albertz has pointed out that in the world of the psalms, we encounter a different type of religion. He speaks of "personal piety" and contrasts this with "official religion." The former focuses primarily on an immediate and intimate relationship between individuals and their deity. It operates on the level of the family and everyday life. The latter focuses on the covenant between God and Israel, which takes place at the level of the people, the temple, and the religious experts. This "official religion" was the main subject of the Bible up to this point. Herein Albertz sees a "socially-related pluralism within the Israelite religion" manifested in "two parallel, non-identical levels of religion."[6] Such an "internal religious pluralism," Albertz believes, is not limited to the biblical religion alone but can in fact "be observed in almost every religion, be it ancient or modern."[7]

We have already examined the religious pluralism that existed in ancient Israel, which prompted us to differentiate between intuitive and intellectual religion. However, we cannot necessarily ascribe this differentiation to social differences. For us, intuitive religion, which Albertz describes as personal piety, is closest to what we can describe as being built on the individual's biological substrate of religious experience. It is tightly interwoven with our innate psychological dispositions and based on our intuitive acceptance of a numinous world inhabited by countless agents. Intellectual religion, on the other hand, is the God-based cultural catastrophe-prevention system that we have already sufficiently explained. It is a product of our third nature.

We've also explained how the monotheism brought back from the Babylonian captivity was intended to silence the intuitive religion rooted in the private sphere of daily life. Why, then, does it suddenly reappear in the psalms, this time with Yahweh as the main protagonist? Before we answer this question, let us look for the expressions of intuitive religion in the psalms. After all, in the Torah we only encounter it in the form of prohibitions. Here, for instance, is a dialog between God and one of his followers as told by the psalms. "But thou art he that took me out of the womb: thou didst make me hope when I was upon my mother's breasts," Psalm 22 reads. "I was cast upon thee from the womb: thou art my God from my mother's belly. Be not far from me; for trouble is near; for there is none to help." Not only can one find a surprising sense of basic trust in these words, but we also encounter a new role for God—that of a midwife. Or does Psalm 22 actually present the notion of Yahweh as a father? This is rather rare in the Old Testament, in which we find a grand total of seventeen mentions of God as a father figure, whereas the New Testament contains at least 260 of them.[8] Assisting with childbirth is a rather unusual role for a war and state god with an explosive temper.

Albertz explains this relationship to God as follows: "Like a small child who based on his parent's loving affection 'preconsciously' grows into a bond of trust with them that can withstand crises and, normally, cannot be dissolved, just like the child finds unconditional consolation and protection when it is afraid, so is the relationship between the individual and his God."[9] Accordingly, the psalms often serve to illustrate how the relationship to God can be explained in terms of psychologist John Bowlby's theory of attachment: the believer feels consoled like a child in the hands of "a protective and caring parent who is always reliable and always available to its children when they are in need."[10] In the language of the psalms it sounds like this: "The LORD is my light and my salvation; whom shall I fear? the LORD is the strength of my life; of whom shall I be afraid?" This yearning for a "safe haven," a "secure base,"[11] also stems from our first nature.

All of this reminds us of William James's classic *The Varieties of Religious Experience*, which even a century later still ranks among the best books ever written about religion. For this work, James drew on interviews conducted by Stanford professor Edwin Starbuck in the late nineteenth century as he explored the American world of religion. To the question of how they experienced God, Starbuck received answers such as the following from a middle-class woman: "I have the sense of a presence, strong, and at the same time soothing, which hovers over me. Sometimes it seems to enwrap me with sustaining arms. God is a personal being, who knows and cares for his creatures." Another woman expressed it differently: "I have often a consciousness of a Divine Presence, and sweet words of comfort come to me." Starbuck concluded that an elementary need "for society, for companionship, for kinship" underlay all these feelings.[12]

William James used these interviews to develop his concept of "personal religion," something quite close to what we describe as intuitive-individual religion. This form of religion centers on "the feelings, acts, and experiences of individual men in their solitude, so far as they apprehend themselves to stand in relation to whatever they may consider the divine." James draws a distinction between this personal religion and "institutional religion," whose main characteristics are "worship and sacrifice, procedures for working on the dispositions of the deity, theology and ceremony and ecclesiastical organization." For James, personal religion is the "primordial thing." As it rests on actual religious experiences, it is "more fundamental" than theology and church, which, as core aspects of institutional religion, feel more like a "secondhand" form of religion to the individual believer. For personal religion, however, the following is true: "The relation goes direct from heart to heart, from soul to soul."[13]

The parallels to the world of the psalms are hard to miss: "I will love thee, O LORD, my strength. The LORD is my rock, and my fortress, and my deliverer; my God, my strength, in whom I will trust; my buckler, and the horn of my salvation, and my high tower." One could of course argue that, given that he interviewed primarily white Protestants, religious teaching molded most of the answers given to Starbuck.[14] Nevertheless, it is remarkable that the psalms continue to touch a chord in people's hearts after more than 2,000 years. Here one is on a first-name basis with the Lord, as it were. You can interact with God as if he were a real person at your side who can help you in the here and now.

BUT THE ANCESTORS AREN'T DEAD

Let us look more closely at this phenomenon from an ethnological-evolutionary perspective. *Homo sapiens* is a hypersocial species. Life together in small groups has always been founded on the duty of the group's members to help one another. The semisedentary Semai of the Malay Peninsula, for example, begin and end every important gathering with a series of reaffirmations: "We are all siblings here; we take care of one another. When I couldn't hunt, you took care of me; when you were sick, I fed you."[15] This confirmation of reciprocal solidarity is basically a form of life insurance. In times when there existed no institutions outside the group, everyone's main focus was avoiding being left out in the cold. Hollywood is quite adept at exploiting this ancient emotion. When the hero is in serious trouble, his own people usually come to the rescue at the very last minute. The feeling of elation that engulfs the viewer in these moments evinces how our first nature simply loves this type of happy end.

This mutual support dominated the group's everyday life. And a lot of evidence suggests that this solidarity did not end with death. Although the hunter-gatherers must also have found death uncanny, it was not the end of everything for them. It was merely the beginning of a different form of existence: "A belief in the immortality (in some form or another) of the dead occurs in all cultures as does the worship (again, in some form or another) of ancestors," argues evolutionary psychologist Lee Kirkpatrick.[16] The ancestors may prefer to remain invisible and unbound by a particular place, but they were by no means completely different from when they were still living people. They were driven by the same psychology and continued to belong to the natural world. They influenced the world of the living, and the living could in turn influence them. The ancestors' world was merely an additional, invisible, and nonphysical dimension of our own social world—no less real when it came to producing consequences in the real world. And as we are anyway "biased toward social explanations," the ancestors always played a role when people needed explanations for unusual events that could not be attributed to a real person.[17]

The ancestors were part of the extended family. They looked after their living relatives and had an interest in the family's well-being and stability. You could therefore count on their support, provided you treated them right. They were there for you when the going got tough. Here kinship psychology and intuitive religion meet, for deceased family members continue to serve as the "attachment figures" they were in life.[18]

This perspective also explains why intuitive religion revolves around aspects of daily life such as birth, illness and death, marriage, and disputes between neighbors. Likewise, we can see why women play a key role in this type of religion.[19] Just as maintaining social connections among living relatives is traditionally women's responsibility,[20] the same is true of maintaining contacts with the ancestors. The ancestors are merely relatives that find themselves in a different aggregate state, so to speak. The concepts, rituals, and magical practices may differ from group to group, but in essence they are very similar as they reflect the same human propensities found everywhere. As guardians of the rules, for instance, the ancestors promote a culture of "unquestionable truthfulness"[21] and help stabilize society. "If you believe ghosts or ancestor spirits might be around and watching," explains Justin Barrett, a pioneer in the cognitive science of religion, "you are more inclined to behave in ways good for your social reputation, thereby making you a more attractive exchange partner."[22]

In the times of the hunter-gatherers, this all resulted in a mix of our first and second natures that was a self-evident, permanent part of daily life. This mix is generally known as household or family religion, in which ancestors universally played an important role in every aspect of life. Ancestor worship also played an important role in ancient Israel, something we look at more closely in our examination of concepts of death in chapter 16. In sum, the belief that we live in a world full of invisible actors was an intrinsic component of basic human psychology. As long as they were our own ancestors, we could blindly trust these agents, but with ancestors from another group or stray spirits, we had to be careful, for these could certainly pose a threat.

GOD'S INHUMANE SIDE

As discussed in a previous chapter, the Israelite elites who returned from exile attempted to eliminate this ancient, intuitive religion in the name of monotheism. Not surprisingly, their effort met with major resistance. The people murmured and murmured. Our first nature clamors for supernatural assistance in times of existential peril. Not even the priests and scholars could completely escape this longing and felt themselves drawn to integrate intuitive religion into the Hebrew Bible in order to satisfy people's inherent psychological needs.

The significant problems that came with monotheism strengthened this pull. With the exception of the prophets, the state god Yahweh did not concern himself with individuals (unless someone broke the rules, of course). As a cultural-protection system, monotheism sought to protect the entire society, and this protection often came at the cost of individual needs. As such, in some instances monotheism could actually greatly exacerbate human suffering. Let us take a closer look at this phenomenon as it pertained to disease.

Taking care of the sick and administering hospitals are the ethical imperatives of today's great religions. This comes as no surprise, for diseases were once believed to be the work of spirits, demons, and, later, gods, which meant that sacrifice, prayer, and exorcism could heal sufferers. "Throughout the ancient world 'health care systems' were integrally related with religion," explains Carol Meyers. "Medical care was a religious activity."[23] The God of the Torah, however, had no comfort to offer the ill. The Bible stipulates the quarantining or even permanent expulsion of the sick. This harsh treatment corresponds to the third-nature logic of monotheism: illnesses were Yahweh's punishments, meaning the ill were not patients but delinquents. They were guilty of some misdeed, so no one should offer them comfort. The illness made the crime visible to all and warned others that by getting too close, they risked losing God's grace.

Since the Torah sought to protect society as a whole, the ill had to be isolated. This makes good sense in scientific terms, for it helped prevent diseases from spreading. For the stricken, of course, it only increased their suffering as they had to endure "isolation and hostile reactions from the community." These are the consequences of the Torah's logic, explains theologian Eckart Otto.[24] Religious scholar Hector Avalos comes to a similar conclusion: "This health care system was not geared toward the needs of individual patients; rather, the aim was to safeguard the community, at least from the perspective of the elite priesthood."[25]

We have already discussed in depth the protoscientific nature of the Torah. In *Illness and Health Care in the Ancient Near East*, Avalos concludes that from a "strictly scientific standpoint," the Torah's health-care system was probably more effective than those of the temples of Asclepius, the Greek god of medicine. Attached to these temples were sanatoriums for the ill, which to us might seem a more humane solution, but they probably did far more to spread disease than the restrictive isolation procedures prescribed by the Torah, which included barring entry to the temple area.[26] The moral dilemma resulting from this collision of social and individual interests continues to trouble us to this day. We only need think of the discussions surrounding mandatory quarantine for persons suspected of having had contact with dangerous infectious diseases such as Ebola.

It is indeed a classic case of mismatch. In order to limit the spread of epidemics and protect society from large numbers of fatalities, isolating the infected makes good sense. To our first nature, however, such actions seem inhumane. The patient's first nature intuitively protests, because in his or her moment of utter helplessness, when succor is desperately needed, the community takes it away. This is equally unacceptable to the families of the diseased, for their first nature prohibits them from abandoning ill group members. How did the Semai put it again? "When you were sick, I fed you." When this sentiment was ingrained in human nature in preagricultural times, however, the prevalence and, more importantly, virulence of infectious diseases were much lower.

Yahweh, the state god and former god of war and weather, had no healing or pastoral skills at his disposal. He didn't need them either, for they were the purview of traditional household religion in which women were mainly responsible for health care. But monotheism created a vacuum when it eliminated intuitive religion. The Torah's charitable prescriptions helped the poor, widows and orphans, and even strangers—any and all, but not the sick. Yahweh's health-care system could be summed up in one short decree: abide by the law!

Yahweh basically considered anyone affected by illness and misfortune a particularly obstinate sinner who had deserved punishment.[27] Thus, the sick, lonely and outcast as they were, were closer to death than life,[28] as reflected in the following lament by a sufferer in Psalm 31: "I was a reproach among all mine enemies, but especially among my neighbours, and a fear to mine acquaintance: they that did see me without fled from me. I am forgotten as a dead man out of mind: I am like a broken vessel."

This was an intolerable situation. The religion required fixing if the priests wished to prevent people from returning en masse to the old gods and practices now officially demonized as superstition. Behind this need was nothing less than a biological imperative. Even today people with severe illnesses leave nothing untried. They visit spiritual healers or embark on a pilgrimage to Lourdes. With a wink, microbiologist David Clark concludes, "As long as scientific medicine is effective, many people today feel little need for supernatural intervention."[29]

We shall see how in response the New Testament began to rethink this approach to the sick (rabbinical Judaism independently did the same and declared care for the sick a moral duty), but the winds of change had already begun to stir in the Psalter. In Psalm 30, for example, Yahweh himself takes a crack at healing the ill: "O LORD my God, I cried unto thee, and thou hast healed me." And in Psalm 41 we encounter the God the people longed for: "The LORD will strengthen him upon the bed of languishing: thou wilt make all his bed in his sickness."

"THOU HAST SMITTEN ALL MINE ENEMIES UPON THE CHEEK BONE"

Illness provides only one example of how monotheism created a vacuum in the existential spheres of everyday life. In chapter 16 we encounter something quite similar in our examination of death. The priests had to fill the void if they were to convince everyone to worship Yahweh alone. They had to transform him into a God for every sphere of life—no aspect of human existence could be strange to him. Because Yahweh in his jealousy prohibited people from making overtures to his supernatural competitors, he had no choice but to expand his own repertoire.

The Psalter accomplishes this. It integrates intuitive religion into the Book of Books, thus enriching institutional religion. Here Yahweh the judge meets Yahweh the philanthropist. This transformation was not the result of some deliberate plan whereby priests intentionally composed new psalms to close the gaps in God's competence. In reality, the Psalter is the product of a process of gradual accumulation. Had this not been the case, we would find clearer traces of institutional religion in the book of Psalms. To give one example: in the laments of the psalms, we encounter no reference to Israelite salvation history. Albertz states, "Never in individual laments does the (arguably entirely natural) argument appear that God should rescue the individual from hardship, having already liberated the people from Egyptian slavery." Similarly, there is virtually no role in the

Psalter for the "complex of sin, God's wrath, punishment, and forgiveness" that dominates the first two parts of the Hebrew Bible.[30]

The psalms' authors often reworked old material. Some of it certainly dated back to an era in which Yahweh still tolerated other gods, whereas they adapted other texts to fit monotheism. For example, it is striking how often the psalms mention unspecified "enemies" that plague the lamenter. "LORD, how are they increased that trouble me! many are they that rise up against me." Originally, these enemies were the "demonic powers"[31] believed to be the cause of disease and social adversity. The believer offers his prayer and implores his God for help: "Plead my cause, O LORD, with them that strive with me: fight against them that fight against me. Take hold of shield and buckler, and stand up for mine help. Draw out also the spear, and stop the way against them that persecute me: say unto my soul, I am thy salvation." Elsewhere, we can read casual lines, such as "Arise, O LORD; save me, O my God: for thou hast smitten all mine enemies upon the cheek bone; thou hast broken the teeth of the ungodly."

Is Yahweh supposed to have offered concrete physical support? Hardly. It doesn't require a great deal of imagination to see how these passages resemble discussions one has with ancestors or tribal gods in a bid to enlist their aid. After all, these supernatural actors are supposed to help when someone is in danger. But now Yahweh himself was set to take over the job, becoming transformed, to a certain degree, into a "superancestor." Our old kinship-oriented psychology molded monotheism into a form that met its needs. Here we see the makings of the concept of God the father who had to step up to the plate and replace the supernatural relatives he had previously cast aside.[32]

Even today we can observe in sports how our monotheistic God still gets treated like the ancestors of the past. We've all seen how—especially right before a penalty kick or a field goal attempt—players as well as fans of both teams send up heartfelt prayers to the heavens aimed at winning over divine assistance. This is impertinent in the eyes of a monotheistic God, of course, but he is a product of our own intellect, and in highly emotional situations—such as when we desperately want our team to win—our first nature could care less about "theological correctness." At times like these, God reverts to being one of the good old spirits that were always at our side when it came to delivering a few punches to our opponents.

As we have seen, Yahweh absorbed aspects of intuitive religion and in doing so strengthened his own being. This way, institutional religion, with its protoscience and its legitimization of morality and power, became enriched by actual religious experiences, which allowed it to tap into the enormous energy reserves of individual-intuitive religion. This was no deliberate transformation. On the contrary, our first nature simply managed to break through and demand its due. Excavated graves reveal that from very early on, the temple elite worshipped Yahweh as a personal tutelary deity, meaning he managed to make his way from the sphere of official religion into that of private, intuitive religion.[33] The lesson from all of this is that new cultural solutions can only have a long-term effect if they do not run directly counter to the needs of our first nature. Here we can discern a pendulum movement: the needs of our first nature will eventually correct a cultural, third-nature solution that is too rational.

GIMME SHELTER

These two very different religious spheres were surprisingly easy to meld together. In both cases the primary interest was protection: the Torah sought to protect society as a whole (the people of Israel), whereas the psalms presented God as a personal protector of people as individuals. In any number of forms, we can hear in the psalms the same plea: "Deliver me, O LORD, from mine enemies: I flee unto thee to hide me." It seems as if God can also function as an effective individual catastrophe-prevention system.

For now we will ignore the contradictions this fusion engendered (we will deal with them in the next chapter) and instead examine this new focus on the individual. God provides his followers with a basic trust, something believers reaffirm to this day. For them this trust is a source of energy. Here God serves as an ideal confidant, stable and always reachable, who satisfies a human need for safety and security. As we hope we have shown here, this works because intuitive religion's basic sense of trust—that feeling of living in a world populated by supernatural beings whom we can ask for support—was projected onto Yahweh. Yahweh's career shift into a personal savior-god is of course highly ironic: he offers consolation for problems that he himself created!

As the vanguard of disenchantment, Yahweh himself had caused the old world of the ancestors to disappear. And suddenly that same Yahweh had to meet the people's now unmet longing for personal support. In Psalm 63 we hear, "O God, thou art my God; early will I seek thee: my soul thirsteth for thee, my flesh longeth for thee in a dry and thirsty land, where no water is." Yet Yahweh himself had turned the land where the springs of intuitive religion once bubbled into a desert. The people found themselves isolated after he drove them out of the paradise of spirits and ancestors. Little wonder that people developed an ever-greater thirst for him, as he was now the only supernatural being to whom they could turn. They had no other choice—slaking their existential thirst at another well of supernatural power would be denounced as superstition or even heresy.

A HYBRID PRODUCT

The addition of the five books of David, the Psalter, to the five books of Moses, the Torah, perfected the cultural-protection system known as religion. This also confirms one of the main theses of our book: the Bible may well be the most ambitious attempt to come to terms with the problems that have plagued *Homo sapiens* ever since our species adopted a sedentary way of life. This attempt takes place at the levels of the individual and of society as a whole: it has both psychological and social consequences. God as a savior works for individuals as well as for a people or nation. God became multifunctional.

This melding of intellectual and intuitive religion is one of the Hebrew Bible's great accomplishments—and also a secret of its success. The result was a hybrid religion that would soon become very attractive in the antique world. The early form of Judaism formed during this time was so appealing, so radically different, that the people enduring repression in Palestine or the difficult conditions of the Diaspora remained true to it throughout their suffering. As a matter of fact, Yahweh even began to attract more and more followers throughout the entire Mediterranean region.[34] This God was responsible no longer only for society but for individual persons as well. "I will both lay me down in peace, and sleep: for thou, LORD, only makest me dwell in safety."

In the next chapter we examine the fact that this hybrid religion would produce a few logical problems of its own. God may be at our side in our time of need, but isn't he also the cause of all the trouble in the first place? Indeed, poor Job experienced this very dilemma firsthand.

15

JOB

How God Got Himself into a Devil of a Mess

Poor Job. His tribulations are proverbial. He was righteous before the Lord, but still he suffered the worst strokes of fate despite being wholly blameless. He received one dramatic message after another. His cattle and asses were stolen and his farmhands slain, his sheep consumed by heavenly fire and their shepherds killed, his camels taken by the Chaldeans and their drivers slaughtered. As Job was still lamenting the loss of his chattel, a final piece of bad news reached him: a desert wind had brought down the house in which his children had gathered to eat and drink—all were dead! Job tore off his clothes, shaved his head, and threw himself on the ground. "Naked came I out of my mother's womb, and naked shall I return thither," he cried. Deep as his despair may have been, however, he never spoke poorly of God: "The LORD gave, and the LORD hath taken away; blessed be the name of the LORD."

As if all this weren't bad enough, Job soon became seriously ill. From the soles of his feet to the top of his head, he was covered in painful boils. His only source of relief was to sit in ashes and scrape himself with a potsherd. Not even his wife had kind words for him, asking if he still wanted to maintain his saintly piety. "Curse God, and die," she told him. But she didn't seem to know her husband that well: "Thou speakest as one of the foolish women speaketh," he replied. "What? shall we receive good at the hand of God, and shall we not receive evil?"

Friends came to comfort him. For seven days and seven nights, they sat together without speaking a word, but then Job could no longer contain his suffering. "Let the day perish wherein I was born, and the night in which it was said, There is a man child conceived," he burst out cursing. "Why died I not from the womb?" Then he would at least have had his rest.

Why did God torment Job so? There must have been a reason for his many pains. His friends pressed him to finally confess his sins, but Job was indignant. His friends tried to outdo one another to find a philosophical explanation. Was his suffering the result of human imperfection? Did God discipline his children like a strict father? Or could it be that the suffering was a means of testing the faithful to find out whether they were virtuous?[1]

Job would have none of this talk. He protested and denounced God: "Though I were perfect, yet would I not know my soul: I would despise my life. This is one thing, therefore I said it, He destroyeth the perfect and the wicked. If the scourge slay suddenly, he will laugh at the trial of the innocent. The earth is given into the hand of the wicked: he covereth the faces of the judges thereof; if not, where, and who is he?" God as a tyrant? Yahweh couldn't let such words stand. From out of a whirlwind he addressed Job directly. God offered no explanation for the man's suffering but instead lashed out at him: "Where wast thou when I laid the foundations of the earth? declare, if thou hast understanding." The morning stars once sang together for God, and all the sons of God shouted for joy in praise of his work. God then went on to talk in great detail about the richness of his creation. "I know that thou canst do every thing," Job answered before admitting, "Therefore have I uttered that I understood not; things too wonderful for me, which I knew not." Job admitted his guilt and repented in dust and ashes.

In the end, God rehabilitated his "servant Job" in front of his friends and richly rewarded him for all his pain and suffering. Job received "twice as much as he had before": 14,000 sheep, 6,000 camels, 1,000 head of cattle, and 1,000 asses. And God gave him seven sons and seven daughters. The story had a happy ending after all. Job lived to be 140 years old and "saw his sons, and his sons' sons, even four generations." He never discovered the reason for all his suffering. And that's a good thing too.

WHAT REALLY HAPPENED IN HEAVEN

Bible readers know from the very beginning of the story that Job owed his suffering to a heavenly bet. Or rather a hellish bet. At any rate Satan was involved in Job's plight. But in Job's days there was no concept of heaven and hell, no notion of an afterlife in which people were rewarded or punished for their actions. Nor was Satan the devil incarnate—he was merely a "lowly supporting player from the heavenly court,"[2] just like the sons of God that Yahweh kept around. "Whence comest thou?" God asked him. After Satan explained that he had been wandering around the earth, God asked, "Hast thou considered my servant Job, that there is none like him in the earth, a perfect and an upright man, one that feareth God, and escheweth evil?" The Bible doesn't mention the scornful glance Satan must have cast in God's direction, for to Satan it was obvious why Job behaved so well. God had richly rewarded his servant; indeed, Job's piety was no great secret. "But put forth thine hand now, and touch all that he hath, and he will curse thee to thy face," said Satan. God agreed to the bet and placed Job in the care of Satan, who was allowed to do whatever he wanted with Job—as long as he remained alive. The source of Job's suffering, therefore, was that God allowed Satan to provoke him. "Thou movedst me against him, to destroy him without cause," God conceded.

Never in his life would Job (or his friends) have guessed he was simply the victim of a perfidious bet. What might Job have said had he known he was a guileless guinea pig in an evil godly experiment? And even though Job was richly rewarded at the end of the bet, the story still leaves the reader feeling more than a little ambivalent. Job's animals, shepherds, and children were all dead—the collateral damage of divine vanity.

THE QUESTION OF THEODICY

In the book of Job we run into the same difficulties of interpretation as in most other books of the Bible. As is so often the case, we are dealing with a text with more than one author, the colorful result of a long composition and editing process that took place from the fifth to the third centuries BCE. Scholars speculate that the story at the heart of the tale of Job has its origins in a legend of a man who endured substantial suffering. As powerful as the impact of Job's story is—most importantly in art and literature, where he continues to serve as the symbol par excellence of the wrongfully suffering innocent[3]—it still remains one of the Bible's greatest mysteries. Professor of theology Konrad Schmid notes the tale's "perspective multi-dimensionality" and stresses that "then as now it remains unclear what the subject of the book of Job actually is."[4]

This confusion may surprise some readers, for whenever the conversation turns to Job, we can be pretty sure that the term "theodicy" will come up. How is it possible that there is so much undeserved suffering in the world of a supposedly well-intentioned God? In light of all the evil in the world, how is it even possible to believe in an omnipotent being at all? The Greek philosopher Epicurus (ca. 341–270 BCE) is attributed with having developed the classical formulation of the problem: "Is God willing to prevent evil, but unable to do so? Then he is not omnipotent. Is he able, but not willing? Then he is malevolent. Is he both able and willing? Then whence cometh evil? Is he neither able nor willing? Then why call him God?"[5]

We need to point out that theodicy—the "vindication of God" aimed at defending the "plausibility of faith"—is a relatively modern phenomenon.[6] It presumes the existence of a single, omniscient, omnipotent, and absolutely benevolent God in whose world evil nevertheless remains.[7] We do not yet encounter the concept of a perfect God in the Hebrew Bible, for God was not linked to the idea of absolute goodness until Greek philosophy percolated into the Judeo-Christian religion much later.[8] Above all, the discussion surrounding theodicy presumes the existence of "senseless" suffering.[9] Only then do questions like the following arise: "If God does exist, why did he create a world with so much pain and suffering? Why does he allow natural catastrophes to happen?"[10] For the God of the Old Testament, questions such as these are insulting. There's no such thing as senseless suffering! All suffering reflects acts of God intended as moral sanctions. He doesn't just send natural catastrophes; he's responsible for everything bad that happens. If we accept the notion of Yahweh as the "god of just retribution," then the question of theodicy becomes irrelevant.[11] The catastrophes *are* God! God didn't mince his words in the book of Isaiah: "I am the LORD, and there is none else. I form the light, and create darkness: I make peace, and create evil: I the LORD do all these things."[12]

And so it is with the story of Job. God was the cause of Job's suffering. He allowed Satan to torment Job with one calamity after another. Yet the important question in our examination of Job is not the question of theodicy, because the question of the source of his suffering is actually unambiguously clear: Job's suffering came from God. The real question, then, is why God tormented him in the first place. Did Job deserve such punishments? The answer is both yes and no. And this is where things really start to get interesting.

FRICTION IN CULTURAL EVOLUTION

Here we find ourselves faced with what theologians like to call the "crisis in the deed-consequence nexus," which is another way of saying that divine justice is not fully served in the here and now, as promised by Torah and the prophets. Too many good people have to suffer while too many bad people live lives of luxury. Reality makes a mockery of the Old Testament's assumption that one need only follow God's law and everything will be fine. The people of Israel obviously continued to suffer. A series of foreign rulers reigned in Palestine, and the conquests of Alexander the Great (356–323 BCE) established Greek culture, Hellenism, as the first global culture in history, a development early Judaism viewed as a major threat.

But that was only the half of it. For us the mysteriousness surrounding the book of Job—the lack of a clear message in the key question of God's fairness—demonstrates that we are dealing with friction in the process of cultural evolution. There's a snag here, a very serious problem, but one to be expected in our anthropological take on the Bible. The previous chapter shows how the psalms welded together two different concepts of religion: the intellectual-institutional religion we know from the Torah and the intuitive-individual religion that originates in our first nature. Any attempt to combine these two distinct forms of logic, these two disparate concepts of justice, is bound to lead to friction. The resultant grating of the gears finds expression in the story of Job, whose wrestling with sudden and undeserved suffering appears in exactly the right part of the Hebrew Bible.

As illustrated by the example of diseases, the two concepts of religion have opposite intentions. Intellectual religion aims to provide society with a cultural recipe to prevent catastrophes. It focuses on the big picture, measuring punishment in generations. Its protoscientific approach rests on a single rational principle: the preferences of the single God. Intuitive religion, in contrast, focuses on the individual human being. It stems from emotions and ancient patterns of perception. Its time horizon is the human lifetime, and the interactions between numerous supernatural beings and forces determine its world.

These two concepts were fused—we see this in the book of Job, where we find both types of religion. But they had not yet merged to form a new religion; they were still cobbled together in makeshift manner. We have become convinced that the book of Job tells the story of the poor man's suffering twice—once from the perspective of intuitive religion and once from the perspective of intellectual religion—for completely different reasons. This is why Job appears in two completely opposite roles. On the one hand we have Job the endurer, who patiently suffers his fate without complaining, and on the other hand we have Job the rebel, who openly denounces God. This may be the very reason why theologians have such a difficult time coming to terms with him. Let us attempt to unravel the two stories.

Job: Take One

Here is intuitive religion's version of Job the way our first nature would tell his story. All the misfortune visited upon good Job couldn't be an accident, could it? Someone must have wanted to do him harm, so why didn't God protect him? Could it be that he, too, had a part to play in all these terrible events? This is the way our minds work: it's all about social causality. In this version we find the resentful Satan who was jealous of Job's fame. And God allowed himself to be riled when he saw an opportunity to take part in a wager. This is the plot at the very beginning of the book of Job. Here God, Satan, and the sons of God are a similar host of supernatural characters to the ancestors, spirits, and tribal gods. The Job in the intuitive version reacts exactly like any hunter-gatherer would who had been dealt a miserable hand: he gets angry and demands to know why God failed to protect him.

And Job was right to be angry. He hadn't done anything wrong; his reputation was spotless. He had always remained true to God, and God had done him wrong for no reason whatsoever. So it was God's fault—Job himself was innocent. This is why God so richly rewarded Job at the end of the story; he wanted to make amends. To our intuition this version of Job is completely plausible. It is thoroughly human in scale, and the emotions and reactions it depicts are part of *Homo sapiens*'s standard repertoire. In chapter 13 we discussed the gods' morality under polytheism, and in this version of Job we find an explanation for misfortune that would have been at home in such a religion. Here it is the result of supernatural beings squabbling among themselves. Job has to suffer because of a vain dispute between God and Satan. Their conflict is fundamentally no different from the competition between Hera, Athena, and Aphrodite to determine who was the most beautiful—a beauty contest that resulted in ten years of war over Troy.

Job: Take Two

The same story told from the perspective of intellectual religion is completely different. Here we find a version of God who is responsible for all of creation, and anyone with the greatness of mind to grasp this profound principle can understand why Job had to suffer. The story is about our inadequacy to grasp God's plan for the world. Even theologians cannot help but scoff that God's only answer to Job is no more than "a three-hour natural history lesson." God's speech tells us everything about "the origin of hail, how the doe bears her fawn, the leap of the locust, the nest of the eagle, about thunder and lightning, the unicorn or the leviathan . . . but he has absolutely nothing to say about Job and his suffering."[13] Here we meet Job the endurer, who can do nothing more than accept that everything apparently is right with creation, even though he himself must suffer terribly. His intellectual capacities are just not big enough to comprehend all the causal relations in this big world, so in the end he can only admit his guilt. Here there is no need for demonic figures such as Satan or human-sounding explanations such as a bet between him and God. In this version we find a monotheistic God whose reference to the relationships between the universe and nature is similar to what we might hear from a modern-day scientist seeking to explain why we have earthquakes or tornados. It might make a lot of sense to our third nature, but it doesn't do much to comfort us.

A QUESTION OF BALANCE

We are not reading the Bible to present new theological interpretations; we merely want to find out what it can tell us about human nature. In this respect the book of Job is a true godsend. The first insight it has to offer comes as no surprise: to which of the two versions of the story can people more easily relate? No doubt it's the one in which Job is the victim of dubious heavenly intrigue. Inferring agents behind events is simply the way our minds work; selection molded this trait as an adaptation to real experiences in our social environment. And this is why it is no accident that we suddenly encounter characters such as Satan and other polytheistic players, even though monotheism was firmly established by the time of the book of Job. Our first nature demands this explanatory structure. It senses that the world is simply too complex and inconsistent to be the product of a single divine force.

The second insight regarding human nature is more remarkable. Precisely because the story is told from two completely different perspectives, we can identify their shared core. In both versions the dominating theme is the law of reciprocity, the natural law of fair exchange that defines cooperation between pairs of individuals in human groups. In both versions of the story it is critical that the balance in the relationship between man and God is eventually restored. At the end, no debt can be left unsettled between the two. And since the Last Judgment did not yet exist, it had to happen in the here and now.

In the intuitive interpretation of Job's suffering, the fault clearly lies with God, for he allowed Satan to draw him into this "cruel experiment."[14] This is why Job denounced him, and it's also why God had to reward Job at the end of the story to pay back the debt. It all takes place with mathematical precision: God pays back double what he took in order to replace the children, farmhands, and cattle *and* recompense Job's suffering. In the end, God and Job are even.

In the second version of the story—the one told from the perspective of intellectual religion—Job's suffering is the punishment of a just God. But this assumes that Job must have been guilty of something. We're dealing here with the God of just retribution, after all. Thus Job must admit his guilt (even though he was unaware of having done anything wrong) so that in this case, too, the relationship between man and God can be brought back into equilibrium.

In both instances human psychology could not imagine a situation with no settlement. It would have been impossible for God to win his bet with Satan without recompensing Job for his suffering. That would have been scandalous. That God punished someone who was utterly innocent would have been even more scandalous. No, in both instances the scales had to be balanced, and Job had to have his happy end. But why? It's simply the way our minds work—we are programmed for the tit for tat of reciprocity in our social bonds. One hand washes the other: this is the foundation of all cooperation. Not even God could escape this basic rule.

EVERYONE SHALL DIE FOR HIS OWN INEQUITY

Reciprocity is therefore the theme at the very heart of Job's story. If the assumption on which we base our reading of the Bible—that mismatch problems drove the human mind to make up stories, which in turn catalyzed the contents of the Bible's stories—is correct, then reciprocity, one of the most self-evident aspects of human coexistence, could only become an issue in the Bible if social relations had become seriously unbalanced in the real world.

Indeed, by the time the Hebrew Bible was written, reciprocity had been hard to come by for thousands of years. Ever since sedentarization, humans had suffered great misfortunes, even though they believed themselves innocent of all wrongdoing. Of course hunter-gatherers also suffered disasters, but the effects of these were limited, and tried-and-tested explanations clarified why they happened. And then sedentarization threw the whole world out of whack: people suddenly found themselves subjected to unprecedented calamities. Neither sacrifices nor rituals could mollify the supernatural beings believed responsible for these acts. As a result we find stories similar to Job's in multiple cultures.[15] All are variations on the same theme: we do everything to honor the gods, so why do they allow us to suffer?

The gods are subject to the same rules of behavior as people, spirits, and ancestors. A divine being who fails to nurture reciprocity—who receives sacrifices without giving something back or absorbs praise without showing his own appreciation in some way—is simply a freeloader. And anyone unleashing his wrath for no good reason is nothing more than a tyrant. Such a deity is no longer worthy of praise, and people will simply switch to other gods.

The situation becomes particularly prickly if there is only one supernatural actor to choose from. In chapter 13, on morality, we showed how monotheism required of God a previously unknown measure of moral integrity. The problem was greatly exacerbated the moment the Bible attempted to integrate intuitive religion and Yahweh confronted new expectations of fairness. Once again, we can practically hear the gears of cultural evolution grinding away.

The intellectual religion of the Torah focused on the people of Israel as a whole. If true to Yahweh, they would be rewarded; if not, they would be punished. The time horizon of reward and punishment could be measured in terms of generations—a fact we tend to overlook today. In the Ten Commandments we read, "I the LORD thy God am a jealous God, visiting the iniquity of the fathers upon the children unto the third and fourth generation of them that hate me; And shewing mercy unto thousands of them that love me, and keep my commandments." We come across similar statements again and again in the Bible. This made it easy for people to find an explanation when misfortune visited good people. The book of Kings showers King Josiah (ca. 647–609 BCE) with praise: "And like unto him was there no king before him, that turned to the LORD with all his heart, and with all his soul, and with all his might, according to all the law of Moses; neither after him arose there any like him." But then the following happens: "In his days Pharaohnechoh king of Egypt went up against the king of Assyria to the river Euphrates: and king Josiah went against him; and he slew him at Megiddo, when he had seen him." So why does Yahweh allow a pharaoh to swat away this model king as if he were nothing more than a bothersome fly? The Bible answers that God is still furious over the disgraceful deeds of Josiah's grandfather Manasseh while he was a vassal of the Assyrians. Not even the grandson's perfect behavior can soothe his divine rage.[16] In biblical times this type of thinking was not as unusual as it would seem today. Blood vengeance, tribal feuds, and even wars had cultivated a belief that people would be held responsible for the misdeeds of their long-departed ancestors. If sanctions can appear generations later, the problem of theodicy never arises. Surely, some event in the distant past justified the divine punishment. In a previous chapter we described such strategies as immunization against refutation (IAR).

Now Yahweh, this heavenly being with a heretofore unknown sense of justice, became transformed into a personal patron god for individual people. He could no longer acquit himself of his duty of meting out justice among the Israelites by dribbling out rewards or punishments for several generations. His justice had to manifest itself within the lifetime of the individual, since—as we have previously explained—notions of a Last Judgment did not yet exist. The prophets Jeremiah and Ezekiel had already proclaimed invalid the old proverb "The fathers have eaten sour grapes, and the children's teeth are set on edge." Now it was known to all that "every one shall die for his own iniquity: every man that eateth the sour grape, his teeth shall be set on edge." This meant that it should now be impossible for someone like King Manasseh to do evil and still reign for fifty-five years before enjoying a peaceful death and equally impossible for his grandson Josiah to be made to pay for his grandfather's misdeeds. From this it is easy to derive the maxim that whoever "hath walked in my statutes, and hath kept my judgments, to deal truly; he *is* just, he shall surely live."[17] Now every individual could enjoy the promise of just retribution God had once made to the Israelite people as a whole—but this put God in a devil of a mess, as the book of Job shows.

The reason is simple. This maxim might sound well and good, but real life just doesn't work like that. Evildoers go unpunished; the righteous go unrewarded. Job laments, "Wherefore do the wicked live, become old, yea, are mighty in power? . . . They spend their days in wealth, and in a moment go down to the grave. Therefore they say unto God, Depart from us; for we desire not the knowledge of thy ways. What is the Almighty, that we should serve him? and what profit should we have, if we pray unto him?" It's just not fair, Job grumbles. God shouldn't punish the children for their ancestors' misdeeds; he should punish the godless directly and so the latter "shall know it."

But we all know there is to be no settlement during our earthly existence. In life as we know it, Job's reward for his loyalty to God at the end of the story is the exception. But if God himself does not adhere to the law of reciprocity, then he is not a dependable partner to his people. This places him in existential danger. He either has to turn to the devil as an excuse, or he has to come up with a place where he can still grant each and every person his or her proper rewards.

AND THE WORMS SHALL COVER THEM

Let us summarize: With increasing misfortune the laments directed at the gods' unreliability grew louder. But things became especially precarious where monotheism reigned—where the one true God had waved the banner of justice and now, as a personal patron god, had to guarantee fairness for all people within their lifetimes. God had to pay people back for the blameless misfortune they had suffered in life. His not doing so could mean only one thing: the individual had deserved it. The unfortunate person had to be a sinner, even if unaware of any transgression. That horrible belief was to play a prominent role in Christianity. The invention of original sin, the presumption that all people are sinners from birth thanks to Adam and

Eve's mistake in Paradise, served to justify all the wrongs of the world from the very get-go. The Christian concept of original sin is one of the most effective IARs of all time.

As we have seen, a great deal needed to be done if the priests and scholars were to prevent the grinding wheels of cultural evolution from pulverizing the new hybrid religion. They had to come up with a way of explaining how all could enjoy God's justice, and hence reciprocity, the fundamental law of human coexistence. As the next chapter shows, the notion of the afterlife would therefore suddenly play an important role in the Bible. What had Job said? It simply cannot be the case that the righteous and the evildoer "lie down alike in the dust" covered with "worms." Death would eventually offer the final hope for God's justness.

16

DANIEL

The Discovery of the Afterlife

Astonishingly we have yet to encounter the subject of death in the Old Testament. We're not referring to killing, of which there is certainly no dearth in the Hebrew Bible. People die in all sorts of ways. For example, Yael hammered a tent stake through the sleeping captain Sisera's temple, Samuel chopped the king of the Amalekites into pieces, and Judith lopped off Holofernes's head. No, here we are talking about death itself and the question of what people expected in the hereafter. None of the books of the Hebrew Bible examined so far appears to consider this a question worthy of special reflection.

This changes toward the end of the Hebrew Bible. In Psalms, Job, and Ecclesiastes, we begin to notice a degree of uncertainty about the subject of death and what happens to people after they die. The book of Daniel, one of the last of the Old Testament's books to be written, suddenly broaches the issue and tells of how the dead will be resurrected. Later, the New Testament will make the question of life after death one of its major themes.

Death is certainly nothing new in evolution, which has always taken place in its shadow. The oldest traces of human burials date back over 100,000 years. We find them among our direct ancestors—modern *Homo sapiens*—as well as our sister species, the Neanderthals. This means that our forebears probably already maintained some conception of death far earlier. If the subject is as important as the evidence suggests, should it not be a major topic of the Hebrew Bible?

No, it shouldn't. The phenomenon of death has always occupied humanity's thoughts. As should be obvious by now, a fundamental presumption of our book is that the most predominant issues we encounter in the Bible are mismatch problems—difficulties created by our sedentary way of life, which were so new and urgent that people had to develop cultural solutions to bring them under control. These cultural solutions may have made the problems more manageable, and in best-case scenarios even made them more bearable, but they certainly could never make them disappear entirely. That is why mismatch problems continue to nag us.

It therefore follows that if the Hebrew Bible didn't find the subject of death worth addressing for so long, this must have been because death was not a new mismatch problem. Our forebears had been dealing with it for hundreds of thousands of years. As this chapter will show, people have intuitive notions of what happens to them after death. Furthermore, all of the cultural practices surrounding death and mourning are an ancient component of our second nature that is deeply interwoven with our first nature. Death has always been one of life's greatest certainties.

That is why it is a massive—albeit widespread—misconception to presume that religion's main function is to take away our fear of death. This is simply not the case. Of course, like all animals, humans possess an instinctive aversion to death and generally tend to do everything in their power to stay alive.[1] Likewise, the loss of close relatives and proven cooperation partners has a deep emotional impact.[2] However, today's widespread "metaphysical" fear of what happens after our biological bodies cease to function is a relatively modern phenomenon.

Dualism is part of our first nature: we intuitively assume that body and spirit are two completely different things. This is why we believe that the spirit—or soul—doesn't die with our bodies. In his experiments Jesse Bering showed that children believe a person's mental abilities somehow survive death.[3] Paul Bloom even goes as far as to state that we observe the world "in terms of bodies and in terms of souls" and rightfully points out, "For most of human history, there was no scientific reason to doubt that the soul can outlast the body."[4] This is why, as we noted earlier, in all of the world's cultures we find a more or less developed belief in an afterlife in the form of spirits and ancestors.[5] Death is not the end; humans never doubted this. Death is no reason to panic.

But in the Ketuvim we see for the first time how death gradually loses its self-evidence. It suddenly developed into an existential issue. We highly suspect that Yahweh is responsible for this outbreak of agonizing about what happens after we die. His motive? There are two, in fact: his jealousy of any form of divine competition and his desire to appropriate the realm of the dead so that he can finally fulfill his promise of just retribution. If religion today is the most important strategy for combating earthly transience, then people look to God for solutions to problems he himself let loose in the world. We cannot leave this unexamined.

POSTMORTAL CACOPHONY

Whereas in other lands gods such as Osiris and Hades ruled over opulent netherworlds together with crews of helpers and maintained sophisticated procedures for the journey into the afterlife, the Hebrew Bible is so tight-lipped when it comes to death that one can only describe this silence as "deafening," to quote William Dever.[6] Nowhere does the Old Testament find it necessary to contemplate, let alone systematically discuss, the nature of death.[7] Scholarly research into the historically existing apprehension of death resembles a jigsaw puzzle reconstructed from texts scattered throughout the Hebrew Bible that mention the phenomenon. For instance, we find a reference to Yahweh's power in the book of Isaiah ("he will swallow up death forever") or a throwaway line in Ecclesiastes about nonexistence ("For to him that is joined to all the living there is hope: for a living dog is better than a dead lion").

In fact, the Bible doesn't even offer an explanation of why people have to die at all. Some scholars refer to Genesis and explain death as a consequence of our being chased out of the Garden of Eden, but this is incorrect. Death was not a punishment God introduced into the world—he had made humans mortal from the outset. From dust we are made and to dust we shall return.[8] It is only in the late-biblical wisdom literature of the Ketuvim that we encounter an utterly confusing cacophony surrounding what happens to people after they die and the role of God in the whole affair.

Some of the psalms deny that Yahweh is in any way responsible for death ("For in death there is no remembrance of thee: in the grave who shall give thee thanks?"). But others express the hope of a transcendent connection to him ("But God will redeem my soul from the power of the grave: for he shall receive me").[9] In the book of Job, death is the "great equalizer" that puts an end to all and levels the differences between the righteous and the sinner.[10] And Ecclesiastes bewails with the deepest sadness, "For the living know that they shall die: but the dead know not any thing, neither have they any more a reward; for the memory of them is forgotten. Also their love, and their hatred, and their envy, is now perished."

In sum, we have a jumble of hopes and fears. Whereas the prophets Elisha and Elijah were able to bring dead individuals back to life, Isaiah and Ezekiel even had visions of a possible resurrection of the entire people of Israel ("Thy dead men shall live, together with my dead body shall they arise"). Bible scholars, however, tend to interpret this as a metaphor rather than a concrete vision of the afterlife. In the book of Daniel we find "for the first and only time" in the Old Testament the idea of an eschatological judgment and the resurrection of the dead.[11] People today hold this to be a particularly Christian notion of the afterlife, but as we will see here, this is also not the case—reason enough to take a closer look at the book of Daniel, which Christian Bibles group with the prophets.

Daniel's Secret Knowledge

Daniel and his friends are carried off to Babylon after King Nebuchadnezzar conquers Jerusalem. At the royal court they make careers for themselves as wise men, but despite all animosity, they remain steadfastly loyal to Yahweh. This is illustrated by the story of the three friends' surviving being tossed into a fiery furnace and the proverbial tale of the lions' den in which Daniel, a victim of slander, is fed to the beasts. His trust in Yahweh proves to be his salvation, however, for an angel of the Lord holds shut the lions' mouths. Daniel survives the night, and the next morning his enemies are served to the lions for breakfast.

The book of Daniel is famous for his dreams and visions, which make it into the only apocalyptic book of the Hebrew Bible.[12] Four monstrous beasts representing the world's greatest empires would rise from the sea and strike terror into the hearts of the people until a heavenly being would appear—who looked "like the Son of man"—and enter into a final struggle for the dawning of God's dominion. On this Son of man, Judaism would pin its hopes that one day a savior from the House of David, the Messiah (the "anointed one"), would appear and save the people of Israel from their misery. The New Testament sees the coming of Jesus Christ as the fulfillment of this prophecy.

Daniel's vision ends with a famous passage:

> And at that time shall Michael stand up, the great prince which standeth for the children of thy people: and there shall be a time of trouble, such as never was since there was a nation even to that same time: and at that time thy people shall be delivered, every one that shall be found written in the book. And many of them that sleep in the dust of the earth shall awake, some to everlasting life, and some to shame and everlasting contempt. And they that be wise shall shine as the brightness of the firmament; and they that turn many to righteousness as the stars for ever and ever.

In Daniel's vision death is not the great equalizer as foretold by Job; nor is it like the end prophesied in Ecclesiastes. A judgment will be held: those who have proven themselves in life will live forever; others will suffer eternal damnation. But all this was to remain a secret: "But thou, O Daniel, shut up the words, and seal the book, even to the time of the end."

Readers could be forgiven for thinking that this command amounts to a virtual declaration of eschatological bankruptcy on behalf of the Bible. First it neglects to offer us an explanation for the most important question in human existence, and then, when it finally does provide one, it classifies the information as top-secret. The fact that we find this collision of conflicting notions of the afterlife in the Old Testament[13] demonstrates that the Bible's authors desperately sought solutions to a problem that hadn't previously existed. After all, in ancient Israel, too, death was in the best of hands.

Hushed Up

The Hebrew Bible does actually address the subject of death in the guise of the proscriptions scattered throughout the Torah. Mourning rites—shaving one's hair and beard or scarring and mutilating one's body, for example—were taboo. Nor was offering food to the dead allowed. Conjuring up the spirits of the departed? Strictly forbidden. Indeed, any contact with the deceased made one unclean.

Admittedly, some actions prohibited by the Torah appear as daily practice in other parts of the Bible. This comes as no surprise: "Societies do not forbid practices that no one would think to perform."[14] A spectacular scene occurs when King Saul goes to the witch of Endor and asks her to summon the spirit of the dead prophet Samuel. "Why hast thou disquieted me, to bring me up?" Samuel complains, but he still finds time to answer Saul's questions. We also find the entire spectrum of mourning rituals, such as the rending of clothing, the sprinkling of the head with ashes, and the loud wailing of mourning women.[15] These same practices existed in countless other cultures. In this form of death and ancestor cult one finds the "basic anthropological components" of what is often referred to as household religion.

This is all true, but the Bible's authors do all they can to "downplay the important role of the dead and their care in ancient Israel."[16] Clearly they did not entirely succeed in doing away with these rites. This leads us to conclude that death was originally the domain of intuitive religion, the practice of which was universal in nature. Monotheism declared war on all of the spirits and other gods and attempted to silence the ancestor cult.

A NEW HOME FOR THE DEAD

These facts provide the backdrop against which we can explain how this transformation of the netherworld came to be, a transformation that over time led to precisely the same metaphysical homelessness that is still with us today. Many contemporary theologians refer to religion first and foremost as a means of coping with contingency, by which they primarily refer to the fear of death. This transformation process is driven by a dialectical interaction between our biological predispositions and cultural transformations. Let us now explore in some detail how death was banned from the realm of the living.

Our Nature

In our chapters on the flood and the psalms, we discuss how *Homo sapiens*'s inborn belief system transformed the departed into spirits and ancestors. "This is represented as the normal outcome of human life, not as a special prize awarded for high morality," Pascal Boyer explains. Accordingly, the souls of the dead "are the most widespread kind of supernatural agent the world over."[17] This is the way our first nature works: people simply cannot imagine being dead. So what could be more natural than to assume that the dead live on as "bodiless souls," asks evolutionary psychologist Jesse Bering. This is our dualistic intuition—and we find it not only in children, as Bering's research had shown, but in adults as well. A convinced atheist, Bering experienced it himself firsthand. After the death of his mother, he was well aware that the coroner had taken away her body, but when a breeze set the wind chimes outside her bedroom window jingling, he immediately knew "she's telling me everything is OK."[18]

These types of age-old intuitions do not require the existence of a special netherworld. For example, hunter-gatherers often believe in reincarnation in which the spirit of the deceased seeks out a new body.[19] But even if they live on as ancestors, the departed do not require some special abode, for they live on invisibly in the natural world, albeit in a sphere that is difficult for the living to reach. The spirits are the same as they were when they were living people. They can be loving and caring, and they watch out for their relatives' best interests. But they can also be moody and vengeful, particularly when it comes to their old enemies. Forgotten ancestors become evil spirits or even demons.

Humans live in a "symbiotic relationship" with their ancestors.[20] People ensure the spirits get enough to eat and in return expect them not to play any mean tricks. But the fear is always there, for the spirits can also send illnesses and misfortune. Thus we find everywhere the self-effacing rituals that are so typical of mourning. Tearing up garments, uninhibited cries, and demonstrative self-mutilation are, of course, credibility-enhancing displays (CREDs) aimed at translating the loss of loved ones into actions that are visible to other people, for it is important to show just how dependable and loyal a partner one actually is. But that's not the whole story: these CREDs are also directed at the dead themselves. The message: Just look at how important you are to me! Look how much I miss you! An ancestor who notices that a living relative isn't mourning for him will certainly try to get back at him somehow.

All of this happens "just like that." It's intuitive. Some of these rituals remain with us to this day—when we "visit" the departed at the cemetery, for example. Even if we might not believe that they live on six feet under, we don't want to leave them alone—we even speak with them.[21]

Death has only recently become something for intellectual-institutional religion to deal with. For most of human history, this type of religion didn't even exist. No sphere separated from the rest of culture was responsible for the beliefs and actions that we would now call "religion."[22] Intuitive religion and culture were one and the same—and death was a completely self-evident part of life.

Ancestor Worship in the Bible

Even though the Bible tries to keep a lid on ancestor worship, it contains a number of striking references to it. The Bible mentions teraphim—figurines of ancestors that stood in the bedrooms of private homes—on fifteen different occasions. Teraphim were important in promoting identity and the "continuing growth and welfare" of the family. Even though the Bible tried to cover it up, in the eyes of "their descendants, the dead were very much alive."[23] Rachel, Jacob's wife, stole her father's teraphim, and Michal dressed up an ancestor figure in a cloak and goat-hair wig to cover up her husband David's escape. As these examples suggest, women interacted with the teraphim; after all, they were the experts in intuitive religion's rituals and the family's cohesion, regardless of whether its members were dead or alive. The dead were therefore in the best of hands. Of course, death wasn't the most pleasant event, especially when it came suddenly or with great suffering, but at least one knew what would come next. Living on as an ancestor was completely in line with the dualistic intentions of our first nature, which views the body and the soul as two different things. The fate of the one is not the same as that of the other.

During his early days as a state god, Yahweh had absolutely nothing to do with this part of daily life. He had always been a "god of life," never a "god of death."[24] For this reason, the departed were not his concern—they were anyway well taken care of. Here we see a properly functioning division of labor. Extrabiblical inscriptions offer evidence of the gradual expansion of Yahweh's duties. They invoke him as the personal tutelary god of deceased individuals. One comes from the grave of a well-off family in Jerusalem. The elites who had dedicated their lives to the cult of the official god Yahweh would certainly not have given him up in death. Here we see how the state god could also serve as a personal patron god, in effect showing how intuitive religion had brought Yahweh into the fold.[25]

Then came the break. While in exile, Yahweh's elites developed a radical form of monotheism and declared war against intuitive religion and all of its actors. From then on, only the one God was permitted. During the Babylonian captivity, the elites were separated from the graves of their ancestors for two or three generations. The traditional funerary cult probably became a reduced part of their experience, likely making its elimination for ideological reasons easier. All of this seems to indicate that Yahweh hadn't originally intended to conquer the netherworld; otherwise we would encounter corresponding texts in the Bible. Emptying the world of the ancestors was an unintended side effect of monotheism.

Sheol

The theological circles working on the biblical texts attempted to erase any hint of "a positive appreciation of the cult of the dead."[26] Thus we find very little information in the Bible concerning the older funerary cult. All the more surprisingly, then, we repeatedly come across a strange and forbidding place in its pages: Sheol, a dreary land of shadows, a place of dust and forgetting. In the King James Version it is usually translated as "the grave" or "the hell." Like in the Greek underworld, Hades, here the dead wander around in a daze as powerless beings in an unending dream.[27] Scholarly literature often depicts Sheol as the accepted notion of the afterlife that existed in Old Testament Israel. This is rather unlikely, however, for it directly contradicts the existing ancestor cult. We concur with biblical scholar Karel van der Toorn, who believes we are dealing here with the biased views of the Bible's authors when they "suggest that most of the Israelites looked upon the dead as bleak shadows without power or influence." This "authorized version of Israel's past is inconsistent . . . with the prominence of the ancestor cult."[28]

When we read of how a patriarch like Isaac or Jacob was "gathered to his people" at the end of his life, this could hardly mean that he entered into a miserable existence in a depressing realm of shadows. Is it really imaginable that God dooms his loyal servants, such as Abraham and Moses, to vegetate in Sheol for all eternity alongside damned sinners like the Babylonian king to whom Isaiah gleefully proclaimed, "The worm is spread under thee, and the worms cover thee"? Is it even thinkable that a God who values justice over everything offers "the same fate for saints and sinners" in the end?[29]

Religious scholars Kevin Madigan and Jon Levenson point out that Sheol only ever appears in the Bible in descriptions of unnatural deaths—and we know by now that these resulted from divine rage. Thus, Sheol was a place of punishment, so those who lived their lives to God's satisfaction didn't end up there. True, the Hebrew Bible mentions no positive alternative to Sheol, "no postmortem Heaven or Garden of Eden to which those loyal to God can look forward." It didn't need to; for reasons by now familiar to readers, the Bible remained silent on the topic. The positive alternative was the well-known concept of ancestorhood. When the people died "old and full of days," they lived on in and with their descendants.[30]

Yahweh's Big Chance

We can conclude there was no targeted colonization of the netherworld. The battle against the old, intuitive religion merely created an afterworldly vacuum. The strange polyphony that we encounter in the pages of the Ketuvim evinces the multifaceted attempts to fill the void. The task was urgent, since it would simply be preposterous for the competence of a universal God to end at the gates to the underworld.[31] In Psalms we even read, "The dead praise not the LORD, neither any that go down into silence." From a monotheistic perspective this was nothing less than scandalous. This situation called for some serious compulsive coherence seeking.

There was one other, extremely delicate issue. The prospect that everyone—whether loyal to God or not—would suffer the same fate after death was nothing less than a disgrace for Yahweh. This makes it easier to understand why the solution proposed in the book of Daniel of a postmortal court presented such an attractive way out of the dilemma—a solution, incidentally, with a promising future ahead of it. The notion that the dead would be resurrected and rewarded or punished according to how they had lived on earth not only meant that the afterworld was now under Yahweh's command but also offered him a way out of the quandary regarding his unfulfilled reciprocity obligations.

We came across this problem in the book of Job, where the option for recompense in the afterworld did not yet exist. Job had to be compensated in the here and now for his unjustified suffering. Earthly experience showed, however, that good people still died all the time without having received their rewards, and bad people could die having avoided punishment altogether—a heavy blow to God's credibility. This meant that divine justice would have to take place some time after death—a brilliant solution as well as a perfect strategy for immunization against refutation. To this day no one has been able to prove that God does not reward or punish in the afterlife. It's little wonder that death would advance to become the strongest bastion of intellectual religion.

Martyrs

We can only rarely trace cultural innovations back to a single cause. At the theoretical level we find ourselves confronting a discursive undertow: certain adjustments were necessary in order to maintain the system's logic. But there is also a historical level, and here, once again, catastrophes turned out to be the driving force. In the centuries that followed the return from the Babylonian captivity, the elites succeeded in establishing monotheism (though without completely eliminating intuitive religion). This did nothing, however, to change the terrible political situation that existed at the time. Things were to get even worse in fact. The Macedonian king Alexander the Great destroyed the empire of the Persians, and Judaea, the name used for the region in Hellenistic and later Roman times, first went to the *Diadochi* state of the Egyptian Ptolemaic Dynasty before falling to the Syrian Seleucids. These two Hellenistic empires warred with each other for all of the third century BCE, during which the Ptolemaic and Seleucid armies marched through Judaea no less than seven times.[32] The apocalyptical visions in the book of Daniel play out against this dismal backdrop.

Even worse things were soon to follow. The Seleucid ruler Antiochus IV Epiphanes conquered Jerusalem on his way back from an Egyptian campaign, plundered Yahweh's temple, desecrated the altar, and installed the cult of Zeus. He suspended Jewish law per decree.[33] "The Jewish religion was to be wiped out, its followers were forced to perform idolatry, to perform acts that were forbidden to them by their religion such as eating pork and other forbidden foods, and make sacrifices to pagan gods," explains historian Aharon Oppenheimer. "The possession of Torah scrolls was punishable by death." Circumcision and observing the Sabbath were also forbidden. This repression resulted in the Maccabean Revolt, during which we find "for the first time in world history . . . a widespread readiness for a martyr's death."[34]

If people are ready to die so that they can remain true to their religion, the issue of justice suddenly becomes crucial. Should the martyrs really end as mere shadows in Sheol together with the evildoers? An absurd idea, for God would certainly reward his most loyal followers. But where? And when? If the elites wanted to prevent the people from turning away from

Yahweh, they would have to make some improvements in the representation of the afterworld.

The Bible is fascinating in that it sometimes feels as if we can see its authors ruminating over the problem. The books of Maccabees are not a part of the Hebrew Bible, but they do belong to Catholic Old Testament canon. They present readers with two tales of martyrdom dating from the time of Antiochus's repression. The first begins as follows: "Eleazar, one of the principal scribes, an aged man, and of a well favoured countenance, was constrained to open his mouth, and to eat swine's flesh. But he, choosing rather to die gloriously, than to live stained with such an abomination, spit it forth, and came of his own accord to the torment." We are meant to take away this message: "And thus this man died, leaving his death for an example of a noble courage, and a memorial of virtue, not only unto young men, but unto all his nation." There is no mention of a reward after death or eternal life for this ninety-year-old man. But was the story convincing enough? Eleazar's loyalty to God was certainly worthy of praise, but he was already an old man who had nothing left to lose.

So the Bible's authors added another story—one about a mother and her seven sons who were tortured as well in an attempt to force them to eat pork. But the seven men, too, all stood firm against the most terrible of pains and eventually suffered a martyr's death. Each and every one of them. The mother, forced to watch it all happen, even encouraged her sons to resist. In the end she, too, was executed. In this story it is not an old man already at death's door but seven young men and their mother who remain true to their beliefs. The main point, however, is that the brothers are sure that "The King of the world shall raise us up, who have died for his laws, unto everlasting life." The dying sons tell the Seleucid king Antiochus to his face that he will be overtaken by God's judgment.

The Bubbling Pot

We understand why toward the end of the Old Testament the subject of death begins to bubble over. Monotheism necessitated a new solution. In light of the repression, persecution, and violence, someone urgently needed to come up with a new answer. A number of fixes were proposed in the centuries before and after year zero. There existed no uniform Judaism at the time (there probably never has); rather, different theological groupings experimented with various solutions in a protean formation process.

The Pharisees, with whom Jesus often argues in the New Testament, believed in resurrection. The Sadducees, who made up the temple aristocracy, rejected it.[35] Ideas imported from neighboring cultures were also entertained. We have already discussed the Egyptian concept of the judgment of the dead, for example. Experts disagree about the possible influence of Iranian Zoroastrianism—also known as Zarathustraism—in which we also find the concept of a divine court of law for the dead where the bad are punished and the good returned to physical life.[36] Greek influences are reflected in the coins found in some Jewish tombs as payment for the ferryman Charon, who transported the dead over the river Styx. In a sweet irony, such a coin was found in the skull of woman in the family grave of the high priest Caiaphas,[37] the very one who handed Jesus over to the Romans. Do we need more powerful proof of just how much uncertainty reigned when it came to the nature of the afterlife? Even the family of Yahweh's top servant on earth was open to foreign influences.

It's hardly surprising that a subject of such tremendous importance would come to play a major role in the New Testament, in which new catastrophes would once again bring the cauldron of cultural evolution to a boil. The reprisals carried out by the Romans surpassed anything the Jews had ever suffered before. Again and again protests and rebellions erupted, and people were crucified by the thousands.[38] But what fate awaited them?

At that time there was still "no systematic, thoroughly developed concept of the afterlife." In the New Testament we continue to find a confusing collection of concepts side by side.[39] A number of questions remained unanswered. Would all of the dead be resurrected? Are body and soul resurrected together, or does just the soul live on? Are the dead resurrected for a limited time or for all eternity? Each of these options had its defenders, but all of them embraced the notion of group resurrection, "for it was not the individual who was resurrected, but the entire group of the righteous in a single, encompassing event."[40] According to some theories of the biblical afterworld, only the martyrs reawakened immediately after suffering a violent death.

As is typical of third-nature products, every concept was endlessly debated. The evangelist Luke tells of a striking encounter. The Sadducees, who did not believe in resurrection, confronted Jesus with the following enigma. Suppose a woman marries and her husband dies, so she remarries, but that husband dies too. She then marries a third time only to lose that husband as well—and so on until she has buried seven husbands. What would happen after the resurrection? the Sadducees asked. "Whose wife of them is she?" One can virtually see the smirking between the lines: A woman with seven husbands! Gotcha! But Jesus has an answer ready: "They which shall be accounted worthy to obtain that world, and the resurrection from the dead, neither marry, nor are given in marriage." If there are no marriages in the afterlife, the problem simply evaporates.

In the writings of the apostle Paul, considered to be the first Christian theologian, we come across two completely different ideas of what happens to people after they cross death's threshold. First Paul draws on the Jewish tradition of physical resurrection as propagated in the book of Daniel. The details are so complicated, however, that Paul later switches to another concept inspired by Greek philosophy in which the soul detaches itself from its mortal shell and ascends into heaven.[41]

And What About Heaven and Hell?

From a modern perspective, this hubbub surrounding the afterworld is unexpected. One would reasonably assume that the Christian Bible had sorted everything out; the dead go either to heaven or to hell after they die. It depends on how they behaved in life. At least the current catechism of the Catholic Church is clear on the subject: directly after death (and not after a Last Judgment at the end of days, as proclaimed in the book of Daniel or the New Testament's book of Revelation), the fate of a person's immortal soul is decided. Either it gets sent down into the "eternal fire" or enjoys "perfect life with the Most Holy Trinity—this communion of life and love with the Trinity, with the Virgin Mary, the angels and all the blessed" called "heaven." If the poor human soul in question is not unblemished, it is sentenced to spend a period in purgatory.[42]

These are, of course, postbiblical additions and therefore fall outside our book's purview, but a somewhat longer look at the history of hell is merited, because it provides such compelling evidence of the great creative power of the compulsive coherence seeking inherent in human thinking. Just as *Homo sapiens* cannot help but explore the darkest corners of space and time, he also cannot help but draw up detailed maps of the places that only exist in the imagination. He has no other choice. As Carl Sagan once said, "Exploration is in our nature."[43]

The point of departure for the Bible's concepts of hell is the prophet Isaiah's much-cited vision in which God proclaims both a new heaven and a new earth. The corpses of apostates were to be thrown into the Valley of Hinnom near Jerusalem, and "their worm shall not die, neither shall their fire be quenched; and they shall be an abhorring unto all flesh." The name Gehenna, used in some versions of the Bible to describe hell, the place of eternal damnation, derives from the Hebrew *Ge-Hinnom.*[44] Jesus of Nazareth invokes the name several times, especially as a place of eternal torment. At the end of the New Testament, John of Patmos describes the fate of the damned in the book of Revelation. Unfortunately he isn't all that consistent. At one point, the book explains how nonbelievers will be destroyed after the judgment of the dead, in another passage we read how they will suffer in hell only temporarily, and in yet another the damnation is eternal. The last of these options is reserved for the real villains of the apocalypse: the devil, the beast, and the false prophet, and they "shall be tormented day and night for ever and ever."[45]

The church fathers Origen (185–254 CE) and Augustine (354–430 CE) attempted to put hell in order. Origen was something of a friend of humanity, for he believed hell was actually a place of purification, and after spending a bit of time there, *all* souls would rise up to be with God. Augustine, who was more of a hard-liner, eventually prevailed: he ratcheted up the notion of eternal agony. Hell not only served as punishment for the deadly sins but was made primarily for original sin, which, as explained by theologian Bernhard Lang, "is the blemish on all human souls that goes back to the Fall of Man in Paradise." Baptism could wash away the original sin, giving people a chance to escape from being sent to hell.

The blossoming of jurisprudence in the Middle Ages encouraged theologians to revisit the idea of hell. Punishment had to fit the crime. It was unacceptable that infants who died before they could be baptized and cleansed of original sin were condemned to rot forever in hell, so theologians came up with minor hells, such as the purgatory described in the catechism of the Catholic Church. Those guilty of lesser crimes received smaller punishments in preparation for a later ascent into heaven. Unbaptized children on the other hand no longer ended up in hell, but landed in "limbo," the "margins of hell," a place of less bliss.[46]

Turning now to heaven, we can keep things short, for, as usual, positive matters receive less attention. In postbiblical times, two main opinions on the subject hold sway. On the one hand heaven plays host to an ideal human world in which all of the blessed live a paradisiacal life together with family and friends. On the other hand we find the concept of God's kingdom of heaven as a mighty cathedral, "where all of the blessed gather together to praise God for all eternity and enjoy his exalted presence."[47] There's no doubt which of these two versions our first nature would prefer—and it's certainly not the never-ending church service.

TO BE CONTINUED

After our tour of heaven and hell, we can conclude for sure that the Bible contains a great deal of uncertainty about what happens after we die. We have witnessed continuous attempts to improve upon and add to notions of the afterworld. In fact the process continues to this day. In *The God Delusion*, Richard Dawkins describes the "hell houses" he encountered in Colorado in which actors visualize the suffering of the damned (with the stink of sulfur included) to prove to visitors that hell is a place best avoided.[48] When on one of his book tours Dawkins claimed during a German talk show that threatening children with hell was a form of child abuse, the Catholic and Protestant bishops in attendance greeted his remarks with smug smiles. They explained that hell is not one of Christianity's greatest inventions and "actually" was no longer a focus of their preaching. Dawkins replied that he was very pleased to hear that German theologians no longer believe in hell.[49]

In Judaism, often misleadingly described as a religion concerned exclusively with life before death, the plurality of explanations remains to this day.[50] When asked what Jews believe about the afterlife, Leila Leah Bronner, former professor of Bible and Jewish history in Johannesburg, replied, "Jewish tradition has not one answer but many," before listing the possible options: resurrection, immortality of the soul, reincarnation, the world to come, and the arrival of the Messiah.[51]

All of the guesswork about what comes after death and all the attempts to describe what heaven and hell might be like make one thing perfectly clear: none of the existing concepts of the afterworld are all that convincing. All of them are brainy, third-nature creations that try to reconcile the notion of a single righteous God with the undeniable injustice of the world. None of these visions of the afterlife fit snugly with our intuition—they all tell us we must believe in them. And being forced to believe in something inevitably calls forth doubt, which is why theologians take great pains to explain how a God of love and compassion can also subject people to eternal torment. For this reason, philosopher Kurt Flasch believes that all "theologies of hell" are doomed to end in an "intellectual fiasco."[52]

The afterworld offers yet another example of the problems facing intellectual religion—and why it still needs theological theories to support it. Intellectual religion remains jealous of intuitive religion, whose predilections have survived for thousands of years. Recall the experience of psychology professor Jesse Bering, when, despite his atheistic rationality, he was simply convinced of his mother's continued existence and the possibility of communicating with the dead. He didn't have to learn this in Sunday school.

The result of all this uncertainty is existential insecurity. The fact that none of these visions of the afterlife was really convincing to our first nature, together with the stigmatizing of our intuition as superstition, created a metaphysical vacuum. And with it grew the fear of what might happen to us after we die. No one knew for sure anymore, and no one could provide more than words. It is an interesting irony of history that God would have to throw himself into the breach. Today his most natural task is to offer comfort in the face of fear of death. He may not offer a convincing solution to the problem of the afterlife, but he lends a sympathetic ear to believers when they turn to him with their fears.

How is it, then, that God can help people overcome the problems that he himself introduced into the world? Because he has been transformed into a hybrid being incorporating the righteous God, the Lord of heaven and hell, and the warm-hearted patron god who listens to the daily worries and existential fears of every single person. The latter must repair the damage caused by the former, the monotheist God of intellectual religion who drove out the trusty spirits and ancestors that had accompanied us for so many thousands of years.

One thing remains clear, however. The idea that God judges the behavior of individuals after their death only appears toward the end of the Bible. We first encounter the notion in the book of Daniel; later it will become one of the New Testament's major themes. This is important, for apparently even most current research into the cultural evolution of religion assumes that the afterlife is an ancient divine feature aimed at maintaining morality in human societies. But God didn't have to hold court in the afterworld to influence our behavior; the catastrophes, seen as his punishments meted out in the here and now, were already keeping people in line.

We can firmly conclude that God is the culprit behind our fear of death—he instilled it in us when his intellectual-institutional religion managed to ruin systems that were working just fine. And he never really managed to introduce functioning solutions to the problems he himself had caused. Once again we are dealing with a clear case of mismatch: this time monotheism created new problems by breaking up the old symbiosis between our basic psychological configuration and our intuitive belief system. Monotheism could only repair the situation by introducing cultural workarounds. It's no wonder that people felt they had been left in the lurch in matters of the afterlife, a theme that will occupy us again. And that brings us to the New Testament.

PART V

THE NEW TESTAMENT

Salvation

At long last we arrive at the New Testament, dominated by one towering figure with magnetic allure: Jesus of Nazareth. For Christians, Jesus is the key to understanding the entire Bible. For us he is a prime example of cumulative cultural evolution. Jesus was a Jew who honored the Torah and never intended to found a new religion. Yet today more than 2 billion Christians worship him as the son of God. We can only understand how this came to pass against the backdrop of the Hebrew Bible.

Our book aims to point out the shared origin of seemingly independent cultural trajectories. We believe this point occurred some 12,000 years ago with the onset of the agricultural revolution, when our species, *Homo sapiens*, began to rely exclusively on its greatest talent, culture, to deal with the new challenges for which biological evolution had not equipped us. We can trace today's rich variety of religions, sciences, cultures, and technologies back to this point.

The New Testament, too, can be seen as a response to the calamities brought on by mismatch problems. Moreover, the coping strategies it proposed were surprisingly old-fashioned—or should we say timeless? The fact that the paths chosen were not truly novel stems from a trait that cultural evolution shares with biological evolution: path dependence. Once a particular direction is taken, reversing directions or making other drastic departures from the current course is difficult. Changes and corrections are sought within the system, and this in turn severely constrains the directions that can be taken. But such limitations obviously do not preclude a system from splitting into two new systems following slightly diverging paths.

What we can explain is such a split around the time of the arrival of Jesus. Because our approach focuses so clearly on the Bible, we run the risk of giving the impression that we regard Judaism as a mere precursor to Christianity. Nothing could be further from the truth. The "Jesus movement" was merely one of many such movements found in early Judaism, such as the Sadducees, the Zealots, the Essenes, and the Pharisees. Christianity would follow its own direction as it developed during the first century CE, but so did the rabbinical Judaism that evolved from the Pharisees' movement at around the same time. Christianity and modern Judaism are therefore sister religions: both sought and found innovative solutions,[1] and both descended from the Hebrew Bible. And as in the past, the era's violence and injustices triggered the changes in the religious concepts by which each of these two siblings sought to deal with its misfortune.

It is important to realize that the two religions' shared point of origin—early Judaism—was already extremely successful across a wide swath of territory ranging "from Rome to Asia."[2] By year zero about 1 million Jews lived in Palestine and between 5 and 6 million were in the Diaspora.[3] Only a great number of conversions can explain these numbers, which attests to the religion's great appeal.[4] Of course, this is not all that surprising: we already noted just what a cultural masterpiece the Hebrew Bible actually was. In early Judaism, too, we find many of the "psychological and intellectual factors that would later make Christianity so irresistible."[5] Christianity's efforts to make the New Testament appear as the fulfillment of the Hebrew Bible were certainly not without reason.

But what made Christianity so enormously successful? Due to the numerous independent influences on the course of history, we may never be able to answer this question fully, but the new religion did experience two great strokes of luck. For one thing, the Pauline wing of the Christ movement survived the catastrophe of the Jewish-Roman war (66–70/74 CE) in the Diaspora, unlike the Sadducees, Zealots, and, probably, Jerusalem's early Christian community.[6] The second stroke of luck was that the Roman emperor Constantine (270/88–337) converted to Christianity and gained the upper hand in the inner-Roman power struggle. The historian Paul Veyne is convinced that "without Constantine, Christianity would have remained simply an avant-garde sect."[7]

What we can explain is why Christianity had the potential to enthrall countless millions from the most diverse cultures over the millennia. This potential has a name: Jesus of Nazareth. Today Jesus even manages to sway nonbelievers. Indeed, the pope of atheism himself, Richard Dawkins, can't seem to escape his magnetic attraction. "Atheists for Jesus" one of Dawkins's essays, is a hymn to the Nazarethan's "super niceness."[8]

But there is more to Christianity's success than Jesus. The religion also has an entire cosmos of heavenly and infernal figures at its disposal: Mary, legions of saints, and sundry devils and demons. The careers of many of these figures were decidedly postbiblical. Even so, they had to follow trajectories already laid out in the Book of Books; hence, we devote a chapter to these characters. They became an essential ingredient of Christianity, and we can attribute their rise to the demands of our first nature.

And finally we examine another bible! Early Christianity spawned the idea that God wrote not just one but two holy scriptures. The second is known as the Book of Nature. This development would eventually produce another momentous split—that between religion and science. For many years they followed parallel paths so close that they are difficult to tell apart. But suddenly, and not all that long ago, they diverged sharply. The knowledge that for so long they shared such similar paths fuels our hope that today's conflict between religion and science may one day disappear. In theory, at any rate. Taking this perspective helps us to clarify religion's current situation and to understand why, in spite of all progress, religion is not about to disappear.

So just who was Jesus? According to our evolutionary approach, his enormous success over the ages and throughout numerous cultures should offer us insights into human nature. It does indeed, and what we find is quite impressive.

17

JESUS OF NAZARETH

God Becomes Human

"THE FACT THAT CHRISTIANITY EXISTS AT ALL IS ITS GREATEST WONder," states German theology professor Jörg Lauster. After all, the Romans made short work of its main protagonist when they nailed him to the cross.[1] This disturbing event cries out for an explanation. In *The Evolution of God*, Robert Wright notes, "Losing your life wasn't part of the Messiah's job description."[2] A messiah is supposed to defeat enemies and liberate the people, not get himself killed before the trumpets announce the decisive battle. We don't want to come across as impious, but once again we can clearly see how the Bible's authors made the best of a catastrophic situation.

Let us therefore examine the marvel of cultural evolution we find in the New Testament. We'll begin by perusing the difficult question of who the "historical Jesus" was before we take up our biblical-anthropological analysis. In the first step we identify all of the issues that can distort our view of the phenomenon of "Jesus Christ." This is a fairly easy task, because by now we have become virtual experts in the field of miracles. We then move on to cultural evolution and the trick it used to get a better handle on reality. Our first nature forced cultural evolution to perform something of a backward somersault: the looming apocalypse meant that demonic powers would finally stage their comeback and, in so doing, iron out some of the kinks that continued to plague monotheism. We ask Jesus why the first shall be last and the last shall be first, as well as why the poor and the meek shall enter heaven. Next we analyze his charismatic essence, which has mesmerized millions from vastly different cultures for centuries. We will see that his message appeals to deep-seated needs of the human psyche that aren't even religious in nature. Once we have completed all of these steps, we will have all we need to show how Christianity developed into a multifunctional and polyvalent belief system. As a religious Swiss army knife, it made a career for itself all over the world. Indeed, Christianity could never have pulled off this feat with just the one God.

THE PROBLEM WITH THE REAL JESUS

Who was Jesus? This is more than merely a difficult question—it's a rather delicate one, too, considering that we are dealing with probably the most important person in the history of the Western world. Whereas for most of the time since his death scholars and theologians have dwelt on his divine qualities, for the past 250 years or so, the focus has been on the historical Jesus. Not a lot of what we know about him is certain. The four evangelists cannot be considered dependable sources, and not only because the gospels attributed to them were written down between forty and seventy years after the events took place. Neither Mark (ca. 70 CE) nor Matthew (ca. 80/90 CE) nor Luke (ca. 90 CE) and certainly not John (ca. 100 CE) was interested in a historically correct version of what actually transpired. For them Jesus was not a historical figure—he was the resurrected redeemer, the eschatological savior. They intended to cast aside any remaining doubts concerning Jesus as the prophesized Messiah; sticking to the facts would have just got in the way.[3]

The evangelists set to work with commensurate enthusiasm and portrayed Jesus in broad, powerful strokes against a fantastic backdrop with a full cast of supernatural characters. People nowadays hardly notice this fact. The oldest of the gospels begins by telling us that Jesus went to John the Baptist, who wore clothes made of camel hair and lived on locusts and wild honey, and had himself baptized. Hardly had Jesus come out of the River Jordan than the heavens opened up and the Holy Spirit descended from above "like a dove" and said, "This is my beloved Son, in whom I am well pleased." But what did the Spirit do next? He drove Jesus into the desert, where Satan tempted him for forty days and angels waited upon him.[4]

Since hardly any extrabiblical sources mention Jesus—other than brief remarks by Flavius Josephus (37/38–100 CE) and Tacitus (55/56–120 CE)—we know very little for certain about Jesus's biography. He was born some time between 6 and 4 BCE in the village of Nazareth and probably worked in construction in the neighboring town of Sepphoris. Jesus did indeed encounter the apocalyptic prophet John the Baptist before moving about rural Galilee and the villages on the banks of the Sea of Galilee as a wandering preacher for several years. By the time he arrived in Jerusalem, if not earlier, he had roused the ire of the authorities, which led to his crucifixion by the Romans for agitation in or around 30 CE.[5]

Theologians, or at least some of them, find the search for the historical Jesus rather unsettling. His human side poses a threat to his divine, Christian side. Over the last few decades, research into his life has helped clarify the extent to which Jesus viewed himself as a Jew, the degree to which Judaism inspired his ethics, and why he never intended to found a new religion. In this regard, statements such as the following by New Testament studies professor Annette Merz are positively explosive: "It is obvious that Jesus had an unusual awareness of his own self; he believed he was God's authorized mouthpiece, just like many prophets in Israel before him. But as the Jew that he was, Jesus himself would have rejected any attempt to assign to him a 'unique' relationship to God, unique in the sense that it exceeded normal human experience of God, as blasphemous."[6] Jesus—the son of God?

"What Jesus preached and what Christianity believes are not the same thing," writes theology professor Jörg Lauster in his cultural history of Christianity. He even concludes, "Jesus is not the founder of Christianity."[7] It comes as no surprise, therefore, that Pope Benedict XVI, in his three-volume book on Jesus, laments the "gap between the 'historical Jesus' and the 'Christ of faith.'" This divide is "a dramatic situation for faith, because its point of reference is being placed in doubt: Intimate friendship with Jesus, on which everything depends, is in danger of clutching at thin air."[8] For this reason, European theologians especially have tended to show little interest in Jesus as a historical figure.

By concentrating on the "Christ of faith," the experts in effect created a vacuum. The public continued to show a great interest in Jesus as a real-life person and desperately wanted the vacuum filled. Conspiracy theories thus spread like wildfire: Could it be that scholarly findings surrounding him were so dangerous that the church stamped them "classified" and locked them away in an attempt to protect its religion's very foundations? There was no limit to the fantastic interpretations of Jesus's past. Was he a revolutionary? The first hippy? The lover of Mary Magdalene? There's hardly a role that hasn't been attributed to Jesus Christ.[9]

The findings of recent research are rather bland in comparison. No other text in world literature has been so thoroughly raked over as the New Testament, complains German filmmaker and author Leo Linder. An "unimagined tangle of interconnections, lines of development, parallels, variations on interpretations, contradictions, and levels of tradition" has come to light, "but strangely all we have left is the shadow of a modestly original wandering preacher whom a horde of infatuated followers post-mortem kneaded into the figure they would have liked to have followed in his lifetime."[10] Does the search for the "real" Jesus not run the risk of destroying the magic surrounding him?

From a biblical-anthropological perspective, the answer is simply no. Here we do not intend to pay tribute to the divine being of Jesus of Nazareth. We agree with Pope Benedict: belief in Jesus's divinity is an "act of faith."[11] Similarly the historically correct Jesus is for us also of secondary importance, as was the historicity of the figures in our examination of Moses and the prophets. Instead, we are fascinated to find in the figure of "Jesus Christ" the perfect product of cumulative cultural evolution. Once we understand how this process came to be, we can understand just how this "modestly original wandering preacher" managed to outdo the competition. Even today in the digital age, he still ranks an uncontested first among "history's most significant people," as demonstrated by an analysis of Google and Wikipedia in 2014.[12]

"PROVE TO ME THAT YOU'RE DIVINE"

As experienced biblical anthropologists, we begin by identifying all the embellishments that were added to eliminate any remaining doubts concerning Jesus's mission. The Bible's authors added them to make one thing clear: Jesus was no charlatan; he was the Messiah (Greek: *Christos*), God's chosen savior of Israel. This entailed a great deal of serious effort, for in Jesus's time there was no lack of charismatic itinerant preachers. For example, Honi the Circle-Drawer could conjure up rain, Eleazar cleverly used an amulet to pull out demons through the nose of the possessed, and the wonder-worker Hanan ha-Nehba had intimate conversations with God just like Jesus—indeed, both called him *Abba* (Aramaic for "father").[13] The evangelists knew they would have to come up with something special if they hoped to prove that Jesus really was in a league of his own.

"Jesus you just won't believe / The hit you've made around here / You are all we talk about / The wonder of the year," sings King Herod in Andrew Lloyd Webber's rock opera *Jesus Christ Superstar*. And historically speaking he isn't far off the mark. Wandering wonder-workers were a big attraction in those days, and many of them really knew how to put on a show. Appropriately, in the song Herod prompts Jesus, "Prove to me that you're divine; change my water into wine." Wonders—as we saw when discussing the Old Testament prophets—are traditional credibility-enhancing displays (CREDs) for counteracting our skeptical cultural immune system. They convince us that we really are dealing with an individual in possession of extraordinary powers. And so, in the first passages of the gospels we encounter a Jesus who was an experienced, well-nigh hyperactive miracle worker. He reminds us of Yahweh who, at the beginning of the exodus story, first had to establish his credibility by proving he was capable of much more than his origins in a burning bush might have suggested.

Most of Jesus's wonders had to do with healing the sick, lame, blind, and possessed, and he did not shy away from applying unconventional methods. In one incident he sticks his fingers into the ears of a deaf-mute, touches the latter's tongue with saliva, gazes up into the heavens, sighs, and says to the afflicted, "Ephphatha," which the Bible translates as "Be opened." And wonder of wonders, the man can speak and hear. Jesus also brings the dead back to life. In other words, he performs the same miracles as Old Testament prophets like Elijah and Elisha, but he does so on a larger scale.[14] Whereas Elisha multiplied the oil in a widow's pot or fed a hundred people with twenty loaves of bread, Jesus made sure that the wine never ran out at the wedding at Cana and fed 5,000 people with only five loaves of bread and two fishes. Particularly impressive in Jesus's repertoire of miracles was his calming of a storm and walking on the water over the Sea of Galilee. To ensure his credibility, every new prophet ideally outperforms his predecessors.

We also find the same Old Testament style of CREDs that served to separate the true prophets from the false. Jesus never took any payment for healing the sick. He refrained from doing anything that could enhance his reputation, and so he forbade his patients and disciples, indeed even the demons he exorcized, from telling of his miracles. He fled from his admirers when their numbers grew too large and played down any speculation that he might be the Messiah. The evangelists wished to clarify that Jesus did nothing for personal gain. Quite the opposite in fact: he proclaimed the Kingdom of God with full knowledge that it would cost him his life. The evangelists sprinkled their texts with Jesus's foretellings of his own death.

The only CRED for which we have historical confirmation is the crucifixion. As explained in chapter 10, this act places an "ineradicable seal" on the authenticity of the mission Jesus believed he was carrying out in the service of God:[15] he was prepared to make the ultimate sacrifice of laying down his own life in order to ensure that mission was fulfilled. Early Christianity was deeply aware of the overwhelming power of this act, and that's why it was so important that Jesus died as a human being, not a god. Dying would have been no great feat for a "divine tourist."[16] We need only think of the Egyptian god Osiris, murdered by his brother Seth but brought back to life by his sister Isis. Human nature only accepts CREDs when they entail true costs.

In the end, however, we find that in terms of miracles, the ones Jesus performed hardly went beyond the Bible's standard repertoire. When we compare the New Testament to the Hebrew Bible, we find nothing truly spectacular, such as the parting of the Red Sea, for instance. Even the transfiguration of Jesus when he meets the prophets Moses and Elijah atop a mountain is strangely lackluster. John the Evangelist felt he had to spice up Jesus's story: "And many other signs truly did Jesus in the presence of his disciples, which are not written in this book."[17] We can assume that the evangelists did not feel compelled to whip up the miracles thanks to the two powerful trump cards they had hidden up their sleeves.

He Is Risen!

Their first trump card was of course the resurrection. The spirits of the dead, such as that of the prophet Samuel, had always been summoned in the past. Elijah and Elisha brought the dead back to life (and Jesus even did the same on three occasions), though ultimately the resurrected would die again some day, just like the rest of us. But a single individual returned from the dead and resurrected in flesh and blood for all eternity? This was unprecedented.[18]

The idea that Jesus had returned from the dead must have appeared soon after his crucifixion on Golgotha. This belief guaranteed that Jesus, unlike all the other failed messiahs who had died on the cross, was never forgotten. People began writing down his words and chronicling his Passion at a very early date, and these writings would later serve as templates for the evangelists. Even in the texts of Paul, who wrote down the earliest account (50–60 CE) of Jesus's life, death and resurrection are the *conditio sine qua non.* He isn't interested in anything else about Jesus's life. In his first letter to the Corinthians, Paul writes, "For I delivered unto you first of all that which I also received, how that Christ died for our sins according to the scriptures; And that he was buried, and that he rose again the third day according to the scriptures. And that he was seen of Cephas, then of the twelve." After listing a few more witnesses for good measure, he adds, "And if Christ be not risen, then is our preaching vain, and your faith is also vain."

Something must have happened that gave rise to the fantastic rumor that Jesus had returned from the dead. We can be sure many doubted the veracity of the rumors. The evangelists, who penned their texts decades after the fact, were well aware of this skepticism and had already assembled their immunization against refutation strategies. Reza Aslan provides us with a list of them. Did the apostles only see a ghost? Well, a ghost cannot partake of fish and bread, can it? Jesus did so. Could it be that the witnesses were suffering from delusion? No, for Doubting Thomas immediately touched Jesus where the nails and spear had been driven through his flesh. Might not his corpse merely have been stolen? No, Matthew made sure armed guards were posted before his tomb. And these guards even saw the resurrected Jesus themselves, but the priests had bribed them to say, "His disciples came by night, and stole him away while we slept." In this manner Matthew also seeks to explain the existence of such rumors: "And this saying is commonly reported among the Jews until this day."[19]

We will never know for sure what really transpired, but the mystery adds to our fascination with the resurrection story. We're dealing with a classic case of what the cognitive science of religion refers to as a "minimally counterintuitive concept." These are concepts that are self-evident yet also include significant but minor deviations from normality.[20] A man lives and dies—so far so good—and then he returns from the dead. This is rather counterintuitive but not to such an extent that the story sounds completely ridiculous. It is actually less counterintuitive than one would think, at least if we assume that belief in people continuing to live on after death is an integral part of our innate intuition. Then the story is as follows: a man lives, dies, and lives on as a spirit, but for a few days at least he remains visible—a perfect minimally counterintuitive concept. Such a story has the potential to spread like wildfire.

This potential owes a lot to timing, for it appeared at the perfect historical moment. It offered innovative answers to a number of pressing questions. As we saw in chapter 16, intellectual religion prohibited belief in individuals' life after death as spirits and ancestors. We also saw how much effort went into filling the resulting vacuum. One such attempt was to suggest the concept of resurrection, which appears in apocalyptic texts such as the book of Daniel. But resurrection was always thought of in collective terms. At the end of the story in the book of Daniel, the entire people of Israel (or at least large parts thereof) will reawaken. But the idea that a single individual could reawaken to eternal life immediately after death could only tentatively apply to certain individual martyrs.[21] In this modified form, however, the resurrection idea rapidly gained in prominence because it met the needs of both intellectual religion (the righteous will be rewarded in the afterworld) and intuitive religion (every individual will continue to live on after death). Thus, the Christian concept according to which the fate of every individual will be decided after death could look forward to a spectacular career. Jesus himself had set the example.

It Is Written

That the evangelists did not put too much effort into the tales of Jesus's wonders has to do with their second trump card in addition to the story of the resurrection. CRED strategy number two was to make powerful use of the aura and authority of old scripture. The evangelists did all they could to present Jesus's mission as the fulfillment of prophecy. Christians transformed the Hebrew Bible into an overwhelming overture to their magnum opus: *The Life and Death of Jesus Christ.*

"It is written": with this magic formula the evangelists lent their texts a legitimacy that grew forth out of the Hebrew Bible. Christians managed to gain recognition mainly by demonstrating that Jesus was the fulfillment of the old prophesies surrounding the Messiah.[22] The evangelist Matthew worked particularly hard at such fulfillment references that he introduces as follows: "Now all this was done, that it might be fulfilled which was spoken of the Lord by the prophet."[23] The best-known and most consequential example was Jesus's fatherless conception and birth to a virgin.[24] Let us read the words of the evangelist Matthew:

> Now the birth of Jesus Christ was on this wise: When as his mother Mary was espoused to Joseph, before they came together, she was found with child of the Holy Ghost. Then Joseph her husband, being a just man, and not willing to make her a publick example, was minded to put her away privily. But while he thought on these things, behold, the angel of the Lord appeared unto him in a dream, saying, Joseph, thou son of David, fear not to take unto thee Mary thy wife: for that which is conceived in her is of the Holy Ghost. And she shall bring forth a son, and thou shalt call his name JESUS: for he shall save his people from their sins. Now all this was done, that it might be fulfilled which was spoken of the Lord by the prophet [Isaiah 7:14], saying, Behold, a virgin shall be with child, and shall bring forth a son, and they shall call his name Emmanuel, which being interpreted is, God with us.[25]

Here we have before us arguably the best-known and most momentous translation error in the history of humankind. In the Greek translation of the Isaiah passage that Matthew refers to, we find the word *parthenos* ("virgin"), but in the Hebrew original the word we read is *almah*, which means not "virgin" but "unmarried daughter," "girl," or even "young married woman." This is of course a simple mistake, unlike Matthew's tearing this Isaiah passage out of its context to use to his own ends. In Isaiah we read of how King Ahaz, who reigned over Judah in the eighth century BCE, was made to hope that the coalition of his enemies would be destroyed before the unborn child of a young woman, to be called Immanuel, grew up to be a man. Thus, no prophet had ever prophesized that in the distant future a virgin would become pregnant.[26] We suppose Matthew had his reasons for this little mystification.

The fact that Mary was pregnant before she married Joseph has always been a source of much speculation—for obvious reasons. In the oldest of the gospels, Mark tells of how the people in Nazareth's synagogue asked, "Is not this the carpenter, the son of Mary, the brother of James, and Joses, and of Judas, and Simon? and are not his sisters here with us?"[27] Son of Mary? In patriarchal societies such as ancient Israel, people were only referred to in this manner when the father was unknown. The evangelists Matthew and Luke, who took on some of Mark's text, corrected the tale: "Is not this the carpenter's son?"[28] or "Is not this Joseph's son?"[29] Could it be that the story of the virgin birth was meant to cover up Jesus's illegitimate origins?[30]

"This much is certain," says Mary biographer Alan Posener, "rumors and slander surrounding Mary and her firstborn son were circulated very early on."[31] Some even suggested that Jesus was the son of a Roman legionnaire. But then Matthew countered the rumors with his powerful story, which transformed a potential scandal into a miracle and at the same time made it appear that a venerable prophecy had been fulfilled.[32] If Matthew had only known of the repercussions!

For church father Augustine, the notion that Jesus was not born as the result of sexual intercourse was crucial, since he based his theory of original sin on the idea that Adam and Eve's guilt was transmitted from generation to generation through that act. Accordingly, if sex played no role in Jesus's conception, then he was born completely free of sin. But this led to another problem: How could Mary give birth to a son free of sin when she herself was conceived in sin? The solution: Mary, too, must have been born of a virgin conception. In 1854 the Catholic Church elevated the notion of Mary's "immaculate conception" to official dogma. And in 1950 the idea that she had remained a virgin throughout her entire life was confirmed, ensuring that she never could have passed original sin on to her son.[33] Once again we see how compulsive coherence seeking is a powerful principle of human thought, a move that contributed to transforming Mary into a "goddess of the hostility to lust" or a "Venus without loins,"[34] the dream girl of a paternalistic church whose elite practiced celibacy.

From very early on, Jews accused Christians of having lifted the idea of a heavenly conceived son from pagan myths. Zeus was known to have impregnated more than a few virgins, for example. "Jews criticized the Christians," writes Kurt Flasch, "that they should be ashamed for spreading such pagan gossip about God."[35] But this unbelievable story also created some practical problems. If Joseph wasn't the father of Jesus, wouldn't this also mean that Jesus's purported descent from of the House of David—artfully constructed by Matthew and Luke to prove that Jesus was the legitimate messiah—was a hoax? And what about Jesus's four brothers and unknown number of sisters, which the Bible mentions in such a manner-of-fact fashion? His brother James played a key role in the original Christian community in

Jerusalem until he was stoned to death in 62 CE.[36] The explanation that Jesus's brothers were children from Joseph's first marriage never gained wide acceptance.

While on the topic of prophecy fulfillment, we would be remiss if we were to omit Christianity's favorite idyll: "And Joseph also went up from Galilee, out of the city of Nazareth, into Judaea, unto the city of David, which is called Bethlehem; (because he was of the house and lineage of David:) To be taxed with Mary his espoused wife, being great with child."[37] Luke then goes on to tell the sentimental story of Jesus's birth in Bethlehem, with its cast of shepherds and jubilant bands of angels. Although in the Gospel of Matthew, Jesus was also born in Bethlehem, this occurred under different conditions. And Mark and John have nothing to say about Jesus's birth. Today's scholars for the most part agree that the birth in Bethlehem was meant to make it look as if Jesus was born in the city of David, as is right and proper for the Messiah. His birth in the obscure village of Nazareth—a town that receives not a single mention in the Old Testament[38]—was nothing short of curious. In the Gospel of John we read of how Jesus's origins provoked sneers: "Can there any good thing come out of Nazareth?"

We now have all the ingredients needed to equip Jesus with the highest possible allure. He was a talented wonder-worker with CREDs that acquitted him of any suspicion of profiteering or fraud. He overcame death and came back to life. And all his acts and aspirations appeared to fulfill Judaism's ancient holy texts. An unbeatable recipe! Nonetheless, we are convinced that apart from the resurrection, these CREDs are secondary. The figure Jesus has a lot more to offer.

"DELIVER US FROM EVIL"

The gospel writers spared no effort to convince people that Jesus was God's messiah, and, indeed, this idea continues to dominate our image of him to this day. This emphasis hampers our understanding of the main issue in the amazing product of cultural evolution that congealed in the New Testament, namely, that we are witnessing an intimate fusion of intellectual-institutional and individual religion. The multifaceted hybrid of these two forms of religion that is Christianity was to become extremely successful.

A fundamental theme of this book is that catastrophes are a principal formative force behind cultural evolution. The intellectual-institutional religion that arose in the Near East as a form of catastrophe prevention was under constant pressure to explain why people still endured one misfortune after another. It sought to bring the people's ideas of the nature and number of supernatural forces acting upon the world into agreement with actual events. And it had to do all this in a way that would not conflict with the intuitions of our first nature. When it came to Jesus, too, catastrophes played a major role. We will see that the programmatic designs of the Lord's Prayer, "deliver us from evil," which Jesus presented to his followers, basically meant "deliver us from all catastrophes."

Let us remember, catastrophes such as droughts, epidemics, and all forms of rampant violence had transformed the spirits into gods, and the double whammy of the Assyrian and Babylonian conquests finally paved the way for a monotheistic supergod. The seemingly interminable wars that followed in Alexander the Great's wake helped advance the borders of Yahweh's empire into the realm of the afterworld. And then the Romans came marching in.

Pompey the Great (106–48 BCE) brought an end to the power struggle–ridden Hasmonean kingdom in 63 BCE. When the Roman general conquered Jerusalem, he entered the temple's Holy of Holies but left the temple's treasures—the same cannot be said of all of the Romans who came later. The country was repeatedly plundered; people were enslaved or crucified. The Romans then placed Herod (73–4 BCE) on the throne as their client king. The Massacre of the Innocents attributed to Herod in the Gospel of Matthew was indeed an invention, but it has some historical justification, for Herod had seven of his own ten sons killed.[39] In light of our examination of the patriarchs and the conditions in the House of David, this does not come as much of a surprise. Incidentally, Emperor Constantine, who contributed so much to the great success of Christianity, also had his own wife and one of his sons killed.[40] The egoism of despots really pushes them to devour their own children.

After the death of Herod in 4 BCE, the kingdom was divided among his surviving sons, but real power remained in the hands of Roman legionnaires. Rebellions were put down with extreme brutality. In one punitive act, governor Quinctilius Varus had 2,000 rebels crucified. His successors would inherit Varus's "fatal lesson that the obdurate Jewish people could only be reined in by means of terror, and mass crucifixions were the most effective means of doing so," as historian Werner Dahlheim explains. The result, however, was a "vicious circle of Jewish rebellion and Roman terror,"[41] which would come to a dark end with the catastrophe of the Jewish-Roman war (66–70/74 CE), the demise of Jerusalem, and the mass suicide of the last Jewish rebels at the fortress of Masada high above the Dead Sea.

The temple in Jerusalem was destroyed—forever. Today only the Wailing Wall remains to remind us of what was once one of antiquity's most impressive structures. Casualty figures soared into the hundreds of thousands.[42] Large parts of the first Christian community in Jerusalem were eliminated; only the Christian communities in the Greek-speaking Diaspora survived.[43] We can be sure of one thing, however: without this Roman terror there could never have been a Jesus Christ or a Christian religion.

The Rise of Evil

Other countries also suffered under Roman occupation, but in ancient Israel's culture—a culture that had placed catastrophes squarely at the center of its belief system and interpreted them as a fundamental means of divine expression—coping with calamity had become second nature over hundreds of years. This system had enabled Jewish culture to get through the worst of disasters. When this calamity-focused system then received yet another overdose of misfortune, it once again found itself subject to fundamental change.

Catastrophes have an impact at two levels. First, they have functioned for millions of years as a decisive selective force. They can eliminate part of the competition and so create dramatically improved conditions for survivors. Calamity served such a function in this instance. From among the countless Jewish groups and sects that existed during Jesus's lifetime, basically only two currents survived the catastrophe of the Jewish-Roman war: the rabbinical Judaism that grew out of the Pharisees, and Pauline Christianity founded in the communities of the Hellenistic Diaspora.[44] We call this, without wishing to sound cynical, the biological side of the evolution of religion, which works in much in the same way as the meteor strike believed to have led to the extinction of the dinosaurs and cleared the field for the rise of the mammals. The massive military defeat of the Jews created an opening for these two surviving currents because the competing religious groupings were largely wiped out.

Second, catastrophes also function as powerful forces of purely cultural evolution—and this is our cultural-protection hypothesis. Every catastrophe was proof that existing cultural risk-management practices were not yet effective enough, and every new disaster forced people to come up with new coping strategies. We see exactly this at work here. The extremely brutal military oppression sparked an infernal fire that brought the cauldron of cultural evolution to a boil. The first belief to falter was the Torah's assertion that people could avoid disasters by following the Mosaic Law and soothe God's wrath by showing obedience.

To understand what happened, we have to remind ourselves how it all began. We have seen how the idea of monotheism resulted from an act of self-defense. Because the defeats at the hands of the Assyrians and Babylonians might naturally have led people to believe other gods were more powerful than Yahweh, scholars and priests came up with the idea that actually only one god controlled the world's fate. Accordingly, he was also responsible for all of the evil in the world and would not hesitate to use a foreign army to punish Israel for its disloyalty.

But we have also cataloged our first nature's difficulties with the notion of a single god. Since our innate intuitions always assume the existence of numerous supernatural actors, people continued to doubt Yahweh's single-handed responsibility for everything. And because catastrophes never ceased to torment them, people felt compelled to conclude that Yahweh was probably not the only moving force behind everything. When disaster struck again—be it a military catastrophe or an epidemic—our first nature whispered that it must be the mischievous work of other gods or resentful demons.

As a matter of fact, the Hebrew Bible had already left the back door open to additional evil forces. Satan bedeviled the pious Job, who had done absolutely nothing to deserve such tribulations. Of course Satan was said to be working on God's orders, so as not to cast doubt on Yahweh's sole responsibility for everything. Nevertheless, we see here none other than unadulterated monotheism's fall from grace: Yahweh's high moral demands on his people were called into grave question when he subjected one of his blameless charges to so much suffering.

We saw how Job's misfortune led to a reframing of calamity. His suffering was (at least in one version of the story) no longer punishment but a test of his faith—one he passed and for which he was richly rewarded at the end. We also encountered the same idea in the book of Daniel: misfortune, whether in the form of an oven or a lions' den, was a trial of faith that Daniel and his friends had to withstand. And the martyrdom that a mother and her seven sons endured in Maccabees was also a test for which they were rewarded in the afterlife.

In all these cases God may have ordered the misfortune, but for Yahweh himself to act as the tormentor would have been unthinkable. He had accomplices for such tasks. In the Hebrew Bible, Satan had not yet become the Lord of Evil; he was just an angel in Yahweh's court who now and then sought to lead humans into temptation. Remarkably, he and other dark beings took on lives of their own—unofficially, at least. Contemporary apocryphal writings conjured up an entire mythology explaining the career of this fallen angel and his mob of demons. Once again our first-nature expectation of consistency was the driving force behind these developments. Beings who do bad things for a good cause simply don't make sense. Anyone causing us to suffer has to be evil. Accordingly, Satan's career made a great leap forward thanks to the same mechanism that once transformed the spirits into gods. Proportionality bias simply demands that big misfortune have a big perpetrator. So inevitably, as the calamities became more intense, God's former agent gradually transformed into his greatest adversary.[45]

The New Testament seethes with such evil forces. Earlier in this chapter we mentioned the Gospel of Mark, the oldest of the evangelists' texts. No sooner was Jesus baptized than he was sent off to wrangle with the devil in the desert. Later Satan would tempt him again with an offer of all the kingdoms on earth if he would only turn to the dark side. Demons, too, were a dime a dozen in those days, and Jesus made a name for himself as an exorcist. He once drove out so many from a single possessed person that the unclean spirits took up residence in a herd of 2,000 pigs. The poor animals threw themselves into the Sea of Galilee and drowned miserably.

Our First Nature Will Be Pleased

For our first nature, the return of the evil spirits was a triumph. Our innate intuitions represent the tangible wisdom that accumulated during hundreds of thousands of years of living on earth. This experience taught us there existed numerous beings who wanted to do us harm or even kill us outright. The return to this animistic way of thinking that causes us to believe in numerous supernatural actors might seem like a step backward; indeed, some saw in these developments a radical revision of the earlier form of monotheism.[46] Christians also had to put up with constant accusations of "superstition."[47] The truth is that we are dealing with the adjustment of institutional-intellectual religion to fit with both existing reality and human nature. After all, the world is too complex for a single force to be behind everything that happens.

As we have seen, in the end our first nature received its due. It had never been happy with the attribution of everything we experience—good and bad—to a single force. According to our intuitions, illnesses and other forms of suffering had always been the work of evil spirits. And now people could once again aid the sick by exorcizing demons—just as they had always done under intuitive religion.

"Although monotheism is usually regarded as a step upward from polytheism, from a medical viewpoint, it was a step backward," explains microbiologist David Clark regarding monotheism's potential for combating disease. "The idea that assorted evil spirits inflicted infections comes closer to the germ theory of disease than later rationalizations. Under monotheism, the victims of disease were thought guilty of secret sins, despite lack of evidence. In contrast, polytheism often regarded the sick as unlucky victims of some passing demon rather than as evildoers. Consequently, treatment of the sick was more humane."[48] Improved care for the sick was indeed one secret of Christianity's success (rabbinical Judaism would independently make a number of improvements in this regard too).[49] This had less to do with the new notion of a merciful God, however, than with the return of the evil spirits. One could heal the sick by casting them out.

"The Time Is at Hand"

Let us briefly review the most important points. Unending repression had weakened monotheism's explanation of the world and allowed the evil spirits to crawl back out of their dark hiding places. Once again people began blaming the evil forces for the misery in the world. The only obvious conclusion was that the evil had to be defeated in order to bring about a new and righteous order. One name for this understanding of the world rolls like thunder throughout history: apocalypse.

We already encountered one example of this outlook in the Old Testament. In his apocalyptic vision, Daniel saw four giant beasts rise from the sea. These represented the world's great empires, and their eschatological battle would bring with it unseen levels of hardship and adversity. But a heavenly being, who looked "like the Son of man," would descend from the clouds, and his dominion would know no end. After a "time of trouble" such as the world had never seen, the archangel Michael would arrive and save the people of Israel. A heavenly court would be held, and the dead would be resurrected—either to eternal life or eternal disgrace and infamy, depending on how they had behaved on earth.

Daniel, however, was made to seal the book in which he had written down his vision "till the time of the end." This is the origin of the word "apocalypse," which is Greek for "revealing" or "unveiling." The apocalyptic visionaries promised to unveil God's secret plan. "In particular, they were convinced that God was very soon to intervene in this world of pain and suffering to overthrow the forces of evil that were in control of this age, and to bring in a good kingdom where there would be no more misery and injustice," writes Bible scholar Bart Ehrman.[50] The apocalypses offered solace and hope in a time when despair threatened to consume everyone.

A deep chasm yawned "between the glorious future envisaged for Israel and the sorrows of her actual condition."[51] People wanted to know whether God really was following a secret plan. Apocryphal texts, such as the First Book of Enoch, the Testament of Moses, the Second Book of Baruch, and the Apocalypse of Abraham, bore witness to this type of crisis management. Indeed, they were very popular in the two or three centuries straddling year zero.[52] Jesus, too, was a first-class apocalyptic visionary who fought both the devil and demons: "But if I with the finger of God cast out devils, no doubt the kingdom of God is come upon you."[53] We shall examine this in detail later. The New Testament ends with one of the most famous representatives of the genre: the revelation of John of Patmos, whose Four Horsemen, Whore of Babylon, and seven-headed, ten-horned dragon haunt the West's collective consciousness to this day. Jesus's Last Judgment follows the destruction of these dark forces. "The time is at hand," promises Revelation.[54]

Seven Bowls of Wrath

Around year zero, apocalypses became intellectual religion's state-of-the-art instruments to wring meaning from the continuing series of calamities. As Elaine Pagels explains: "The Book of Revelation reads as if John had wrapped up all our worst fears—fears of violence, plague, wild animals, unimaginable horrors emerging from the abyss below the earth, lightning, thunder, hail, earthquakes, erupting volcanoes, and the atrocities of torture and war—all into one gigantic nightmare. Yet instead of ending in total destruction, his visions finally open to the new Jerusalem—a glorious city filled with light. John's visions of dragons, monsters, mothers, and whores speak less to our head than to our heart."[55]

The Four Horsemen of the apocalypse, who ride across the face of the earth in John's vision, personify war, hunger, plague, and death. And the seven bowls of wrath contain the worst of plagues. John's vision pushes what Jeremiah described as "the sword, the famine and the pestilence" to unseen heights. We can therefore understand an apocalypse as a type of infernal drama in which catastrophes play the main role. No longer God's punishment as they were in the Torah, they have become the ultimate trial before God transforms the world into something new.

The catastrophes, then, are the birth pains that precede the arrival of God's dominion. Jesus says so quite clearly: "And when ye shall hear of wars and rumours of wars, be ye not troubled: for such things must needs be; but the end shall not be yet. For nation shall rise against nation, and kingdom against kingdom: and there shall be earthquakes in divers places, and there shall be famines and troubles: these are the beginnings of sorrows."[56] And he continues, "But in those days, after that tribulation, the sun shall be darkened, and the moon shall not give her light, And the stars of heaven shall fall, and the powers that are in heaven shall be shaken. And then shall they see the Son of man coming in the clouds with great power and glory. And then shall he send his angels, and shall gather together his elect from the four winds, from the uttermost part of the earth to the uttermost part of heaven."[57]

The catastrophes portended the end of the world as we know it. For the apocalyptic visionaries, the severity of the catastrophes became a yardstick for just how soon salvation would arrive. God would make his appearance just as everything couldn't possibly get any worse. Every time great despair visited the people, they readily interpreted it as a sign that the triumph of good over evil was at hand.[58] "But he that shall endure unto the end, the same shall be saved," promised Jesus.[59]

"I Came Not to Send Peace"

Could the people have interpreted the terrors visited upon them by the Romans in any other way? Wasn't it obvious that Satan and his earthly minions had sunk their claws into the world? Wasn't the battle raging in Judea one between God's people and the devil's legions? Wasn't defying the Romans a way of pleasing God? The apocalyptic interpretation of events on the ground was so attractive because it meant the people were no longer damned to suffer catastrophes as God's punishment. Now it was possible to strike back.

We see here a fundamental transformation of religion as a cultural-protection system. Intellectual religion, tasked with finally bringing catastrophes under control, was being radically reformulated. It is important that we try to understand Jesus's actions against this background. Researchers have long debated the self-image of the historical Jesus in this apocalyptic setting. What did he believe his role really was? And how did he feel about violence?

It is hard to avoid the idea that propagating the dominion of God under Roman rule was basically a call for open rebellion against the Romans. Didn't the Romans execute Jesus on Golgotha as a rebel leader, just as the sign on the cross—"King of the Jews"—indicated?[60] Wasn't Jesus a zealot who hoped to provoke God's intercession by waging a violent struggle against Rome? There is certainly some truth to this suggestion. Simon, one of the twelve apostles, was nicknamed "the Zealot," after all. And didn't Jesus himself claim, "I came not to send peace, but a sword"?[61] In *Zealot,* Reza Aslan sums up Jesus's attitude toward violence as follows: "There is no evidence that Jesus himself openly advocated violent actions. But he was certainly no pacifist."[62]

Theologians have a different view. They tend to believe Jesus was no "political-revolutionary zealot." He refused to promote violence, and his ragtag band of followers showed none of the signs of a typical guerilla unit.[63] Nevertheless, it's important to realize that the gospels tend to paint too "gentle" a picture of Jesus. For purely opportunistic reasons, later depictions of Jesus portray him as less radical and less political than he probably was in real life. The gospels were written after the Jewish-Roman war had ended in a crushing defeat for the Jews, and their authors could not afford to allow Jesus to appear as an anti-Roman revolutionary. Accordingly, the evangelists worked hard to absolve the Romans of Jesus's condemnation (prefect Pontius Pilate washed his hands of guilt) and instead pinned the blame on his Jewish opponents.[64] Even more importantly, they had to dehistoricize Jesus. Had he been only another Jewish zealot executed by the Romans, what significance would he have half a century later when there was no longer a Jewish polity and Christians were attempting to recruit new members among the non-Jewish population in the Roman world?

As already discussed, the gospels present Jesus as Christ, as God's messiah, who would bring about the Kingdom of God. Yet the Hebrew Bible presents no programmatic tradition of the Messiah, and the picture is accordingly vague.[65] Traditionally, these images drew on notions of the return of a king from the House of David who would liberate Israel from foreign occupation. Thus Jesus was mocked as the "King of the Jews" when condemned as a political rebel. In the apocalyptic thinking, the tradition of the Messiah gets melded together with the book of Daniel's divine "Son of man" who will bring salvation. But the historical Jesus never referred to himself as the Messiah, and he was actually quite hesitant when confronted with this title.[66] Instead, Jesus himself used the term "Son of man." Whether he was actually speaking of himself remains unclear, however, and if he was, was he speaking apocalyptically or literally, as in, "No, no. I'm not the Messiah. I'm just a son of man."[67] But this, too, can be interpreted as a CRED. It simply wouldn't have been fitting for the real Messiah to brag about his status. Way too many candidates were already busy doing that.

Jesus was just as loath to describe himself as the son of God. Many people do not realize that in the Jewish tradition, "son of God" did not have the same literal meaning that Christianity would later attribute to it. "Son of God" was the traditional term used to describe kings in ancient Israel, and a number of people appearing in the Hebrew Bible, such as David, for example, were described as such. In other words, the term does not refer to a heavenly father-son relationship. This interpretation would arise only after early Christianity found itself confronting pagan notions of sons of the gods.[68] The notion of "God the father" was also not as unusual as one would think, for it too was a standard expression in the vocabulary of early Judaism.[69] Once again, we can attribute no statements with any certainty to Jesus in which he clearly explains his relationship to God. Most likely he never said anything about it at all, for if he had, someone would have written it down.[70]

From the perspective of cultural evolution, all of this is secondary. What matters is that scholars all agree that Jesus spoke with a great deal of authority stemming from his conviction he had a divine mission. He appears to have been convinced that something was wrong with the world, that a new one was needed, and that he, Jesus, would play a key role in establishing this Kingdom of God.[71] In fact, his teacher, John the Baptist, was already an apocalyptic visionary who believed the end of the world was nigh. John propagated baptism as a means of washing away all sins in preparation of the coming Great Tribulation. Jesus transformed this message into something more radical: the Kingdom of God was not just close; it had already dawned![72] According to the Gospel of Luke, Jesus saw Satan fall from the sky—before long, evil's rule on earth would be abolished. "Verily I say unto you," Jesus spoke to his followers, "That there be some of them that stand here, which shall not taste of death, till they have seen the kingdom of God come with power."[73] The historical Jesus saw himself as a warrior in the final battle against demonic forces. His apocalyptic message even made its way into the Lord's Prayer, which millions of Christians today still recite: "Our Father which art in heaven, Hallowed be thy name. Thy kingdom come."[74]

Apocalypse Now

Jesus Christ—and we mean not the historical Jesus but the cultural product of the New Testament—is the centerpiece of an apocalyptic worldview. He is the champion intellectual religion sent in to do battle with evil. This is not the only Jesus we find in the gospels (we will meet the second Jesus—the Jesus of intuitive religion—later), but Jesus Christ essentially dominates the New Testament's morality. We can only understand this morality against the backdrop of the apocalypse, for only then does it become clear that Jesus sows as much confusion as he kindles admiration because there are actually two sides to the New Testament's morality. His morality is deeply rooted in our primate past and has always come to our aid when we felt ourselves confronted by diabolical enemies.

The basic apocalyptic assumption is that the decisive moment is at hand, and we must therefore focus on what really counts. Only within this framework can we make sense of the high demands Jesus made of his followers. We are dealing here with a massive shift in the principle behind the Torah, which states that individual behavior provoked catastrophes as a form of divine punishment. And because God favored collective punishment, everyone had to follow the rules down to the letter. But in Jesus's time, evil was held responsible for catastrophes; as a consequence, people now had to focus on behavior that helped combat this evil. Of course God remained interested in moral behavior, but now the moral agenda served to strengthen the solidarity and resistance of the community. And so, any behavior that might aid the evil enemy was now forbidden. Only in this way could the powers of darkness be defeated.

Jesus himself stressed that he did not wish to abandon the Torah, as illustrated by his Sermon on the Mount: "Think not that I am come to destroy the law, or the prophets: I am not come to destroy, but to fulfil."[75] Nevertheless, he significantly modified the Torah's message. For one, he weakened some of the rules governing formal behavior. This was true of the purity laws ("Not that which goeth into the mouth defileth a man; but that which cometh out of the mouth, this defileth a man"[76]) as well as rules governing the Sabbath ("The sabbath was made for man, and not man for the sabbath"[77]). When the decisive moment is at hand, mere adherence to the law is not enough. When the time is nigh, everyone must stand with the righteous and be prepared to fight to the end. At that time the wheat is separated from the chaff (a saying that goes back to another apocalyptic visionary, John the Baptist). To prepare people for the coming emergency, Jesus therefore tightened the Torah's laws dealing with the people themselves. Let us listen in on Jesus's Sermon on the Mount, perhaps the most famous moral sermon the world has ever heard:

> Ye have heard that it was said by them of old time, Thou shalt not kill; and whosoever shall kill shall be in danger of the judgment: But I say unto you, That whosoever is angry with his brother without a cause shall be in danger of the judgment: and whosoever shall say to his brother, Raca, shall be in danger of the council: but whosoever shall say, Thou fool, shall be in danger of hell fire. . . .
>
> Ye have heard that it was said by them of old time, Thou shalt not commit adultery: But I say unto you, That whosoever looketh on a woman to lust after her hath committed adultery with her already in his heart. . . .
>
> Ye have heard that it hath been said, An eye for an eye, and a tooth for a tooth: But I say unto you, That ye resist not evil: but whosoever shall smite thee on thy right cheek, turn to him the other also. . . .
>
> Ye have heard that it hath been said, Thou shalt love thy neighbour, and hate thine enemy. But I say unto you, Love your enemies, bless them that curse you, do good to them that hate you, and pray for them which

despitefully use you, and persecute you [the commandment to hate one's enemy is *not* in the Torah].[78]

Jesus's preaching is so drastic that even theologians have asked themselves, What kind of person could live that way?[79] or Can this ethic be practiced at all? Does it not ask too much of people?[80] In light of this "unbelievable radicalization of the daily rules of life," Christianity has always doubted if the faithful can ever fulfill the demands Jesus set forth in the Sermon on the Mount.[81] Occasionally, theologians posit that we are dealing here with an ethic of transition.[82] After all, Jesus did expect the Kingdom of God to arrive quite soon.

We take a similar view. Jesus saw that the moment of truth was at hand, and now it was up to the people to prove themselves. He thus laid out the ethics of the final struggle. Now, on the eve of an apocalyptic eschatological battle, everyone had to stand with those they could trust blindly—people who had complete control over their emotions and would not abandon their friends or collaborate with the enemy for the sake of greed. This principle is deeply etched in the human psyche. When the going gets tough, you have to be sure you can count on the one standing next to you. Now it was of the utmost importance that everyone stand tightly together and work as a team. Jesus forbade precisely the types of behavior that the Hebrew Bible had already recognized as *asabiya* sapping: fighting among brothers, lust for women, and escalation of violence. Jesus even spoke out against lust for wealth in his Sermon on the Mount: "Ye cannot serve God and mammon."[83] Luxury, as Ibn Khaldun argued, is the surest way to destroy *asabiya.*

The Two Faces of Morality

In this apocalyptic moment of truth, one important detail often does not garner much attention. Human morality is full of double standards—and it has always been that way.[84] Nowhere is this as obvious as in conflict situations. Some readers may feel we have painted a rose-colored view of our hunter-gatherer past, but this is partly because we have so far concentrated on within-group life, the social setting in which the psychological preferences of our first nature took form. But in doing so we have neglected one important factor: hunter-gatherer groups were always competing with other communities for limited resources. If they wished to come out ahead, they had to make sure to maintain cooperation within their own macroband, while simultaneously remaining prepared for conflicts with other communities. As a result human morality acquired two faces.

On one side there is the friendly face directed inward toward the group, the inner morality, our main focus up until this point. It strives for equality and reciprocity and is forgiving. It guarantees smooth and enduring cooperation. The second face of morality, our outer morality, is rather ugly. A separate set of criteria governs treatment of outsiders. Outer morality aims to keep outsiders at arm's length unless they are deemed either beneficial or, conversely, a threat. "There seems to be a special, pejorative moral 'discount' applied to cultural strangers—who often are not even considered to be fully human and therefore may be killed with little compunction," writes Christopher Boehm in *Moral Origins.*[85]

Here we can see age-old mechanisms at work. Primatologist Jane Goodall first described how chimpanzees deal with chimps from other groups in a warlike manner. The process is one of bloodcurdling cruelty: when the males jointly patrol the borders of their territory and suspect the presence of a small number of members of a neighboring chimpanzee community, they intrude into the foreign territory and attack the animals they find there—provided they outnumber them by a factor of three or more. Anyone witnessing these attacks would come away with the impression that the males intended to kill the outsiders.[86] Goodall introduced her own term to describe the behavior these animals display when on the attack: "dechimpization."[87] The attacking chimpanzees behave as if they were dealing with prey. They do not treat the "enemy" in the same manner as they would normally treat other chimps but instead behave as if they were dealing with another species entirely—one they might normally tear to pieces.[88] They aim to expand their group's territory, and, because their neighbors are pursuing the same goal, offense really is the best defense—as long as it offers the possibility of eliminating enemies with little risk. Two neighboring chimpanzee communities can never truly be at peace with each other.

Homo sapiens behaves in a similar fashion. Hunter-gatherer groups never conducted wars as we think of them today, for they simply did not have the logistical wherewithal to do so; "raiding warfare," however, an unending series of ambush attacks basically like those seen among chimpanzees, was quite common. This type of conflict, spurred on by tit-for-tat acts of revenge, can produce appreciable per capita casualty rates.[89] To maintain this type of adaptive behavior, human nature requires special mechanisms that prevent the same vicious behavior from surfacing within the group. Natural selection's solution resembles that found among chimpanzees: the enemy is dehumanized and demonized. If the human characteristics attributed to all members of the in-group can be denied to outsiders, it is no longer necessary to treat them as people. Inhibitions against killing no longer apply. "Those people are our enemies. Why shouldn't we kill them?" the warrior argues. "They're not human."[90]

Outer morality, which helps groups demonize enemies, is enlisted as soon as the group feels threatened or comes under attack. The dehumanization program then dependably kicks in, although not automatically. Whereas that distrust of strangers is always present, sometimes trade with other groups can bring substantial benefits, and selection therefore favored among our forebears a more flexible approach than that of the chimpanzees for dealing with outsiders.[91] But as soon as things go wrong, we can always fall back on the logic of friend versus enemy. Recent history has shown how, even today, normal people can display dehumanizing behavior that culminates in torture and genocide.[92] This, too, is a part of our first nature.

Thou Shalt Love Thy Neighbor as Thyself

We can apply this knowledge of humanity's double morality to Jesus's ethics. Let us begin with the friendly morality directed at the Israelites themselves. Here we can see how everything serves to preserve *asabiya*, the social force that holds groups together. For Jesus it was clear which of the commandments was the most important: "Thou shalt love the Lord thy God with all thy heart, and with all thy soul, and with all thy mind." This is the greatest of the Torah's commandments, Jesus told the Pharisee when asked which of God's laws were most important. One other is equally important: "Thou shalt love thy neighbour as thyself."[93] As we saw in our discussion of Ibn Khaldun in chapter 11, religion is a powerful means of binding anonymous societies together. Whereas small groups of hunter-gatherers could make do with the golden rule,[94] larger, truly anonymous societies needed the love of God as social glue to maintain cohesion. It therefore comes as no surprise that this was the first commandment Jesus mentioned. Without a common obligation to a moral God, society would descend into anarchy. But Jesus also added the rule about loving one's enemies in his Sermon on the Mount. He was doing all he could to preserve all-important *asabiya*.

We should be aware that Jesus was not speaking universally. "The oft-repeated commandment to 'love thy neighbor as yourself' was not Jesus's invention. It comes directly from the Torah and is meant to be applied strictly in the context of internal relations within Israel," explains Reza Aslan. "To the Israelites, as well as to Jesus's community in first-century Palestine, 'neighbor' meant one's fellow Jews."[95] Neither the Torah nor Jesus demanded that people love out-group "strangers."[96] The same is true for the commandment to "love your enemies" and "turn the other cheek," for these also targeted exclusively the Israelite in-group. In other words, we're dealing with pure inner morality aimed at promoting peaceful coexistence.[97] The upcoming apocalypse required maximum social cohesion.

The Bible drops enough clues to support the thesis that Jesus was only interested in the fate of the Jews. When he sent forth his apostles, he told them, "Go not into the way of the Gentiles, and into any city of the Samaritans enter ye not: But go rather to the lost sheep of the house of Israel."[98] "Jesus'[s] mission was for Israel," confirms theologian Jürgen Roloff. "A universal plan for all of humanity was, in light of all that can be gleaned, foreign to him."[99]

Nor was Jesus the first to suggest the idea of loving one's enemies. Classical philosophers such as Socrates (469–399 BCE) also encouraged people, particularly the powerful, to avoid seeking revenge, to counter evil deeds with good, and even to promote the enemy's welfare. The Torah, too, includes commandments that encourage helping one's enemies.[100] These types of appeals aimed to break the vicious circles of violence that threatened post-hunter-gatherer societies everywhere. And it's equally evident that Jesus's notion of loving one's enemies wasn't meant as a universal moral precept. Had that been the case, the entire apocalyptic struggle would have made no sense, for evil enemies had to be vanquished, not embraced. Only then would the Kingdom of God be established. And whoever decided to take the other side was damned for all eternity. "In other words, 'love your enemy,' like 'love your neighbor,' is a recipe for Israelite social cohesion, not for interethnic bonding."[101]

One can only win out against a powerful enemy if one is prepared to stamp out disputes within one's own ranks, making it impossible for the enemy to sow discord. It's not enough to swear love and loyalty to God and one's own neighbors. One must also prevent all actions that could endanger *asabiya*, which is why the New Testament denounces the same mismatch sins that the Old Testament deemed so objectionable. Thus the separation between inner and outer morality is artificial; they are the two sides of the same coin. Every internal action undertaken to boost *asabiya* also serves as a strategy aimed against competing groups. We have often seen such commandments up until now, but in the New Testament they became much more radical, for—to put it in no uncertain terms—Jesus felt he was at war.

In the case of Jesus we also find the opposite: behavior that is anything but stern and particularly admired today as the "human" side of his morality. Here we refer to things such as his unconditional advocacy for the poor, the sick, and the weak and his plea that we avoid judging others. What did Jesus say? "Why beholdest thou the mote that is in thy brother's eye, but considerest not the beam that is in thine own eye?"[102] We find an impressive example of the forgiving side of Jesus's morality in the story of the adulteress. The scribes brought her to the temple in which Jesus was teaching. "Master, this woman was taken in adultery, in the very act. Now Moses in the law commanded us, that such should be stoned: but what sayest thou?" Jesus was silent and wrote on the ground with his finger. When they pressed him, he spoke the one sentence that has never lost its potency. "He that is without sin among you, let him first cast a stone at her." The scribes departed and left the woman behind. "Woman, where are those thine accusers? hath no man condemned thee?" Jesus asked her. "No man, Lord," she replied. "Neither do I condemn thee: go, and sin no more," Jesus commanded.[103]

Even if the story's historicity is subject to debate, it matches perfectly with Jesus's biography.[104] The story itself is phenomenal, given how nicely it fits the old spirit of the hunter-gatherers. We need not mention how adulteresses, even today, are treated in patriarchal societies when caught in flagrante. In the times of the hunter-gatherers, however, women enjoyed a great deal more sexual freedom than they have in most so-called civilized societies throughout history. At the same time, Jesus cast out a fundamental legal principle according to which a defined crime meets with a defined legal consequence, and he replaced it with a hunter-gatherer moral principle aimed at promoting compensation and reconciliation within the group. Forager morality knows no rigorism, for it is all about peaceful conflict resolution and emotional reconciliation. Unlike in modern societies, interpersonal relationships in traditional societies had much greater existential importance and were generally meant to last for a whole lifetime.[105] In traditional societies, people depended much more on one another, and for this reason they had to be prepared to forgive. After all, who doesn't make a mistake now and then?

Wailing and Gnashing of Teeth

Whereas compensation and reconciliation were at the very core of inner morality, Jesus's outer morality was merciless. He divided the world into good and evil, friend and enemy, in a reflection of his own apocalyptic views: "He that is not with me is against me."[106] Anyone not on the side of God deserved no mercy. Jesus told the parable of the weeds sown by "an enemy" in a wheat field. The servants wanted to weed out the field, but the master forbade them from doing so. To avoid damaging the wheat, they were to let both grow. When it was time to harvest, he told the reapers, "Gather ye together first the tares, and bind them in bundles to burn them: but gather the wheat into my barn." The apostles wanted to know what Jesus meant with this story, and Jesus said,

> He that soweth the good seed is the Son of man; The field is the world; the good seed are the children of the kingdom; but the tares are the children of the wicked one; The enemy that sowed them is the devil; the harvest is the end of the world; and the reapers are the angels. As therefore the tares are gathered and burned in the fire; so shall it be in the end of this world. The Son of man shall send forth his angels, and they shall gather out of his kingdom all things that offend, and them which do iniquity; And shall cast them into a furnace of fire: there shall be wailing and gnashing of teeth. Then shall the righteous shine forth as the sun in the kingdom of their Father. Who hath ears to hear, let him hear.[107]

It would have been unthinkable for Jesus to step in and prevent the children of the wicked from wailing and gnashing their teeth in eternal agony. Loving one's enemies was never meant to go that far.

All those people who were not among the children of God's dominion were therefore to meet the worst of all fates. And who could these people be other than the pagan oppressors? Because the New Testament was written during Roman times, its authors could not openly identify the Romans themselves with Satan, so the evangelists demonized those Jews who opposed Jesus: the traitor Judas Iscariot, the high priests, and the scribes. According to the evangelists, they were all under the devil's influence.[108]

Another group was also clearly associated with evil: the rich. Indeed, they were the profiteers par excellence. We examined the issue in our chapter on Adam and Eve and encountered it again and again throughout the Hebrew Bible. The invention of private property and the resulting "positive feedback loop" in which "the rich get richer, and the poor get poorer"[109] represents one of the greatest threats to *asabiya*. The elites were also to blame for the massive hardships that led to enormous social problems in antique societies. They often joined forces with the Romans, and quite a few owed their success to them. For all of these reasons, Jesus condemned the rich in the strongest terms.

In the Sermon on the Mount, Jesus preached, "Lay not up for yourselves treasures upon earth,"[110] for it is impossible to serve the two masters of God and mammon. "It is easier for a camel to go through the eye of a needle, than for a rich man to enter into the kingdom of God," Jesus said.[111] The parable of the rich man and poor Lazarus underscores this message. After a life of misery, when Lazarus dies he finds himself in the good hands of Abraham, whereas the rich man is left to rot in the flames of hell. Finally, it's important to remember that Jesus was even willing to resort to violence over this issue, as when he expelled the moneychangers from the temple. He accused the priestly elites of having transformed the holy shrine into a robbers' den.[112] "Ye cannot serve God and mammon"[113]: Theology professor Annette Merz sees in Jesus an "absolutely uncompromising position" when it comes to money and property.[114]

Here we can hear our exploitation-abhorring first nature calling out. In some instances it even supports the ultimate sanction; hunter-gatherer groups are not afraid to kill those whose "behavior is threatening to everyone else's welfare."[115] Here again we encounter the old truth that anyone wishing to win in conflicts with out-groups has to press for egalitarianism. Everyone has to benefit equally from the common cause, for only then will everyone stand firmly together. Thus people enriching themselves at others' expense simply cannot be tolerated. Here we are reminded of the book of Joshua, where the pillager Achan was stoned to death on God's orders (see chapter 11).

We can also see the true nature of Jesus's outer morality in the details. Take the story told by Mark and Matthew. Jesus visited the region of Tyre, where a woman caught word of his arrival. She threw herself before him and begged him to rid her daughter of an evil spirit. Jesus refused. The reason: "The woman was a Greek, a Syrophenician by nation." He spoke to her: "Let the children first be filled: for it is not meet to take the children's bread, and to cast it unto the dogs." Dogs? Even if we claim that Jesus was only speaking metaphorically, his dehumanizing words are typical of those spoken since time immemorial when it comes to excluding outsiders. Jesus did help the woman in the end—but only because of the answer she gave. And what were her words? "Yes, Lord: yet the dogs under the table eat of the children's crumbs."[116] The woman had accepted her own dehumanization and displayed submissiveness.

We do not intend here to put Jesus on trial. We only wish to show the origins of his moral principles. And these origins should by now be perfectly clear: tolerance and reconciliation within the in-group and implacability and a tendency to dehumanize toward the out-group. These are the age-old components of the double standard in human morality, that easy-to-activate logic of friend versus foe. And this is exactly what made Jesus's morality so successful—and Jesus into a fine example of the species *Homo sapiens.*

Paradise Found

The beatitudes with which Jesus began his Sermon on the Mount rank among the Bible's most famous passages: "Blessed are the poor in spirit: for theirs is the kingdom of heaven," Jesus began.[117] According to Luke, the real poor belonged in God's kingdom.[118] And Luke lets Jesus add, "But woe unto you that are rich! for ye have received your consolation. Woe unto you that are full! for ye shall hunger. Woe unto you that laugh now! for ye shall mourn and weep."[119] Here we once again encounter the logic of reciprocity so characteristic of humanity. Since the rich have helped themselves to material resources beyond all measure in the here and now, they will pay dearly in the Kingdom of God. Balance will once again return to the world.

The beatitudes are the program behind the apocalypse. In Matthew they continue as follows: "Blessed are they that mourn: for they shall be comforted. Blessed are the meek: for they shall inherit the earth. Blessed are they which do hunger and thirst after righteousness: for they shall be filled."[120] Robert Wright speaks of the principle of "reversed polarity" at the heart of the apocalyptic visions: "Someday the oppressed will rise to the top of the heap, and the oppressors will find themselves at the bottom."[121] According to author Leo Linder, Jesus "turns everything upside down, or downside up."[122] From the perspective of cultural evolution, Linder is right on the mark, for everything Jesus talked about represented an attempt to bring things back to the way they had once been and to reintroduce a hunter-gatherer morality based on equality and justness. Of course, no one knew in those days that they were dealing with an age-old human legacy: it just felt right.

If Jesus was to set the world to rights, the mismatch phenomena of injustice, inequality, and oppression first had to be eliminated and all of the catastrophes that came about in the wake of sedentarization had to be made to disappear. The same was true of patriarchy. We have seen how forgiving Jesus was when it came to the adulteress, and we soon shall see just how important a role women played in the Jesus movement.

The apocalyptic vision, and thus Jesus's dream of the kingdom of heaven, represents a return to the world of our first nature, to Paradise, where equality really meant something and no one could monopolize property and oppress his fellow human beings, where everyone was fair to everyone else. And we shouldn't forget that this paradisiacal worldview inevitably implies the existence of hell—the place where enemies, evildoers, and exploiters would go to suffer for all eternity. That, too, is part of the dream of our first nature.

THE SECOND JESUS

Eschatology is the study of the end of everything—the end of the world and the end of history—as described in apocalyptic visions.[123] Today one group of theologians believe they have discovered a "noneschatological" Jesus in the New Testament, a version of him uninterested in the apocalypse. Instead, they believe, he was inspired by the Greek philosopher Diogenes of Sinope (ca. 410–323 BCE), who, according to legend, lived in Athens in a big clay jar. According to their theory, Jesus was a Jewish cynic, an outsider who goaded his contemporaries with his alternative lifestyle and paradoxical words of wisdom. A nonconformist who celebrated a "life of untethered freedom," he represented the philosophical ideal of a "self-absorbed existence independent of external influences."[124]

In the United States, this view was held primarily by representatives of the group known as the Jesus Seminar, such as John Dominic Crossan or Burton Mack. In Germany theologian Bernhard Lang dedicated a book to the theory.[125] Even so, the idea has never really caught on. Its critics claim the group assigns too great a Hellenistic influence to Jesus—something that does not sit well with his own deep-seated Jewish identity and the key role of Jewish teaching in his mission.[126] We mention this debate here because it shows that the gospels present not just the apocalyptic Jesus but also a second one with a completely different character (although we believe Hellenism did not inspire this second version).

Jesus's apocalyptic interpretation of the world emerges from the same impulse that inspired the Torah: humanity's desperate attempts to deal with catastrophes. The only novel element in Jesus's case was his vision for how to accomplish this. In both instances we see how intellectual religion mainly served to thwart catastrophes. The apocalyptical Jesus—the Jesus we have examined so far—must accordingly be understood as a representative of this phenomenon. His morality aims at protecting society from evil. This Jesus, then, isn't interested in the worries and needs of individual people. Doesn't this suggest that a second, intuitive Jesus must be buried in the pages of the gospels, one who attends to the needs of each individual human being?

In the Old Testament we observe how in the Psalms the personal protective God of salvation joins the monotheistic Yahweh, the God of the Torah. Features of intuitive religion addressing individual needs for personal spiritual succor therefore complement intellectual religion's concept of a God interested in morality and protecting the community. Whereas the Old Testament merely juxtaposes these two concepts, the New Testament fuses them together. Nevertheless, it is still easy to pick out the intuitive Jesus. He seems very familiar to us. In fact—to put it in more provocative terms—he appears as the perfect hunter-gatherer, an idol of our first nature.

A Longing for Community

We do not intend to romanticize Jesus, and we certainly do not claim that everything happened just as the gospels describe. But we do believe that the Jesus figure created by the evangelists overall is a product of cumulative cultural evolution that used tried-and-tested means to strike particular chords in our hearts.

Think of Jesus and his twelve disciples making their way through rural Galilee: Is it really so absurd to imagine them as a group of hunter-gatherers fallen out of time? The New Testament actually offers quite some evidence in favor of this view. Many have focused on the symbolism of the number twelve: one apostle for each of the legendary twelve tribes of Israel. But their number is of just as little interest to us as the eschatological claim that the twelve apostles would be enthroned alongside the Son of man and rule over God's people.[127] In numerous passages the Bible reveals that Jesus's group consisted not just of his disciples, but also attracted many people, including women.

Jesus's circle was egalitarian, a "zone free of oppressive power structures."[128] Although Jesus was indeed its undisputed leader, he did everything possible to avoid solidifying his rule. Hunter-gatherers, too, placed their trust in leaders in times of emergency, but as soon as the danger had passed, the group reabsorbed this person, who was expected to avoid giving the impression he might seek to hold on to power. "Ye know that they which are accounted to rule over the Gentiles exercise lordship over them; and their great ones exercise authority upon them," Jesus told his followers. "But so shall it not be among you: but whosoever will be great among you, shall be your minister: And whosoever of you will be the chiefest, shall be servant of all."[129] If you ever wondered why Jesus washed the feet of each of his disciples in the Gospel of John, now you know the answer.

Again and again the New Testament stresses the strong bonds within Jesus's group. Its *asabiya* is valued above the ties of family and biological kinship. "There came then his brethren and his mother, and, standing without, sent unto him, calling him. And the multitude sat about him, and they said unto him, Behold, thy mother and thy brethren without seek for thee. And he answered them, saying, Who is my mother, or my brethren? And he looked round about on them which sat about him, and said, Behold my mother and my brethren! For whosoever shall do the will of God, the same is my brother, and my sister, and mother."[130]

Many theologians view Jesus's disparagement of the family as one of his more troubling sides.[131] Indeed, Jesus seemed to act like a modern-day sect leader who attempts to dissolve his followers' original family ties. Families embody the traditional world, and precisely these chains were to be broken. Anyone wanting to stand alongside Jesus when God's kingdom finally arrived had to renounce family and property and cast aside his old second nature. In his own life Jesus exemplified how he expected all of society to live in the future. Belief in God can bring even total strangers together into one great family, a perfect example of fictive kinship.

Jesus and his followers didn't hold steady jobs, and he demanded they give up all possessions. They made a name for themselves as healers and exorcists. "Take no thought for your life, what ye shall eat, or what ye shall drink; nor yet for your body, what ye shall put on," Jesus told them. "Behold the fowls of the air: for they sow not, neither do they reap, nor gather into barns; yet your heavenly Father feedeth them. Are ye not much better than they?"[132] They celebrated shared meals—often with individuals from the edges of society—and Jesus was consequently accused of keeping bad company: "Behold a man gluttonous, and a winebibber, a friend of publicans and sinners."[133] It is absolutely no accident that the Eucharist would become one of Christianity's central institutions. Here the community celebrates together, just like in the days of the hunter-gatherers: no one eats alone. Eating is a social event that strengthens bonds among the group's members.

Why do people find themselves so drawn to this type of dropout culture? From the *Lebensreform* movements of the early twentieth century to the more recent hippy movement, people again and again find themselves attracted to alternative forms of living together. Early Christianity fascinated many of today's Christians, who believe it remained faithful to the original goals of the Jesus movement. With the arrival of Jesus, "we see the emergence of an alternative lifestyle—one that posed a constant challenge to society," writes Bernhard Lang. "It is difficult to turn away from its allure." According to Lang, Jesus and his disciples were living a social experiment.[134] But they weren't really trying out anything revolutionary at all, for Jesus was merely returning to the roots of *Homo sapiens.* Seen against the long history of humanity, Jesus was actually something of a hyperconservative.

Ever since the days of Sigmund Freud, discontent with civilization (*Das Unbehagen in der Kultur*) has become something of a catchphrase. From an evolutionary perspective, patriarchal societies' focus on material property and neglect of social relationships explain this discontent. With the arrival of such societies, life became materially richer but socially poorer.[135] But humans evolved as hypersocial beings, and for hundreds of thousands of years our survival depended on a tightly woven network of social relationships. This was our only source of life insurance, so it comes as no surprise that we felt and continue to feel discontented in our new surroundings. The fascination we have for a sharing and caring collective, like the one that gathered around Jesus, stems from our first nature's longing for the times when our psychological preferences evolved, a longing left largely unsatisfied in our anonymous, materialistic societies.

The Revolutionary Role of Women

The prominent role that women played in the Jesus movement also fits with what we would describe as the "back-to-hunter-gatherer-culture" trend. Not only do we see a "remarkably varied inclusion of women and their world," but women also belonged to Jesus's wider circle of followers. They were responsible for material support and played a decisive role in the events at the crucifixion and Jesus's resurrection. Some women even remained active as wandering preachers after Jesus's death.[136]

Mary of Magdala is the most renowned of these women. Luke tells of how Jesus drove seven evil spirits out of "Mary called Magdalene." Like other women, Mary Magdalene followed Jesus from Galilee to Jerusalem; she didn't flee with the other disciples after his arrest but remained to witness the crucifixion. And all of the evangelists agree that Mary Magdalene discovered Jesus's empty tomb. According to three of the gospels, she was the first to encounter the resurrected Jesus, who tasked her with announcing his return to the disciples. "She thus certainly played an important role in the Jesus movement, one comparable to that of Simon Peter," explains Jürgen Roloff, who concludes that the importance of women in Jesus's circle "was unusual in the culture of the day with its patriarchal society, if not completely revolutionary."[137]

As explained in chapter 9, we must view the Hebrew Bible as a kind of minority report.[138] Normal women are virtually invisible in the Old Testament, and those women who do make an important appearance are "not representative of women in ancient Israel."[139] The Bible only includes them because they stand their ground for the people of Israel like men would. Judge Deborah went to war against the Canaanites; Yael killed an enemy commander; Esther, the consort of a Persian king, saved her people from a pogrom; and Judith, whose book is considered apocryphal, cut off the head of Assyrian general Holofernes. These stories all serve to bolster the patriarchal message: Yahweh will help even weak women achieve great things if their belief is strong enough.

Ruth the Moabite would become famous for remaining true to Yahweh's people after the death of her husband. This case was the exception rather than the rule, because the Hebrew Bible generally views foreign women as a threat. After all, they might encourage their husbands to get involved with foreign gods (Solomon being the best-known example). Hence the book of Ezra demands that mixed marriages be dissolved, and the book of Nehemiah forbids them entirely. This confirms the observations of a number of religious scholars: women are the actual bearers of religion, meaning they determine religious praxis in their respective households. In ancient Israel, too, they were the real "ritual experts" of intuitive religion.[140] Incidentally, the same remains true today. "The greater religiosity of women, demonstrated in consistent research findings over the past 100 years, is one of the most important facts about religion," writes religious psychologist Benjamin Beit-Hallahmi.[141]

Apart from these few noteworthy appearances, women in the Old Testament only appear as combatants in patriarchal family disputes or as victims of violence offering men an excuse to wage war. The rape of Dinah provided her brothers with the justification to slaughter the residents of the city of Shechem, even though they were prepared to convert. And then there is the tale in Judges of the infamous deed at Gibeah, where a woman was given to the enemy to be raped for one whole night. The next day, the consort took a knife and "laid hold on his concubine, and divided her, together with her bones, into twelve pieces, and sent her into all the coasts of Israel"—all in order to rally troops for a retaliation campaign.[142]

In contrast to the stories of the Old Testament, the gospels contain various antipatriarchal elements. Jesus himself refused to reject ostracized women, such as prostitutes, and we have already discussed the incident with the adulteress, whom he would not judge for her behavior. And when Jesus was anointed in Jerusalem, as was proper for the Messiah, it was a woman who carried out the act. All this adds up to a major piece of the puzzle in our argument that we are witnessing the comeback of our first—our hunter-gatherer—nature. As we have explained, in hunter-gatherer groups women enjoyed much more freedom than they did in the patriarchal societies of the sedentarized world. We could also describe what we see in the Jesus movement as the emancipatory inclusion of women.

No One Is Lost

For Jesus, every single person counts. We hear this from him again and again. He packages this message in parables such as the one about the lost sheep:

> What man of you, having an hundred sheep, if he lose one of them, doth not leave the ninety and nine in the wilderness, and go after that which is lost, until he find it? And when he hath found it, he layeth it on his

> shoulders, rejoicing. And when he cometh home, he calleth together his friends and neighbours, saying unto them, Rejoice with me; for I have found my sheep which was lost. I say unto you, that likewise joy shall be in heaven over one sinner that repenteth, more than over ninety and nine just persons, which need no repentance.[143]

Stories such as these enable Christianity to appear as the "religion of love."[144] The prodigal son, who had squandered his inheritance in foreign lands, returns home out of desperation, repents, and is joyfully accepted by his father. A readiness to display forgiveness is at the heart of hunter-gatherer morality. In small groups, everyone depends on everyone else, so the group cannot afford to give up on errant individuals so quickly. "This is partly because they feel for them as fellow human beings, and partly because they're practical people who understand the need to have as many hunters as possible in the band," explains anthropologist Christopher Boehm.[145]

Jesus's engagement on behalf of the socially disadvantaged and his healing of the sick also form part of this morality. These acts therefore represent more than CREDs cleverly inserted into the gospels by the evangelists. The Jesus movement scored points by demonstrating solidarity in a time of great suffering. The message was clear: we won't give up on anyone; the community will stick up for each and every one of its members. Jesus appears here as a personal counselor and protector, exactly what our first nature has always longed for. He leaves no one behind and drives out our demons. This Jesus is the representative of the God we find in Psalms, not the God we encounter in the Torah.

"Ecce Homo"

Time to sum up: The double Jesus we encounter in the New Testament is a hybrid accumulation product of cultural evolution. Intellectual religion revolutionized its matrix for explaining the world in an alluring, apocalyptic fashion, and in so doing it created a world of good and evil. In this world the first Jesus Christ appears as a hero in the struggle against the forces of darkness. The second Jesus we meet in the gospels corresponds to our age-old psychological needs. Even popes have used the word "friendship" to characterize this side of Jesus's double morality.[146] And so, alongside the eschatological Jesus, we also meet Jesus the friend.

Whereas the former is the incorruptible arbiter of the Last Judgment, the latter remains the "loving friend of sinners," prepared to forgive every misdeed.[147] But both aspects of Jesus fuse to form a single figure manifested in the form of a suffering mortal man, and this is the secret to Jesus's success. John, the last of the evangelists, relates a characteristic scene in which the Roman prefect Pontius Pilate, preparing to sentence Jesus, hesitated and started looking for a way out—although this could hardly have really happened because the real Pontius Pilate was well known for his summary judgments. In any case, the biblical Pilate finally presented the bound Jesus, wearing a crown of thorns and a purple robe, to the people. And then Pilate spoke his legendary words: "Ecce homo"—"Behold the man!"

These words depicted the Jesus of the gospels as a mere human being—vulnerable, pitiful, and destitute. Paul Veyne believes that early Christianity's greatest capital was the fact that Jesus "was no mythical being living in some fairytale time" but a real person.[148] The evangelists portrayed him as an individual in desperate need of help. This, too, helped draw people to him. Sympathy is after all a part of our basic psychological makeup.

An additional asset of the Jesus movement was its ability to maximize *asabiya's* potential. On the one hand the movement's apocalyptic understanding of the world activated the age-old psychology of friend versus enemy. Ideals of egalitarianism and forgiveness welded the believers together, and the radical nature of the movement's ethical demands functioned as a costly signal. Those prepared to live up to such demands must have really meant what they were preaching. This increased the group's attractiveness and ensured, as the Acts of the Apostles attest, the cohesiveness of Jesus's followers even after his death and despite all the reprisals they were made to suffer.

On the other hand the animosity the group displayed toward a devilishly experienced enemy was an extremely effective social glue. "So long as the Christian movement remained a persecuted, suspect minority within Jewish communities and within the Roman empire," writes Elaine Pagels, "its members . . . no doubt found a sense of security and solidarity in believing their enemies were (as Matthew's Jesus says of the Pharisees) 'sons of hell.'"[149] This would prove fatal once the majority, including the powerful, had adopted Christianity, because from then on the new religion was in a position to brand its enemies as Satan and send them straight to hell.

For most of its history, Christianity's success arose from this double potential. On the one hand it was a religion of love that stood up for the poor, the weak, and the sick. On the other hand it was a religion of hatred that demonized its enemies. Its followers believed the powers of evil lurked around every corner in the form of pagans, Jews, heretics, and witches—and we're not even mentioning all manner of crusaders and their holy wars. We only wish to point out the manipulative potential that arises when God blesses our deeply rooted friend-versus-enemy psychology and the battle against a supposed evil suddenly becomes a means of doing away with a demonized enemy in a bid to enter into the kingdom of heaven.

Jesus and his disciples had arrived on the scene in order to destroy Satan's evildoing. The decisive factor we wish to stress here is that people's psychological makeup makes them highly susceptible to such an "apocalyptic matrix."[150] We cannot fundamentally attribute this susceptibility to religion; religion merely enhances it. Believers and nonbelievers alike have a soft spot for those ready to wage war against deepest, darkest evil. Just look at the great epics of our time, such as the *Lord of the Rings* trilogy, the *Star Wars* films, and the various *Harry Potter* books, in which sworn bands of fictive brothers (and sisters) face off against the insidious forces of darkness.[151] Be it Frodo and his companions battling against Sauron and the Orcs, Luke Skywalker and the Jedi standing up to Darth Vader and the Empire, or Harry, Hermione, and Ron giving their all in the struggle against Lord Voldemort and the Death Eaters, for our first nature this is as thrilling as it gets.

18

WHEN JESUS STAYED IN HEAVEN

The Birth of Christianity

JESUS WAS CRUCIFIED. THIS CAME AS A BIG SHOCK FOR HIS FOLLOWers: a messiah should not die such an ignominious death. Nevertheless they found it remarkably easy to recover from the blow by interpreting it to fit with Jesus's wider, apocalyptic message. Jesus did, after all, return from the dead. Didn't that mean that his own life and death represented a completed apocalypse of their own? Jesus fought against devils and demons and refused to be led astray by priests or Romans. He suffered the most terrible of agonies upon the cross and then went on to triumph over death itself. If this wasn't an "apocalyptic sign," then what was? Surely the Kingdom of God was at hand.[1]

But God's Kingdom did not come. Years turned into decades, and God had yet to deliver humanity from evil. In the meantime, the Romans brought destruction upon Jerusalem. Early Christianity took shape in the worst of times: the Romans occupied Palestine and ruthlessly suppressed repeated rebellions. The extent to which the Romans also persecuted the early Christians remains unknown; some argue that the situation was not as bad as commonly suggested.[2] But one question definitely kept nagging

Jesus's followers. Had he been wrong about the apocalypse? Had they been duped by a charlatan?

Our knowledge essentially comes from the few successful religions that endured, so we cannot exclude survivorship bias, but it is striking how in the Bible established concepts were usually not abandoned, even though they no longer squared with reality, but merely modified. The failure of the Kingdom of God to materialize—a deeply disturbing turn of events, to be sure—therefore initiated two mutually reinforcing processes: the individualization of the hope for salvation and a deification that not only elevated Jesus but also created a panoply of heavenly beings. Even though this analysis leads us well beyond the pages of the New Testament, it is still worth taking a look at how human nature influenced the cultural evolution of Christianity.

SALVATION BECOMES PERSONAL

Jesus failed to return, and the apocalypse never happened. Did this mean that early Christians had to abandon all hope of salvation? Fortunately, the concept of the apocalypse itself suggested a way out because it contained two closely interwoven ideas that made it possible to simply shift the focus from one to the other. The idea that the apocalyptic transformation of the world would befall the living during their lifetime was combined with the idea that the dead would be resurrected at the same time. Apocalypses were thus based on the idea of collective salvation of the living and the dead at the arrival of the Last Judgment at the end of days.

If the apocalypse was not to happen in the here and now, then what could be more natural than the belief that judgment would only be passed over the dead? In this model, salvation takes place in the afterlife, individually and at the moment of death. This was an easy inference, because people already believed this was what happened to martyrs after they died anyway. And wasn't Jesus himself dead for only three days, which would suggest that judgment happened right after death?

It was certainly an attractive idea, for it also ironed out another kink in the original notion of the apocalypse. People had always been a bit uncertain about what happened to their souls while they awaited judgment at the end of days. Where and how would they spend the years or even centuries until this day finally arrived? As born dualists, people could never really imagine what this temporary form of nonexistence was supposed to look like.

The formulation of the new concept of salvation was a continuing search for a satisfactory solution of the fate of the deceased's soul that would press on into the Middle Ages. Individual judgment gradually replaced the Last Judgment guaranteed in the Bible: "Each man receives his eternal retribution in his immortal soul at the very moment of his death, in a particular judgment that refers his life to Christ," explains the catechism of the Catholic Church.[3] This not only has the great advantage of being irrefutable (no one has been able to prove otherwise) but is quite appealing to our first nature.

In the end, Christianity decided to postpone apocalyptic hopes for salvation until the afterlife and transformed heaven and hell into the cosmos in which God's justice is realized. To accomplish this, however, it needed the right personnel. Luckily for Christianity, a number of deification processes were already under way.

HIGHER AND HIGHER

"Only Judaism and Islam are strictly monotheistic in principle," wrote Max Weber.[4] Paul Veyne speaks of Christianity's "doubtful monotheism," for with its "three supernatural objects to be worshipped"—God, Christ, and Mary—it is, "quite literally, polytheistic."[5] Indeed, Christianity is often accused of "re-mythization"[6] or even a "primitive relapse into mythical religion."[7] But calling these changes the "re-enchantment of the world"[8] would be more appropriate, given that we can view them as a reaction to the "disenchantment of the world" produced by intellectual religion. This re-enchantment is the second of the reactions to the delay of the Kingdom of God.

We have seen this phenomenon again and again in this book. Cultural-evolutionary solutions of an intellectual nature are hard to sustain, since our first nature tends to resist them. We already noted how people had their problems with strict monotheism and how our first nature's "theologically incorrect" tendencies tend to diversify divine forces. Christianity allows them finally to run free, even if—officially, at least—it maintains the fiction of the one and only God.

Once again we are dealing with a process of hybridization. A thoroughly moral God called a host of additional supernatural actors into action: Jesus and Mary, thousands of saints and angels, and the devil and his demons. There's something for everyone. Let us take a closer look at the individual characters comprising this divine Christian cosmos. What helped them achieve divine status? And what consequences did this heavenly division of labor have for God himself?

Jesus: God and Man

What happened to Jesus after his resurrection? According to the Acts of the Apostles, he ascended into heaven on a cloud after forty days (the reverse of the Son of man's route in the book of Daniel). Years later Stephen had a vision in which he saw Jesus standing to God's right in heaven. The Jews, seeing this as nothing less than blasphemy, stoned Stephen to death, making him Christianity's first martyr: Saint Stephen. When Jesus appeared to Saul of Tarsus—a persecutor of Christians later known as the apostle Paul—on the road to Damascus, Saul was blinded by a bright light from above. Clearly, Jesus had achieved the status of a heavenly being.

The title "son of God" certainly helped. According to Jewish tradition, this was just an attribute of the Messiah, not a claim to divine descent. The pagan world, however, which was quite familiar with the notion of divine children, saw things differently and transformed him into the true son of God. Appropriately, the Gentile Christian Luke greatly embellished the details of Jesus's virgin birth and divine paternity in his gospel.[9]

A human became transformed into a god and ascended "higher and higher" into the heavens. Bart Ehrman couldn't have summed it up better when he wrote, "Jesus went from being a potential (human) messiah to being the Son of God exalted to a divine status at his resurrection; to being a preexistent angelic being who came to earth incarnate as a man; to being the incarnation of the Word of God who existed before all time and through whom the world was created; to being God himself, equal with God the Father and always existent with him."[10]

For three or four long centuries, a fierce debate raged within the Christian Church about how exactly this had happened. Philip Jenkins, who wrote a book on the topic, fittingly referred to the struggle as the "Jesus wars." We do not wish to delve into this highly complex debate here, but in essence the key questions were as follows: Was Jesus a man or a god? If he was only a human, then what's all the fuss about? If he was a god, doesn't the tale of his suffering lose some of its credibility? As theologian Rudolf Bultmann (1884–1976) succinctly puts it, "Anyone who knows he will rise again after three days won't think death is such a big deal!"[11] Thus

Christians had to prove that Jesus was both god and human. They had to square the circle, so to speak.

In the end the doctrine of the Trinity won out. It drew on Greek philosophy to show how the divine being (*ousia*) could be imagined in terms of three persons (in the form of the Father, the Son, and the Holy Ghost), whereby—and here we must introduce yet another paradox—both their consubstantiality (*homoousios*)—in other words, their being of the same substance—and their individual distinctness were ensured. The commitment to the Holy Trinity may have been an impressive achievement, explains Jörg Lauster, but it also pushed Christian theology to the very limits of "what was to be achieved with philosophical concepts."[12] Jesus himself had to be saved for the sake of the Christian community—regardless of the costs.

Officially, Jesus could not be allowed to become a second God (hence the complicated argument concerning consubstantiality). Instead a human was needed—albeit one who was more than your average person. Without Jesus the man, Christianity would never have achieved anything. He provided God with a human face and helped iron out all of our first nature's problems with the monotheistic God of the Old Testament. People could readily imagine Jesus and even depict him: Jesus arguably became the most portrayed figure in the world. We humans desire to depict our gods. These portraits show Jesus as a human being—one with whom one can communicate.

We can turn to Jesus—we're talking about the hunter-gatherer version discussed in the previous chapter—with our problems. He is, as none other than Pope Benedict XVI said, a friend. He assists individuals in times of hardship and illness and fends off evil spirits. And as a healer he introduced a degree of sorely missed medical compassion into the kingdom of heaven.

But even Jesus could not escape the internal momentum toward deification. He moved ever farther away from the people, became transformed into Christos Pantokrator, and was worshipped as the supreme ruler of the world.[13] And the more divine he became, the greater our need for gods with a human face grew.

Mary, Mother of God

The oldest biblical references don't even mention Mary by name. In his Letter to the Galatians, Paul merely wrote, "But when the fulness of the time was come, God sent forth his Son, made of a woman."[14] The evangelists, who did not set down to write until decades after Paul's letters, only noted the bare minimum when it came to Jesus's mother. They made no mention of Mary's appearance and remained absolutely silent about the circumstances surrounding her own birth and death. Legends appeared to fill in the gaps, but the question of Jesus's divine nature ultimately brought Mary to the attention of intellectual religion. After all, she had given birth to a god, so wasn't it appropriate to refer to her as *Theotokos* ("God bearer"), as she was known to the common people? In 431 CE the bishops assembled at the Council of Ephesus decided the answer was yes. Mary advanced to become the "Mother of God," and shortly afterward it was decided that she too had ascended to heaven, even if the Bible makes absolutely no mention of this. As recently as 1950 Pope Pius XII made it official when he announced the dogma of the assumption of the Virgin Mary into heaven.[15]

Mary also found herself facing so much work because she filled probably the greatest gap in biblical religion: she lent a female face to the divine. Finally women had an adequate counterpart, someone with whom to discuss all of their worries and hardships, woman to woman. It's no coincidence that Mary's main role was mother, whether she was offering Jesus her breast or lamenting her dead son. In times when medicine barely existed and infant mortality was extremely high, people needed divine succor more in no other aspect of life. And they needed it from someone who knew what she was talking about.

Institutional religion certainly had some catching up to do. All of the pagan religions had mother goddesses, from Isis to Artemis. "The success of the Christian mission," explains historian Klaus Schreiner, "depended on the church's providing an offer of salvation to pagans who were willing to convert that did not offer less than their old religions."[16] Mary became a decisive factor in this offer. In ancient Israel, too, mother and fertility goddesses were very popular, and we have already discussed Yahweh's alleged consort Asherah, as well as the various terracotta figurines used in Israelite household worship. The female cult might have been forbidden, but in the realm of everyday religion, female deities continued to play an important role.[17] With the veneration of the Virgin Mary, Christianity in effect legalized the people's need for a goddess.

Even men can turn to this godly mother in times of hardship. And she reacts just as any mother would: she offers understanding, forgiveness, and solace. As Jesus could no longer respond to individual calls for aid, protection, and mediation as he once had—thanks to his advancing heavenly career—Mary increasingly took up the slack. To this day millions of pilgrims continue to make their way to Guadalupe in Mexico, Lourdes in France, Częstochowa in Poland, and Altötting in Germany to seek Mary's help with their worries and illnesses. And their prayers seem to be working; thousands of votive tablets at these pilgrimage sites attest, "Mary helped."

As Swiss theologian Josef Imbach explains, Mary is the "accomplice to the disadvantaged."[18] She puts mercy before justice. Indeed, legends tell of how she has helped pregnant nuns and unfaithful wives. She does exactly what the hunter-gatherer Jesus of the New Testament did but the divine Jesus is no longer allowed to do. It is Mary who will lift the scales at the Last Judgment so that people's sins do not weigh too heavily.

Saints and Angels, Devils and Demons

Inevitably, Mary eventually shared the same divine fate as Jesus. Even though not officially a goddess, she still fulfilled that function.[19] As her heavenly career took off, people felt the need for additional divine figures to whom they could turn. Thankfully such figures had always been there.

Max Weber noted that nowhere under monotheism "was the existence of spirits and demons permanently eliminated; rather, they were simply subordinated unconditionally to the one god, at least in theory."[20] Belief in such beings is usually dismissed as primitive superstition resulting from the low religious aptitudes of "the common man." But this explanation is just too simple because such a belief is the result of human nature claiming what is rightfully its own—and it is also what Christianity would eventually provide. At the end of the day, our first nature received all the supernatural actors it had always desired: angels, devils, demons, and legions of saints, both male and female.

We have already encountered the angels in the Hebrew Bible, where they appear as members of the heavenly court. In Christianity, too, they would perform a number of different services. The apocalypses enabled the evil spirits to make a comeback, only now they were officially declared fallen angels.[21] And not even hell remained unaffected by the differentiation that occurs in all established cultural systems. Originally the angels punished the damned,[22] whereas the devil himself suffered in a "lake of fire and brimstone," "day and night for ever and ever."[23] Angels as torturers? This was unacceptable, for good must do good, and evil must do evil. And so, hell became Satan's realm, and he became the Dark Lord, roaming the world of the living to strike terror into people's hearts.[24]

Fascinatingly, ancestor worship would also eventually sneak into Christianity in the form of the martyred saints. Those who passed God's scrutiny and achieved martyrdom would receive a place of honor close to God.[25] This is the well-known logic of reciprocity: outstanding performance must be outstandingly rewarded, and so the souls of the martyrs were promoted to angelic beings in the afterworld.[26] But if this was so, why shouldn't people be able to turn to them as mediators? After all, the martyrs were certainly familiar with human needs and emotions, so surely people could count on them to put in a good word with the boss. Later, individuals who lived exemplary lives of piety could also posthumously achieve sainthood.[27]

The *Martyrologium Romanum*, the book of all of the saints and blessed of the Roman Catholic Church, contains around 7,000 entries. There is more than one saint for every day of the year and just about every conceivable situation. Aid and advocacy are certainly assured regardless of what may happen. One randomly chosen example: Clare of Assisi, who died in 1253, is the patron of Assisi, washerwomen, embroiderers, glassmakers, glass painters, gold workers, the blind, telegraphs, telephones, and television. She intercedes against fever and eye problems and during difficult births and works for good weather.[28]

In this manner Christianity offered a legitimate second chance to the old belief in spirits and ancestors—and for its own good, too. Our first nature of course offered no resistance to a belief system that in no way opposes human intuition. This enabled Christianity to tap into the enormous power of intuitive religion. Psychological tendencies such as the Hyperactive Agency Detection Device give people the feeling that something must be out there pulling all the strings, even if we can neither see nor hear it. Lest anyone continue to harbor any doubts that a belief in spirits belongs to *Homo sapiens*'s basic psychological makeup, we would like to point out that such beliefs are not only found in all traditional societies. Some readers may be surprised to learn that, according to recent surveys, 75 percent of Americans and 66 percent of Germans believe in angels.[29]

SO WHY DO WE STILL NEED GOD?

As a teenager, biblical scholar Bart Ehrman was a born-again Christian. He tried hard to convince his Jewish girlfriend to take Jesus into her heart too. She thought about it for a while before answering, "But if I already have God in my life, why do I need Jesus?"[30] In light of everything discussed in the previous sections, we would like to turn this question around and ask, If we have Jesus, Mary, and countless saints, angels, and demons, why do we really need God? After all, we have a well-organized and gender-differentiated cosmos of gods and spirits at our disposal to help us with the challenges of everyday life. A horde of demons can explain away all the evils in the world. And regiments of good spirits are ready to do battle against the forces of darkness at the command of a female and male deity.

In the Torah, God personally sent catastrophes into the world. Now he had others to do this for him, above all the devil. The latter had taken it on himself to carry out brutal actions. This made the devil into a kind of obverse of God, solely responsible for all the evil in the world.[31] God was not about to miss out on the opportunity that this transformation provided. He withdrew from daily life and began concentrating on his core business. He was the God of creation who had called everything into existence, and he was the guardian of morality. And now—thanks to the devil—he could become entirely good, compassionate, and abstract. In doing so he completed the journey he had begun when the Babylonians destroyed his temple in the sixth century BCE. He had reached transcendence. He could finally become perfect.

Here we see the last phase of God's enrichment. "Christianity began as a Jewish messianic sect," says historian Werner Dahlheim. "But the world in which it was to grow was ruled by the Greek mind and the Greek language." Dahlheim sums this up so succinctly that we quote him verbatim: "Christians only knew the Old Testament in its Greek translation, the Septuagint; the evangelists, the writers of the epistles, and all of the other authors of the first two centuries wrote in Greek; Christianity's spiritual leaders all thought in Greek; and it was in Greek or strongly Hellenized metropolises—including Rome and many other cities in Italy, Spain, and southern France—that its missionaries carried out their works; and it was there that the decision concerning the future of the new religion was made."[32]

The following transpired: The Jewish God and the unprecedented levels of morality he expected of individuals melded with the idea of absolute goodness as propagated by Greek philosophers such as Plato and the Stoics.[33] Only then could our modern understanding of God's greatest characteristics come to be: he became omniscient and omnipotent—something the Hebrew God had never been, as we have observed in detail in previous chapters. Yahweh was neither unchanging nor all-powerful nor all-knowing. Israeli philosopher Yoram Hazony writes, "The source of the equation of God with perfect being is quite clear in Greek thought: We find it in Xenophanes, Parmenides, and Plato." He adds, "An immutable perfect being is a Greek conception of what God must be like, not a biblical one."[34]

God's high moral demands and the downright mathematical beauty inherent in the idea of a single power who created heaven and earth absolutely fascinated Greek thinkers. This God provided a more convincing form to the rather "bloodless" idea that the gods represented the absolute foundation of good. Conversely, this Hellenic impulse gave God the decisive push forward that finally landed him in the realm of transcendence.[35] In this Jewish-Greek hybrid being, morality and philosophy joined forces in a seminal alliance.

Admittedly, the Jews themselves were also influenced by Greek thinking. Jewish philosopher Philo of Alexandria, a contemporary of Jesus (ca. 20 BCE–50 CE), sought to combine Jewish wisdom with Greek philosophy. He took up the Greek concept of Logos, or the eternal reason that governs the world and gives it structure. According to Greek thought, this universal force could be described as Zeus, God, or providence.[36]

Christianity enthusiastically adopted this idea. The Greek-educated Christians saw Jesus as a personification of divine Logos. The last gospel, the book of John, said as much. It begins with the words, "In the beginning was the Word, and the Word was with God, and the Word was God." In the Greek original, however, we find the word "Logos" in the place of the rather nebulous "word." Hence the next sentence should read, "And the Logos was made flesh, and dwelt among us, (and we beheld his glory, the glory as of the only begotten of the Father,) full of grace and truth."[37]

It is hard to overestimate the impact of this turn of events on Western culture. According to philosopher Werner Beierwaltes, despite all their differences, "Greek philosophy and Christian theology found a form of symbiosis that was certainly full of tensions, but also productive and mutually illuminating and sustaining." Logos, or reason, would become a "personal trait of the Christian God." He would become a "pure, absolute being"[38] who would prove irresistible to our third nature. And so God became a purely intellectual principle and lost all contact with perceptible reality. He didn't need it—that's what Jesus, Mary, the devil, and the rest of the heavenly players were there for.

Religious psychologist William James—somewhat derisively—summed it up in a nutshell: "by mere logic," God had become a "metaphysical monster." Beginning with the Aristotelian premise that God is the first cause, systematic theology provided us with "God's list of perfections." He is not only necessary but also absolute; he is the one and only, unchanging, unlimited, immeasurable, omnipresent, eternal, all-knowing, and all-powerful. But according to James, these are just "pedantic dictionary adjectives, aloof from morals, aloof from human needs," nothing more than "an absolutely worthless invention of the scholarly mind," which, when it comes to alleviating human hardship, is of very little help.[39]

On the other hand, this development explains why the very anthropomorphic image of God as the father, as the good shepherd—the clichéd old man with the flowing beard—became so thoroughly established in everyday Christianity. It went along perfectly with patriarchy, especially since the loving father figure could easily turn into the stern *pater familias,* who would spank his unknowing children to teach them valuable lessons. This image also fitted nicely into the claims to authority of worldly and religious rulers.[40] And for the notoriously theologically incorrect first nature of the believers, finally, this is simply the true trinity of Christianity: God the father, Mary the mother, and Jesus the son.

RECIPE FOR SUCCESS

Our detailed dissection of Christianity has made it clear that we are actually dealing here with a double religion. On the one hand it harbors its own animistic-polytheistic cosmos, which speaks to our first nature and fulfills the needs of intuitive religion. It allows us to make images of Jesus and the whole panoply of heavenly beings, drive out evil spirits, make pilgrimages to the Virgin Mary, venerate wonder-working relics, and implore the saints to help us score the goal. On the other hand Christianity also incorporates the intellectually stimulating concept of a monotheistic God who is the most powerful force in the universe. He is pure reason, the pride of our third nature, the inviolable guardian of morality, and creator of the universe. He has come to move in mysterious ways, and his plans will come to fruition not in this world but another one. This, at least, is how Augustine, the most influential church father of Christianity, formulated it in his *De Civitate Dei* (*The City of God*).[41]

This dual makeup is the true reason for Christianity's immense success. Drawing on a wide variety of distinct, independent ingredients, cultural evolution created a religion of enormous range that offers something for everyone. It satisfies both our intuitive and our intellectual needs and offers enormous potential in terms of popular religiosity as well as the philosophical leanings of the elites. First there is the notion of God as an abstract, transcendent being (the one we encounter in pure contemplation), which of course continues to fascinate many a natural scientist. Then there is the wondrous world of supernatural beings who help us out in every aspect of life and whose assistance we can invoke in a wide variety of ways. Their presence allows us to explain every event.

For the sake of completeness, however, we would like to include three other ingredients that belong to early Christianity's recipe for success. First of all, the Christian God was particularly attractive to the rulers of the day. Although Paul Veyne referred only to Emperor Constantine when he wrote the following, his words could easily apply to every Western Christian ruler: "If he wanted to be a great emperor, he needed a great God. A gigantic, caring God who passionately desired the well being of the human race aroused far stronger sentiments than the crowd of pagan gods who lived for themselves. And this Christian God revealed a no less gigantic plan for the eternal salvation of humanity. He involved himself in the lives of the faithful, demanding that they observe a strict moral code."[42]

Once more we come across the amalgamated character of the Christian notion of God. We can trace his love and mercy back to the hunter-gatherer Jesus of the gospels, and his benevolence draws on the needs of intuitive-individual religion. But Yahweh is also the creator of heaven and earth. He can part the seas and has moral aspirations never seen before in the ancient world. This is the legacy of intellectual-institutional religion. And this God also places earthly rulers under his authority—for the good of society no less.

Another successful ingredient was the addition of the apocalyptic matrix, which included the integration of our age-old psychology of friend versus enemy. This was extremely useful, for it helped ensure that the people closed ranks. Although *asabiya* was already on good footing, thanks to the Bible's high level of morality, the apocalyptic message further increased solidarity by forcing charity, love, and support for the poor and the weak. It also gave religious and worldly authorities a new weapon: they could now depict enemies not just as barbarians but as demons and witches to be sent straight to hell.

The final component to the success formula was not all that novel, and we have already discussed it, but it unfolded to the utmost in Christianity. Contemporary evolutionarily inspired research into religion refers to this ingredient as "supernatural monitoring" and "supernatural punishment," and it can be summed up as follows: "Watched people are nice people."[43] An omniscient God who monitors people's actions and judges each and everyone at death can only exhibit his full power when the concept of the afterlife includes a place where he can mete out godly justice. This device makes the system immune to refutation, since everyday experience provides ample evidence of how, in the here and now, gods can be pretty arbitrary in their punishments and rewards. With its heavy focus on the afterlife (culminating in the medieval byword *memento mori*—"remember that you will die"), Christianity developed the perfect monitoring system.

This is what we mean with our claim that Christianity reached previously unseen levels of success because it really was the Swiss army knife among the classical religions. The enrichment process of cultural evolution had produced a powerful hybrid product whose application could offer a great deal of support in every aspect of life.

EXPLOSIVES INCLUDED

Christianity is a religion for everyone and everything, but it's also full of tensions. This should come as no surprise, because fusing the spheres of intellectual and intuitive religion was no easy task. The resulting amalgam is full of explosive potential. We will now analyze three examples, not so much to expose Christianity's inconsistencies as to illustrate the cumulative nature of its cultural evolution. For us, this is just one more piece of evidence showing how we can only really understand religions if we separate their cultural-intellectual components from the intuitive elements anchored in our basic psychological makeup.

The first two problems result from intellectual religion's excessive coherence seeking. As William James explained, God was showered with attributes such as omnipotence, infinite goodness, and eternalness as the consequence of pure logic. With these kinds of issues, it is always helpful to turn to the catechism of the Catholic Church, which lends aspects of faith the requisite authority: "God is the fullness of Being and of every perfection, without origin and without end."[44] But reconciling a perfect being with an imperfect world has caused even the cleverest of thinkers a great deal of headache.

The first dilemma is obvious. Theologians describe it as follows: "Christianity conveys God and the world in Christological terms from the very beginning in that God entered into the world in the form of Jesus Christ as its savior."[45] This means that God became human with the arrival of Jesus. However, it is logically impossible for an immutable and eternal being to appear as a human being on one unique occasion at one particular time in history. Immutable is immutable.

The second dilemma has its origins in the same logic and is particularly difficult to come to terms with. We are referring to theodicy here. If God is perfect, then why isn't the world perfect too? Why is there so much suffering in a world overseen by an all-powerful and infinitely good God? We examined this dilemma in our chapter on Job and explained how such questions did not arise until after the Bible was written. Of course, theologians have always debated the origins and causes of the existence of evil, but for most of history, they believed it to be either a manifestation of God's punishment or the work of the devil. Only with the arrival of pure monotheism in its Hellenism-inspired form—with its notion of an all-powerful and infinitely good God—did the origins of evil become a real problem. If God is responsible for everything, then that must really mean everything! Intuitive religion, however, with its numerous supernatural actors, has no problem with this inconsistency, of course. There are enough devils and demons to explain away even the greatest of the world's horrors.

Both of these problems were homemade, so to speak. They arose because Christians adopted the claim to perfection introduced by Hellenistic metaphysics, inserted it into their own intellectual religion, and then applied it to a nonperfect world. They had to do this, because the biblical God had always claimed to be following a mission of salvation in the real world. But, once again, the God of the Bible was not, and had never claimed to be, a perfect God.[46] As previous chapters have shown, God underwent a massive transformation over the course of his career. How else could Jack Miles have found the material to write his thrilling life story of God for which he even won the Pulitzer Prize—for biography![47]

The Old Testament, too, suffered as a result of this postbiblical perfection craze. Many, if not most, believers—and even nonbelievers—open its pages expecting to find a perfect God. And every inconsistency, every flaw in the text offers grounds to doubt God's omnipotence.

HE DIED FOR OUR SINS

Let us now turn to the third problem that arose as a result of this fusing of the spheres: the question of why Jesus had to die. "Christ died for our sins," says Paul in the New Testament.[48] This is the basis of the fundamental Christian principle of redemption that by giving up his life, Jesus took all of the sins of the world upon himself and, in doing so, saved humanity. But what does this really mean? How could this work? And shouldn't this salvation involve a final judgment, as the Bible repeatedly stresses, rather than a wholesale acquittal for everyone?

Even some theologians have problems coming to terms with Paul's idea. "How can the guilt of one man be expiated by the death of another who is sinless—if indeed one may speak of a sinless man at all?" asked Rudolf Bultmann, one of the most influential Protestant thinkers of the twentieth century. And combatively he added, "What primitive notions of guilt and righteousness does this imply? And what primitive idea of God? The rationale of sacrifice in general may of course throw some light on the theory of atonement, but even so, what a primitive mythology it is, that a divine Being should become incarnate, and atone for the sins of men through his own blood!"[49]

Bultmann therefore called for "demythologizing the New Testament proclamation." Here he meant nothing less than cleansing Christianity of its "world of spirits and miracles"[50] belonging to the sphere of intuitive religion. The result would be a highly rational theology—intellectual religion in its purest form, which is actually quite typical of some forms of European Protestantism. It is hard to avoid the irony of the situation: one of the main reasons for the much-lamented unintelligibility of the idea of redemption lies in the fact that these events had already been "demythologized." Let's see how this might have worked.

The notion of humanity's redemption rests on the idea of sacrifice. Remember the economy of sacrifice: a gift functions to offset a debt or, if given in advance, indebts the recipient to the giver and obliges him to give or do something in return. This principle of reciprocity is a fundamental law of human nature; not even God can disavow it. So let us examine this logic as it applies to the crucifixion. Who was actually making a sacrifice?

The catechism of the Catholic Church maintains, "First, it [Christ's death] is a gift from God the Father himself, for the Father handed his Son over to sinners in order to reconcile us with himself."[51] After all, the evangelist John wrote, "God so loved the world, that he gave his only begotten Son."[52]

If God really sacrificed his most precious resource—his son—then, from a monotheistic perspective, we can conceive of two options. Option one: God owed humanity a debt, so he had to sacrifice his son to pay it off. This idea sounds blasphemous even to agnostic ears. Option two: God was making a sacrifice in advance. He sacrificed his son in order to place humanity under obligation. This, too, is hardly an acceptable solution. Which god has to buy humanity's love with a human sacrifice? It's pretty clear that the idea of God sacrificing his only son for humanity makes very little intuitive sense.

What if humanity sacrificed Jesus? John the Baptist spoke of Jesus as a sacrificial lamb "which taketh away the sin of the world."[53] The Torah introduced the concept of the scapegoat with which the people of Israel could rid themselves of their sins each year, but the people compensated for their sins by sacrificing their own goat—they didn't kill the son of the being supposed to forgive their sins and thereby cleanse them. Their debt would have been immeasurable if they had. So it seems that we must reject this possibility too.

That leaves only one option: Jesus sacrificed himself to pay off humanity's debt. The catechism of the Catholic Church subscribes to this line as well: "At the same time it is the offering of the Son of God made man, who in freedom and love offered his life to his Father through the Holy Spirit in reparation for our disobedience."[54] But this sacrifice doesn't make much sense either. If God accepted the sacrifice of his own son, then he was the one taking a loss, not humanity. Consequently, the people's debt to God still stood. In a strictly monotheistic setting, it is hard to see how Jesus's sacrifice could ever be made to make sense. But it turns out this was actually never the intention! Which brings us to the "demythologization" of the crucifixion.

Jesus's death only begins to make sense if we take into account the Bible's apocalyptic view of the world and introduce a third power into the mix: the devil. "The biblical texts spoke of ransom money that had to be paid in order to free humanity from the devil's rule," explains philosopher Kurt Flasch. In other words, God had to pay with his son's life to ransom humanity from Satan. The church fathers—including Irenaeus (d. 202 CE), Origen (185–253/54),[55] and, most importantly, Augustine, who adhered to the notion of original sin—took a similar view. For Augustine this idea was perfectly logical: ever since Adam and Eve's lapse in the Garden of Eden, humanity had been under the devil's thrall, and the act of procreation ensured the continuous passage of the sin from generation to generation. When he killed Jesus, however, the devil killed the only person he could not lay any claim on, for Jesus was born of a virgin, meaning he was untouched by sinful libido. He was pure. Under this logic, God had to send his only son as ransom to free humanity from the devil's chains. God's powerful sacrifice liberated humanity from evil and proved his love of mankind.[56]

Our first nature finds this rationale very easy to accept. The story derives from a world of numerous gods, devils, and demons, something with which Christianity has no issues. On the other hand, for representatives of monotheistic theology, the avant-garde of intellectual religion, it remained a "primitive" notion, for it threatened the very monotheistic principle. Satan could not be allowed to appear as a kind of "anti-God." Neither should God be forced to pay ransom to the devil. That would raise doubts about his omnipotence. And so the devil had to be airbrushed out of the picture to save monotheism.

Might it not have been better if the theologians had done away entirely with the sacrificial function behind Jesus's death? From the perspective of intellectual religion, the answer is probably yes, but as we have already stated, nothing makes for a more powerful credibility-enhancing display than death. Nothing is as convincing to our first nature. Christianity simply couldn't do without Jesus's crucifixion. And what else could Jesus possibly have died for? After all, his death on the cross was a historical fact. So the church fathers provided a reason. Once again—at risk of sounding cynical—we have to give them credit: they proved their talent at spinning the worst possible catastrophe into an advantage.

THE PERILS OF INSTITUTIONALIZATION

Ironically, these contradictions would play a highly productive role in Christianity's further development. The theological efforts needed to prevent this volatile mixture of intuitive and intellectual religion from blowing up actually led to a new wave of rationalization.

The making of Christianity certainly was a paramount intellectual and institutional challenge. It had to deal with the awkward facts of Jesus's death and the failure of his predictions to come true: the Kingdom of God never did appear, after all. Christianity also had to make a name for itself as a new religion. It had to move away from Judaism and defend itself against claims that Christians had simply fallen for the ruse of a fraudster. At the same time, it had to absorb aspects of Greek thought and hold its own against competing religious doctrines such as Gnosticism, which posited a radical dualism between spirit and matter. Most importantly, however, it needed to stifle the profuse proliferation of alternative ideas within its own ranks. From its very beginnings, contests over true doctrine dominated Christianity. In Paul's letters we can see the apostle's anger over competing missionaries. The process causes even some of today's theologians to resort to martial terminology. Jörg Lauster described the "seething turmoil" that characterized the first three or four hundred years after Jesus's death: "The birth of a religion knows no tolerance and no dialog, rather it is characterized by domination and exclusion."[57]

This was a golden opportunity for theologians—those teachers who were, in the words of Tertullian (ca. 150–220 CE), "gifted with the grace of science"—to assert themselves.[58] As early as the second century CE, theological schools sought to create a compelling Christianity capable of holding its own in debates with outsiders. Hellenistic philosophy offered the rules of engagement, making it possible for Christian theology to become "a legitimate heir to classical philosophy."[59]

In addition to these intellectual challenges, early Christianity also had to organize itself institutionally. A hierarchy of offices and a network of institutions sprang up, including an elaborate system of caring for the poor and the sick, and church dignitaries from the entire Mediterranean region met at councils. The "masterpiece" that is the Christian Church arose within a few hundred years,[60] and it owed its success in no small part to the infrastructural advantages offered by the Roman Empire.

Obviously these intellectual and institutional achievements came at the cost of Christianity's intuitive and individual aspects. Charisma doesn't mesh well with bureaucracy. The anarchic and egalitarian hunter-gatherer Jesus of the gospels often fell by the wayside. The Bible itself offers a nice example of the inevitable consequence of the institutionalization process.

Whereas joint meals played an important role in the days of Jesus, things began to change once institutionalization got under way. The apostle Paul received a number of complaints from congregations in which things were starting to get out of hand at the Lord's supper. Some of the participants were already drunk before the rest of the guests had even arrived. This left Paul no other choice than to order, "And if any man hunger, let him eat at home; that ye come not together unto condemnation."[61]

Institutions exist to control social behavior.[62] Hierarchy soon took the place of Christianity's original enthusiasm and egalitarianism, and the "principle of authority"[63] was applied with particular stringency. Women bore the brunt of these changes. They had played an important role in early Christian congregations, particularly in the area of spiritual guidance, but increasing professionalization meant the church became more and more of a men's club.[64] The notion that sin entered into the world thanks to Eve's blunder in the Garden of Eden—an idea that Paul and Augustine espoused with particular passion—meant that the concept of women's subservience to men due to their own weakness and corruptibility became deeply rooted in Christian doctrine.[65]

Another major victim of the wave of institutionalization was the Bible itself. For over 1,000 years, the Good Book was a work in progress, and every era had added something of its own and adapted the notion of God to fit its needs. With the rise of Christianity, this adaptation process came to an abrupt and definitive end, and by the turn of the fifth century, the biblical canon was more or less firmly established.[66] Exegesis replaced the Bible's effervescent and wondrous storytelling. Ever since, theologians have filled entire libraries with their writings, but only their colleagues read them.

THE HIGH PRIESTS OF THIRD NATURE

Christianity's experience was not unique. In fact, it is the expected trajectory of cultural evolution, the process we normally refer to as "progress." According to Max Weber, "rationalization and intellectualization" mark this process. As a result, people no longer have the relevant knowledge of all aspects of life.[67] Karl Marx chose the term "alienation" to describe the result of this process, whereas Sigmund Freud referred to it as our discontent with civilization. People find themselves at the mercy of the experts, those high priests of our third nature.

The same thing happened to Christianity. Theology became the "property of specialist elites, whose bodies of knowledge were increasingly removed from the common knowledge of the society at large." The "'lay' member[s] of society" no longer knew exactly how their religious world was constructed; they only knew which experts could offer the appropriate explanation.[68]

In our view, the true importance lies in the fact that the institutionalization of Christianity after it became the Roman state religion in 380 CE greatly strengthened the already dominant sphere of intellectual religion.[69] Christianity developed into a two-tiered religion with a small and knowledgeable clerical elite set against a great mass of uninformed laypeople. Nowhere is this more apparent than in the stubborn and excessively long adherence to Latin as the liturgical language. According to religious scholars Justin Barrett and Pascal Boyer, this clerical autocracy played a major role in turning people toward "theological incorrectness." We encountered the same phenomenon in the story of the exodus with its murmuring people. Our first nature simply bristles at such abstract concepts.

That Christianity didn't collapse under all of this theological weight can be put down to the fact that, thanks to its intuitive core, it could draw on its resilient foundation in our first nature. We have already discussed its rich collection of supernatural agents. Christianity gained further strength from its strong community structures, its *asabiya*-enhancing factors such as care for the poor and the sick, its focus on God's love, and the opulence of its rituals and religious services. This rich intuition-pleasing mix was the source of its emotional power, which fulfilled the individual needs of laypeople and clerics alike.

And finally, we should also add at least one other reason for how this thoroughly explosive religious amalgam managed to avoid blowing itself to kingdom come: Christianity was absolutely ruthless when it came to persecuting potential threats to its much-touted religious unity. Indeed, the bonfires of the Inquisition would burn brightly for centuries.

But we do not intend to offer a history of the church. We only wish to stress how intellectual-institutional religion, whose genesis we witnessed in the Bible, was elevated to near perfection in the Christian Church. The most powerful expertocracy in all of history ascended to power—and the sciences owe much more to this development than most people today are aware. This is the topic of our last chapter.

19

THE BOOK OF NATURE

God's Second Bible

HAD IT NOT BEEN FOR CHRISTIANITY, EUROPE WOULD HAVE NEVER witnessed the blossoming of culture and science that played out over the course of the Middle Ages and the modern period. The greater the intellectual and institutional the challenges faced by the church, the more complex its educational institutions had to become and the greater the requirements for the education of clerics. At a time when hardly any other educational institutions existed, the church's educational system represented a first-class engine of cultural change. Without the monasteries, cathedral schools, schools run by religious orders, and, later, universities, Europe would have never witnessed the blossoming of literature, philosophy, and the arts—let alone science.

The sciences are all religion's children. Or to put it more precisely, they are the legitimate heirs of intellectual religion, something that requires a bit of explanation. After all, today people often view religion as the epitome of irrationality. The great majority believe that science had its beginnings and witnessed its first golden age in ancient Greece before disappearing in the Middle Ages due to Christianity's hostility toward science. According to this understanding of scientific history, not until the sixteenth and seventeenth centuries did the likes of Nicolaus Copernicus (1473–1543), Johannes Kepler (1571–1630), and Galileo Galilei (1564–1642) ring in the scientific revolution that would lay the foundation for the modern world. Science and technology would come to reign supreme and force religion into retreat. And once science could offer an explanation for everything, religion would disappear altogether.[1]

But all of this is mere myth. Over the course of the past few decades, modern historians of science have determined that it makes very little sense to speak of a perennial conflict between the sciences and religion or of religion beating a retreat in the face of triumphant science. First of all, claims John Henry, there "can be no doubt that religion and theology played a major part in the development of modern science."[2] And second, the secularization thesis—the idea that religion would completely disappear in modernity—has certainly proved incorrect, because religion, particularly when viewed on a global scale, is more alive than ever before.[3]

In just a few broad strokes, we would like to sketch out how, from the perspective of cultural evolution pursued throughout this book, the sciences evolved out of religion. This exercise can also demonstrate the usefulness of differentiating between intuitive-individual and intellectual-institutional religion and help to complete the cultural-protection hypothesis. Finally, it allows us to point to the fundamental role the Bible played in this process.

THE GENESIS OF SCIENCE

Science and religion are neither "natural kinds" nor "cultural constants." The two developed into distinct concepts only over the past three centuries or so; hence, we must be cautious about analyzing the more distant past from our modern perspective.[4] "The fundamental weakness of the conflict thesis is its tendency to portray science and religion as hypostasized forces, as entities in themselves," wrote British historian of science John Hedley Brooke as early as 1991. Rather, we should view them as "complex social activities" with "the same individuals often participating in both."[5]

Throughout this book, we have repeatedly pointed out that one cannot speak of clearly delineated spheres of science, religion, and politics in the biblical era. We suggested instead thinking in terms of a "primordial cultural soup" from which the individual spheres slowly evolved. Intellectual religion developed as a cultural product aimed at protecting societies from the unprecedented catastrophes, diseases, and social violence that arose in the wake of humanity's adopting agriculture. Its heroic attempts to explain the world at a time when *Homo sapiens* still lacked any and all scientific understanding enabled intellectual religion to help our species overcome the worst and so master the challenges of living in ever-larger societies. Our analysis of the Torah's 613 commandments and prohibitions demonstrates that these religious laws were equally protoscientific and protosociopolitical in nature. They strove to protect society from catastrophe, just as science-based technology does today.

Thus it is simply wrong to locate the birth of the sciences in ancient Greece alone. Science can trace its roots back to all of those places where intellectual religion used empirical observation of the world around us to develop rules aimed at protecting people from misfortune. The hypothesis that the gods were responsible for everything that happened on earth was arguably one of the most successful heuristic instruments in the history of the world. In this respect, the Torah deserves a place of honor in the history of the sciences.

What happened next reflects the momentum of cumulative cultural evolution. Religious institutions became more complex and differentiated according to the needs of society. "Intellectualization" and "rationalization" were the characteristic processes,[6] even in the Christian Church. We have already discussed how Christianity produced a theological elite whose members enjoyed a comprehensive Greek education. As the church became ever larger and more integrated into the state, it produced more and more experts to address questions concerning law, medicine, astronomy, architecture, and mathematics. All of these experts had trained as clerics, and all of them fulfilled their duties for the glory of God. The church's relationship to the Greek—and thus pagan—scientific tradition wasn't always easy, but was certainly very fruitful. The fault for the demise of the sciences during the so-called Dark Ages rests not with Christianity but rather with the fall of the Roman Empire and the confusion brought on by the great migrations that followed. On the contrary, Christian monasteries often served as cultural life rafts, helping to save the knowledge of antiquity in those difficult times.[7]

THE TWO BOOKS OF GOD

Our book is about the Bible. And although the canonization of the Old and New Testaments in around 400 CE meant that new texts were no longer added and old texts were no longer edited, our job is not yet done because work began on a new bible soon afterward. From our perspective, it is anything but a coincidence that we find in Augustine of Hippo's writing the fascinating suggestion that God had written not one book but two: in addition to the Holy Scripture, he also created what Augustine called "the Book of Nature."[8] Augustine believed that we could read God's message not only in the Bible but also in the abundance and diversity of the real world. The idea remained quite popular until well into the modern era, and even today many believers find God in nature rather than the pages of the Bible.

Make no mistake: this was a momentous act. We have already explored the Torah's basis in an empirical examination of the world and how it sought to reconstruct God's order. But because it aimed to uncover the actions that pleased God or aroused his wrath, the Bible never really extolled nature as a source of knowledge worthy of systematic investigation. But now things began to change, initiating a new phase in the cumulative cultural evolution of Christianity.

Plato, Aristotle, and the Stoics all sought to "explain existence in its entirety by means of reason."[9] If God is indeed Logos, the principle that generated the ordered universe, then the world must reflect this reason in its material manifestations. This was Christianity's fundamental premise in setting out to read the Book of Nature and uncover the order of God's creation down to the last detail. This way, scholars could integrate Greek natural philosophy into their own belief system and adapt to their own purposes a cultural-protection system that was as elaborate as it was seminal.

In chapter 13 we saw how science—or, more accurately, natural philosophy—blossomed in ancient Greece. The Olympic gods had proved unreliable companions, and as a seafaring people, the Greeks had a pressing need to develop effective methods of combating storms, floods, and other natural forces. Astronomical, mathematical, medical, and natural scientific study accumulated and soon allowed for evidence-based means of controlling nature. Since there was less and less need for the gods to serve as the possible causes of social and natural adversity, they increasingly withdrew and became the guardians of world order.

The idea that there are two books of God—the Bible and the Book of Nature—owes its success to the fact that it led to two distinct groups of experts. On the side of Holy Scripture, there were the theologians, for whom a binding canon was of the utmost importance. Its existence allowed them to draw on the sacrosanct authority of the Bible in any dispute and thus ensured their hegemony at a time when Christianity threatened to splinter into countless sects. From this moment on, the Bible revealed God's word and was therefore binding on each and all. But because this God also was Logos, the Bible thus had to represent a perfect step taken by a perfect God. Thus generations of theologians were forever busy trying to explain all the gaps, inconsistencies, and mistakes in the Bible—in addition, of course, to occupying themselves with questions of faith and sacred rites. At the same time they sought to master the challenges posed by the knowledge they gleaned from nature. After all, the Good Book had to make good sense and be free of inconsistency.

On the other side, that of the Book of Nature, we have the experts who sought to uncover nature's secrets. They had taken on the task of protecting society from all forms of catastrophe through examination of the natural world, including medicine, astronomy, natural philosophy, and, when it came to the social world, the nature of law. They argued that the better they were able to reconstruct the divine order that held together nature and the universe, the greater the level of peace the people would enjoy.

Behind the notion of the two books of God, we therefore find a process in which the differentiation of expert knowledge led to a certain division of labor. Intellectual religion had generated a number of tools to master the challenges of an increasingly complex world—tools that were effective because they could go beyond revealed biblical truths. This source of power allowed Christianity to master the challenges of an ever more complex world.

"THINKING GOD'S THOUGHTS AFTER HIM"

Although the myth of the eternal conflict between religion and science might suggest otherwise, even the protagonists of the scientific revolution of the sixteenth and seventeenth centuries were themselves religious. Science historian John Henry stresses the "importance of religious concerns to the leading thinkers in providing a general motivation for, and in shaping the precise details of, their natural philosophies"; he includes among these contemporary thinkers Johannes Kepler, Francis Bacon, René Descartes, Robert Boyle, Gottfried Wilhelm Leibniz, and

Isaac Newton, to mention only a few.[10] Newton, for instance, believed himself "one of God's chosen and viewed space and time as God's emanations."[11] According to Henry, "All in all, there can be little doubt of the importance of religious devotion in motivating and shaping early modern science."[12]

Nevertheless, problems soon arose, for God's two books did not harmonize all that well with one another. Johannes Kepler, who discovered how the planets traveled in set orbits around the sun, described himself as "God's priest of the Book of Nature" and believed that he, "by discovering the pattern which God had imposed on the cosmos, was 'thinking God's thoughts after him.'" Kepler had in fact originally wanted to become a theologian (positions as court astronomer being rather rare), but he had little interest in teaching what the Holy Scripture had to say about nature: "The Bible is not a textbook on optics and astronomy."[13]

Here we find we have to stand up for the Bible. The Good Book cannot help that its scientific knowledge is stuck at the level of the ancient world. The Bible had always proven itself a true living medium, for it had always adapted itself to meet the challenges of a given era. The educated elite had continuously worked on it from about 800 BCE to 200 CE. As new stories were added and older ones were reworked, the Bible was a constantly updated product of cumulative cultural evolution. Nevertheless, this important fact is insufficiently embedded in our collective consciousness.

With its canonization in around 400 CE, the Bible became frozen in time and thus could no longer be adapted to fit changing insights. Whereas it managed to remain relevant over the centuries when it came to problems relating to our first nature, the model of the physical world it presented grew increasingly outdated. And while researchers continued to improve their reading of the Book of Nature, Genesis's story of creation still reflected the worldview of antiquity. Inevitably, the rift grew wider.

With fateful consequences. The Bible experts represented the church, which in the meantime had grown to become the most formidable institution in the West. They stood on the side of power. This meant they always had to maintain their authority vigilantly and viewed any suggestion of the Bible as flawed as a threat to their authority. This stance would intensify with the arrival of the Reformation in the sixteenth century, when the dispute over true belief peaked and, in the end, led to the breakup of the church. The Inquisition kept a jealous watch over the correct interpretation of the Good Book.

GALILEO, TOO, READ THE BOOK OF NATURE

This brings us to Galileo Galilei. As everyone knows, the Inquisition put him on trial for his support of Nicolaus Copernicus's heliocentric theories, according to which the universe and all the planets of the solar system revolve around the sun and not the earth. Most historians of science now agree that this case was not the great showdown between science and religion that scholars have made it out to be for so many years.[14] But the Galileo trial does show how the unity of the two books of God had begun to crumble, which, ironically, both Galileo and his opponents in the church were actually keen to prevent. As a result, contrary to popular myth, Galileo's trial actually represents the last major attempt to save the time-honored union of science and the Bible.

The name Galileo Galilei remains forever closely connected to the term "Book of Nature." His formulation of it in his *Il Saggiatore* (*The Assayer*) has become legendary: "Philosophy is written in this great book which is continually open before our eyes—I mean the universe—but before we can understand it we need to learn the language and recognize the characters in which it is written. It is written in the language of mathematics and its characters are triangles, circles, and other geometrical figures, without which it is humanly impossible to understand a word of what it says. Without these, it is just wandering aimlessly in a baffling maze."[15]

This sounds like pure science, but Galileo believed in the truth not only of the Book of Nature but also of the Bible. He firmly adhered to Augustine's two-book theory. In his famous letter to Benedetto Castelli, Galileo wrote, "Holy Scripture and nature proceed alike from the divine Word—Scripture as dictated by the Holy Spirit, and nature as the faithful executioner of God's commands."[16] For the believer Galileo, God had undeniably written both of these books, and neither could be flawed. The latter notion would prove increasingly problematic.

In his letter to Castelli, Galileo refers to the Old Testament book of Joshua. At Joshua's request, God stopped the sun so that the Israelites would have enough time to avenge themselves on their enemies. But how was that supposed to work? If Copernicus's theory was correct, the sun remained in a fixed position in space, and the earth and all the people on it revolved around the sun. So how can we read that God stopped the sun in its tracks? He certainly wouldn't have misled his own people when it came to his own universal truth.[17]

Fascinatingly, both the Catholic Church and the astronomer recognized the conflict between God's two books. Neither side could live with this discrepancy, for both firmly believed that God had produced the Bible and nature, and, as Galileo wrote in his letter to Castelli, "It is impossible for two truths to contradict each other."[18] So the two sides of the conflict agreed on the fundamental assumptions. But they only differed in their solutions—even though these, too, were strikingly similar. Each side sought to manipulate the other's book. The church insisted that the biblical worldview with the earth at the center of the universe was correct and argued that the calculations and telescopes must have been mistaken in their reading of the Book of Nature. Galileo accepted that the Bible did represent the truth, but he refused to acknowledge its right to serve as the ultimate authority when it came to matters of nature, because it often used metaphors and its interpreters could therefore make mistakes. As Galileo wrote, "It is agreed that Scripture, in order to be understood by the multitude, says many things which are apparently and in the literal sense of the words at variance with absolute truth."[19] In fact, Galileo himself tried to develop a correct exegesis that could harmonize the story of Joshua with the heliocentric worldview gleaned from the Book of Nature.

Where exactly did the conflict lie, in the sphere of science or that of religion? For Galileo, there was "no conflict between religion and science," writes German science historian Klaus Fischer. For the astronomer it was a "conflict between good and bad science," and he criticized the latter as relying on "a few of the Bible's popular formulations of astronomical events" to support its own outdated worldview.[20] For the Inquisition, however, the real conflict was between good and bad biblical exegesis. The church's problem, as encapsulated by John Henry, was "Galileo's insistence upon entering into public discussion of biblical interpretation (to show how Copernicanism could be made compatible with various biblical statements), at a time when the counter-reforming Catholic [C]hurch was trying to restrict free interpretation of Scripture."[21] Galileo's conviction and prison sentence (soon commuted to house arrest) were, in essence, the collateral damage of the religious wars raging at the time. They did not represent an attempt to make an example of science.

Both the church's and Galileo's interpretations of the conflict were correct in a way, for science and religion had not yet gone their separate ways, in line with our thesis that science arose as the legitimate heir of intellectual religion. At the time of the astronomer's trial, both sides still believed that both the Bible and the Book of Nature represented the truth and that the experts on the opposing side must be reading them incorrectly—it couldn't possibly be God's mistake, after all! From that point on, however, this conciliatory position became increasingly untenable. Thus Galileo's case marked the decisive split. Science thereafter followed its own path in seeking coherence and left the Bible behind.

THE COMMENCEMENT OF HOSTILITIES

The church started out by compromising and tempering its positions when it came to astronomy and other issues,[22] but it soon found itself increasingly on the defensive. The sciences began demonstrating amazing insights into nature's forces and how they relate to one another. At the same time, the new discipline of historical Bible research began to fan the flames of an unholy suspicion: the word of God was in actuality the work of man. And since only one of the two books of God could be right, scientists and philosophers began to free themselves from the Bible. Still, they remained thoroughly religious, just not traditionally so. During the Enlightenment a number of new concepts began to develop, such as deism, pantheism, and physicotheology, all based on nonbiblical attempts to harmonize God and the Book of Nature.

Not until the nineteenth century did the modern notion of the sciences emerge. Indeed, many historians believe that it only makes sense to speak of science from this point onward, whereas earlier scientific endeavors could best be described as "natural philosophy" or "natural history."[23] At any rate, cultural evolution had made coexistence between the natural sciences and intellectual-institutional religion impossible. It was the time of the ultimate divorce, and henceforth the two sides would have to clearly define their positions. They accomplished this by defining themselves as nonoverlapping. The sciences began to view themselves as professionalized disciplines emancipated from any religious influence, which (at least from their point of view) implied that religion was incompatible with a scientific approach.[24]

With the arrival of the astounding technical discoveries of the nineteenth century, an unshakable belief in the power of progress became widespread. In one of his lectures, physiologist Emil du Bois-Raymond (1818–1896), who was also the president of the University of Berlin, excitedly asked, "What now can check modern civilization? What lightnings can ever shatter this tower of Babel? It makes one dizzy to think of what mankind is destined to be in a hundred, a thousand, ten thousand, a hundred thousand or more years hence. What is there to which it may not attain?"[25] The natural sciences began to celebrate themselves as the best cultural-protection system the world had ever seen!

The new message of salvation was as follows: just like Prometheus had brought fire to humanity, science would lead humanity into a new era in which all problems would be solved. Authors such as Andrew Dickson White (1832–1918) and John William Draper (1811–1882) offered fitting narratives to accompany this message. In works such as *A History of the Warfare of Science with Theology in Christendom* and *History of the Conflict Between Religion and Science*, they depicted science's victorious crusade against superstition. Thanks to the heroic sacrifice of great thinkers science's ultimate victory was finally within reach. Giordano Bruno (1548–1600) and Galileo Galilei were transformed into martyrs of the new religion of science, and the notion was born of a "great war" between science and religion that had been raging for centuries.[26] Of course science was also struggling to wrest power, money, and the social prerogative of interpretation—the "baton of authority"— from the church, which was extremely reluctant to relinquish them.[27]

A powerful narrative emerged that has continued to have an impact to this day. In the war of the sciences versus religion, scientists began to win battle after battle; as a result God only managed to survive as the god of the gaps, finding refuge in the remaining unexplored dark corners that the light of science had yet to illuminate: before the big bang and after death. The inescapable conclusion is that the current conflict between science and religion is not a flare-up of an eternal conflict. It is a legacy of the nineteenth century.

"STEALING GOD'S THUNDER"

The greatest blow to the church came from a former theology student who himself once "did not . . . in the least doubt the strict and literal truth of every word in the Bible."[28] Charles Darwin's (1809–1882) theory of evolution was the final straw that made it impossible for anyone to look to the Bible as a dependable source of scientific knowledge. Not only had responsibility for the wonderful variety of life on earth been torn from God's own hands, but now humankind had been degraded from the "crown of creation" to a mere "ape."

It seemed it was only a matter of time before science could provide an answer for everything. Then, to paraphrase Friedrich Nietzsche, God would finally be dead, and all religion would disappear. This was the birth of the theory of secularization, according to which a modern, enlightened humanity would leave religion behind as a backward worldview. But nearly 150 years later, despite breathtaking scientific progress, religion is undoubtedly still alive and, in fact, doing quite well. How can this be?

All theories predicting religion's demise grow out of a simplified understanding of religion and science. On the one hand they fail to recognize that scientific progress only poses a threat to intellectual-institutional religion, this cultural system for explaining the world that arose at a particular point in history. The religious core, however, intuitive-individual religion, has nothing to fear. It is a part of the human condition and is remarkably impervious to logic.

On the other hand such theories also find themselves wedded to the traditional idea that the scientific worldview had its origins in ancient Greece—that Greek philosophers at some point stopped looking to the gods as a source of explanation and instead began to seek out natural causes for the events they witnessed, giving birth to science as we know it today. As a result, to the traditional view, science has been a kind of antireligion from its very beginnings. As we have just seen, it would not actually become so until the nineteenth century.

We think our anthropological approach can prove its mettle here. As explained in previous chapters, thanks to the unique process of cumulative cultural evolution, *Homo sapiens* was uniquely able to make radical changes to its way of life in order to come to terms with the earth's ecological changes. We have called the cultural-protection system with which humans mastered these new challenges—even though they had no idea what was really behind all of the epidemics and natural and social catastrophes that plagued them—intellectual-institutional religion. This system attributed the misfortunes that befell them to divine actions.

According to our cultural-protection hypothesis, this system functioned to combat the challenges of ever more complex societies and avoid calamity. The constant pressure of new catastrophes forced intellectual-institutional religion to set great store by empirical evidence, something that eventually led to a process of rationalization, which gradually but inevitably gave rise to science. And in the end, science found it could manage without God as a heuristic aid and protect humanity all on its own.

Science, too, is nothing more than a cultural-protection system. The study of the world is not an end in itself, for it fulfills the age-old task of identifying the forces behind disasters and illnesses, the laws they follow, and how we can protect ourselves from or even prevent these problems, so people can live in peace at last. And, indeed, science has gotten increasingly better at the latter. Its accomplishments range from the lightning rod to the tsunami early-warning system, from penicillin to MRI scans, from inorganic fertilizer to solar-powered water treatment plants. The title of Philip Dray's book on Benjamin Franklin's discovery of the lightning rod, *Stealing God's Thunder*, reflects just how threatening this change must have been to the church.[29] The sciences robbed the gods of their oldest and most important credibility-enhancing displays: catastrophes.

The cultural-protection system provided by intellectual religion gave way to another, more efficient system provided by science. That intellectual-institutional religion nonetheless continues to persevere—even though its original task of explaining the world by means of the Bible has been made obsolete—is due in large part to the fact that it has become solidly entrenched in the second nature of most Western societies and continues to have powerful institutions at its disposal. And, of course, its dead-set determination to defend itself against any attempt at refutation is a part of its cultural DNA. Moreover, intellectual religion never exists in a pure form. It is thoroughly fused with intuitive religiosity—the powerhouse of religious belief that is deeply rooted in our biology and highly oblivious to all forms of scientific progress.

BELIEF IN THE FACE OF ALL REASON

"When asked what they would do if scientists were to disprove a particular religious belief, nearly two-thirds (64%) of people say they would continue to hold to what their religion teaches rather than accept the contrary scientific finding, according to the results of an October 2006 *Time* magazine poll."[30] The numbers are probably even higher, as some of those questioned might have feared appearing narrow-minded if they admitted to remaining loyal to their religion despite all common sense.

The numbers are fascinating for two reasons. For starters they show the high value people place on loyalty and dependability—our first nature, once again. We don't give up on our own group's values just because someone claims to have a better argument. But—and we find this truly interesting—these numbers also show that cognitive knowledge and intuitive beliefs originate in two different mental domains. Even if people know for sure that there isn't the slightest proof of the existence of supernatural beings, our third nature still cannot silence the instincts of our first nature. Something inside us continues to believe that spirits and other invisible agents pervade the world. Our intuitive certainties are immune to the power of rational arguments.

We can use Max Weber's notion of a "religiously 'musical' person" to illustrate how our three natures interact when it comes to religion.[31] Indeed, musicality and religiosity have striking similarities. Both have a genetic foundation in human psychology; the strength of their expression, however, depends not only on our genetic makeup but, to an even greater extent, on our socialization in the culture we grow up in. Someone born into a musical environment who learns to sing and play the piano from an early age will probably make for a passable musician, even if, genetically speaking, she isn't all that musically endowed. And the more in tune the musical culture is with our first nature, the deeper music becomes anchored and the more likely it is to become a solid part of our second nature. If our third nature were to come along and prove music to be nonsense, we might agree with the argument in principle but would nonetheless still whistle a tune or snap our fingers to the beat now and then. And people would continue to do so even if they never came in contact with music again. Hundreds of generations would have to pass before musicality disappeared entirely from our genetic makeup.

The same argument holds for religion. Christianity continues to attract over 2 billion followers today. We have shown just how wide its range truly is. In addition to its intellectual and theological facets, it offers everything our first nature requires—in the form of Jesus, Mary, the saints, and the demons—to make it resonate in our souls. Powerful church-based institutions also permeate our society, and this helps keep intellectual religion alive. And with all this in mind, we now take a brief look at the current state of religion.

THE BIRTH OF SPIRITUALITY

We've examined why religion hasn't disappeared in the face of scientific progress. The contrary is true, in fact: thanks to the bustling development of religion all around the world, sociologists have even come to believe that we can finally lay the secularization theory to rest.[32] But it's still worth taking a closer look at the situation today, because it actually confirms our argument. The religions booming today are those with a significant intuitive component, whereas those dominated by intellectual-institutional aspects see people leaving in droves in search of more spiritual alternatives.

Around the world, the charismatic and Pentecostal movements display an enormous ability to mobilize believers. These movements focus on "baptism in the Holy Spirit, auditions, speaking in tongues, and enthusiastic spirituality"—elements that are a part of intuitive religion's repertoire. Religion is quite strong in places where people are open to a number of heavenly and demonic actors, where they have no reservations about opulent rituals and magical practices, where women can play a major role, and, above all, where community and the collective emotional experience are of central importance. Such churches often impose a rigid morality on and demand great sacrifices of their members. They also tend to separate themselves strictly from the world at large.[33]

The situation is rather different in Europe—where secularization theory became so popular in the first place. There one can observe a dramatic decrease in church attendance due to the high levels of religious intellectualization and institutionalism, a characteristic especially typical of liberal Protestantism. European Christianity also never managed to achieve the transition from the monoculture of territorially and bureaucratically organized national churches to a system of religious pluralism.[34] Such churches have precious little to offer people's first nature and simultaneously have a hard time competing against modern knowledge systems. They are simply no longer needed as a means of explaining the world, and as a substitute for philosophy, they speak to an ever-smaller segment of the population. In many places they are at risk of becoming museum-like institutions of the inhabitants' second nature, visited only on special occasions such as Christmas, weddings, and funerals.

But even in places where people are leaving the traditional churches, defectors are not necessarily becoming less religious—another impressive piece of evidence for the strength of first-nature religion. As early as the 1960s, sociologist of religion Thomas Luckmann cast doubt on the power of the secularization thesis when he noted how people were seeking out alternative means of experiencing religion away from established institutions. What began as the New Age movement has in the meantime metamorphosed into a colorful hodgepodge of esoteric offerings ranging from a belief in angels to ascetic ecstasy, from astral projection to feng shui. All possible longings for an authentic experience of supernatural powers are sure to be satisfied. Whereas Luckmann chose the term "invisible religion" to describe such alternatives, today they have become "highly visible forms" and even reached the status of "popular religion."[35]

In the United States, too, where the old mainline Protestant churches are losing members, the people turning their backs on organized religion aren't embracing secularism. The trend is away from "denomination" toward "individual mysticism," explains American sociologist of religion José Casanova, which has led to a new category: "spiritual, but not religious."[36] Ever attentive, scholars have already begun correcting their own terminology: the standard reference book on the topic is now titled *Handbook of the Psychology of Religion and Spirituality*.[37]

As science and technology have continued to weaken the intellectual side of Christianity, our first nature has resorted to spirituality to recoup the lost territory. Nowhere is this transformation as clear as when it comes to the afterworld, which many see as God's last bastion. And here God finds himself in dire straits indeed.

BRINGING DEATH BACK TO LIFE

In chapter 16 we saw how monotheism banned the intuitive certainty of our belief in spirits and ancestors and the attendant mourning rituals by declaring them primitive superstition. In doing so, it created a vacuum when it came to death. Whereas Judaism refrained from regulating the afterlife and allowed various coexisting conceptions, Christianity decided to fill this vacuum and to settle on a single option: the notion of the judgment of the dead. This decision may not have been fortuitous. As historian Paul Veyne notes, "Christianity, for its part, offered a test more likely to scare off a newcomer than convert him: would the outcome be salvation or the eternal torments of Hell?"[38]

Even dedicated Christians found this notion counterintuitive: Charles Darwin initially believed in the Bible's literal truth and even entertained becoming a man of the cloth. However, Christianity's postulates about the afterlife, not his revolutionary notion of evolution, robbed him of his faith and turned him into an agnostic. As he wrote in his autobiography, "I can indeed hardly see how anyone ought to wish Christianity to be true; for if so the plain language of the text seems to show that men who do not believe, and this would include my father, brother and almost all my best friends, will be everlastingly punished. And this is a damnable doctrine."[39]

The absolute nature of such a punishment clashes with our reciprocity-based feelings of justice and encourages our first nature to rebel. Thus, beginning with the Enlightenment, Protestant theologians in particular started arguing that Jesus himself never really preached eternal torment in hell.[40] Indeed, as noted by sociologist Hubert Knoblauch, the European churches began to avoid the topic altogether: "The churches under the influence of the Enlightenment increasingly withdrew from interpretations of death and widespread ritualization." During the second half of the twentieth century, the catchphrase "tabooization of death" would increasingly rear its head.[41]

Official religion's neglect of death is problematic for those growing up in a culture that propagates a belief in heaven and hell. If doubt is thrown upon this belief, the question arises as to how we can live a good life when our actions will merit neither reward nor punishment. Doesn't that make everything random, arbitrary, or—as sociologists would have it—contingent? And above all, what is the point of death? And what happens after we die?

This development is quite revealing: people are beginning to question the traditional second-nature solution offered by Christianity. But they still feel they must reject the ancient first-nature options as superstitious. So, here we have a nice challenge for our third nature to come up with new answers. Unfortunately, rational approaches cannot produce any authoritative statements about life after death. In fact, from a scientific perspective, death, if anything, is the end. As a consequence, people have what is called the contingency experience: death appears to be random and meaningless. It turns into nothingness, something we cannot grasp. That scares us.

Ironically, in this context we often encounter the claim that religion can help us cope with this contingency (or, to use the impressive and ponderous German term, *Kontingenzbewältigung*). Religion, after all, provides answers, offers consolation, and has tried-and-true rituals at the ready—all of which science-based thinking patently fails to do. Many therefore argue that the world cannot live without religion, especially when it comes to death.[42]

It's a common claim, but that does not make it true. The real reason for the crisis is that the very solutions traditionally proffered by intellectual religion have lost their credibility. Of course, the churches may respond by reformulating their notions of the hereafter. In fact, such a movement is afoot: as noted earlier, many German clerics have done away with hell altogether. But for an institution claiming to peddle timeless truths, this is a risky strategy. Philosopher Kurt Flasch may well be right to complain that the "churches have terrified people for over a thousand years with images of hell, and now they reassure us that the Bible never stipulates it."[43]

The contingency experience is in fact nothing but a cultural artifact and far from a cultural universal. It results from the failure of our second and third natures to produce a convincing answer. This fiasco provided our first nature with the opportunity to reinstate our ancient intuitions. These had never really disappeared, of course; they were merely branded as theologically incorrect.

Sociologists are indeed noticing a trend in Europe, according to which "outside the churches" a new culture of death is establishing itself with "its own rituals, forms of experience and interpretations."[44] Belief in reincarnation and transmigration is booming. Books about near-death experiences make it onto the best-seller lists, increasing numbers of people wish to be buried in natural settings instead of cemeteries, and guardian angels have taken over the position vacated by our ancestors. In short, our animistic belief in the immortality of the soul is making a full comeback.

But even people for whom this is all too esoteric and who never believed in heaven or hell cannot so easily escape our old psychology. "No sane persons literally deny their mortality, of course," argues primatologist Frans de Waal, "but many of us act as if we'll live forever."[45] And atheist psychologist Jesse Bering—who was sure his deceased mother had contacted him from the afterworld—admits, "It's this cognitive hiccup of gross irrationality that we have unmistakably inherited" from our ancestors.[46] This irrational delusion may even be adaptive: it brings us peace and prevents us from spending all our time worrying about our mortality. It allows us to secretly hope, in some unarticulated way, that Shakespeare's Hamlet was right when he said, "There are more things in heaven and earth, Horatio, Than are dreamt of in your philosophy."

Death only rears its head as a serious problem when a crisis forces our third nature to focus on it and we try hard to imagine what it will be like not to exist anymore. That's when many of us panic. But third nature cannot offer any answers—apart perhaps from the one supplied by Richard Dawkins in *The God Delusion.* He begins by citing Mark Twain: "I do not fear death. I had been dead for billions and billions of years before I was born, and had not suffered the slightest inconvenience from it." Dawkins later adds, "I shall be just as I was in the time of William the Conqueror or the dinosaurs or the trilobites. There is nothing to fear in that."[47]

AND WHERE DO WE GO FROM HERE?

Let us return to life before death, the here and now. We have seen just what an amazing hybrid religion Christianity was for a long part of its history. It both met the needs of individual people and offered cultural protection and cohesion to society at large. But the declaration that the Bible was finished meant that a gulf inevitably developed between the world and the Holy Scripture. Those scholars who dedicated themselves to the study of the Book of Nature found it increasingly easier to get by without God, and by the nineteenth century at the latest, the fissure had become so great that it shattered Christianity's hybrid construct. Science was able to free itself entirely from God, and work on the cultural-protection system was ceded to the scientists and secular institutions based on social science.

In our view, we are still in a process of transformation that has continued ever since. So where do we go from here? We are certainly not futurologists, and we can only forecast future tendencies from the perspective of cultural evolution. First of all, when religion itself created the improved tools that could help reveal the world's workings, it made its own model for explaining the world obsolete. One might think the potential for conflict with the sciences would thus no longer be an issue—in theory, at any rate—and at least in the case of Europe and Christianity, this has actually largely happened.

Second, we have also seen how religion served to regulate morality in a bid to win the gods' favor and protect society from catastrophe. Although our morality is much older than religion, religion took over the moral reins at a time when the egos of the smaller and larger despots threatened to destroy society. In doing so, biblical religion created the foundation for human rights and democracy. Where secular institutions became firmly established, they no longer need religion to legitimize these values. So, although its moralizing function was crucial at one time in history, religion has accomplished its moral mission. This does not mean, of course, that religious institutions may no longer act as moral watchdogs. But it does mean they do not own any privileged claims to normativity that the rest of society is not allowed to use.

We can therefore also discern a striking third development. Now that religion's cultural-protection function is no longer needed, thanks to the more suitable institutions its intellectual side helped create, intuitive religion now has free rein. Theological doctrine has a much smaller role to play, and as religion's intellectual component weakens, so does its compulsive search for rational coherence. The shared experience and the spiritual high are far more important now. This development allows for diversification—a bit like in music, which has a wide range of styles, from pop to classical. Of course there will always be believers who appreciate religion for its venerable theological system of teachings or who hope to harness these to understand the subtle interplay of nature's forces. In sum, religious belief is clearly becoming more pluralistic, also in Europe.

Even now it is already possible to see how the religious congregations with the greatest influx of newcomers are those whose spiritual offerings appeal to our first nature. Here people find happiness, feel protected by a higher power, and acquire a sense of security. They feel like part of a real community, just like our hunter-gatherer ancestors once did within their vibrant groups. Viewed in this light, the religion of the future might develop into a sanctuary for our first nature, an anti-mismatch space, a temporary paradise where people can recuperate from the demands of our third nature. It could become a place where they find a moment of respite from their discontent with civilization.

And why not? We have an extremely well-developed ability to create a second reality in our minds, as primatologist Frans de Waal explains. It's easy for us to block out one reality for the sake of another. This is why "we fall for romances, rivalries, and deaths in movies while at the same time well aware that the actors are just acting." Jonathan Gottschall has pointed out that stories are our most powerful "virtual reality" technology, one we can use to simulate alternative future scenarios. This is why we love going to the theater or the cinema, reading novels, and immersing ourselves in computer games. Religion, too, ranks among these "dual realities," according to de Waal. For this reason he compares those "neo-atheists" who only care about the facts of empirical reality to "people standing outside a movie theater telling us that Leonardo DiCaprio didn't really go down with the *Titanic.* How shocking! Most of us are perfectly comfortable with the duality."[48]

This brings us to the fourth trend. The traditional churches do not exactly approve of these developments. They regularly criticize the "spiritualization" and "individualization" of religion and argue that religion should not degenerate into some facultative lifestyle choice—a kind of wellness offering. True religion involves the whole human being: a commitment to God encompasses all aspects of life. "Religion light" simply won't do.

This attitude may at least partly reflect our age-old psychology. We have to keep the group together! Our first-nature fear of free riders pushes us to raise the costs of membership. We cannot allow people to come and go as if religion were just another yoga course. This explains why we are now witnessing a boom in uncompromising churches as the liberal ones continue to decline. The churches that demand a lot from their members in terms of commitment and money are seeing the greatest influx.[49] They insist on the literal truth of the Holy Scripture. If, in this day and age, someone claims to believe that the world was created on Saturday, October 22, 4004 BCE, at exactly 6 p.m. (as extrapolated from the Bible by Irish archbishop James Ussher in 1624),[50] he is proving one thing in particular: You can trust me. I am absolutely loyal to the group! This, too, is a part of our hunter-gatherer soul.

But a danger lurks here—this is the fifth trend—and not just because the world has long since changed and we no longer roam around in small bands. Wherever we encounter vigorous in-groups, the old friend-versus-enemy psychology is never far away. After all, it is an intrinsic part of our first nature. As a result, outsiders can quickly find themselves demonized. The members of such in-groups are susceptible to conspiracy theories, for if one is prepared to believe in a supernatural actor (and our first nature believes there are a lot of them), then there is a danger of seeing them at work everywhere—particularly in the big bad world outside the group.

Religion can easily become a double-edged sword—and this is a main reason it has been so successful. Inwardly it welds the group together, and externally it shuts the group off from the outside world. Throughout its history, Christianity has thoroughly understood this point. We can read it in the Bible. Jesus the friend, who stands graciously beside us in the face of hardship, and Jesus the apocalyptic hero, who battles against the forces of Satan, are one and the same. In the modern world, however, only one of these two is acceptable.

We are confident that religion will continue to accompany *Homo sapiens* and that the intuitive part of Christianity will become stronger. But we feel it should not insist that it has a monopoly on prescribing our behavior. That was religion's charge for a critical phase of our species' history, and overall it acquitted itself of the task rather well, especially in the absence of viable alternatives. But if there is one lesson to be learned from cultural evolution, it is that this was not its mission for all eternity. For the Bible, this is good news. It no longer has to be the perfect scripture of a perfect God.

EPILOGUE

We have finally reached our destination. Our journey through the pages of the Bible led us from Genesis to the apocalypse. We were on the scene when Adam and Eve were cast out of Paradise, and we witnessed how humanity tried to cope with its post-Paradise stress disorder. Recovery was only partial at best. Happy moments were few and far between, and at its end, the Bible had the world set to end in a fiery inferno. According to John's book of Revelation, it was high time for a new earth, a new heaven, and a heavenly Jerusalem.

Eden and Armageddon are the alpha and omega of the Bible, and between the two, people had to cope in a world for which our species was not made. Life in a world for which we were not made: this is the Bible's true subject matter. And as we have shown, the Bible is right. Ever since our ancestors adopted a sedentary way of life, they've had to find their way in an environment for which they did not have enough biological adaptations. It's there for all to read in the Book of Books: people did their best to survive catastrophes such as droughts, epidemics, oppression, and wars. Brothers became enemies, women lost their freedom, and big shots let their egoism run wild and enriched themselves beyond all measure.

That which theologians have traditionally interpreted as proof of mankind's sinfulness (if only Adam and Eve had obeyed God's will, we would have been spared this terrible fate!) really reflects its struggle with the most radical shift in behavior that any of the planet's species has ever endured. But people overcame these challenges with flying colors, for they had wagered everything when they played their ultimate trump card: the ability to engage in cumulative cultural evolution.

There was a price to pay for their efforts, however: mismatch. We have used this term to describe the wide divergence between humanity's innate psychological makeup and the new ways of life it encountered. We continue to pay this price for cultural progress even to this day. We don't feel really at home in an increasingly incomprehensible and anonymous world full of unknown challenges. We have described this as our discontent with civilization or, more appropriately in this context, a sense of longing for Paradise.

Viewed in this light, the Bible acquires new significance. Far too few are aware that it is the product of almost 1,000 years of work—and 1,000 years of experience made it into the true Book of Books. Its stories tell of how people tried to come to terms with all of the calamities they had to face, how they experimented with diverse strategies, and, in doing so, how they produced entirely new problems. We can thus see the Bible as humanity's own diary in which it put to paper all of its attempts to master life in a new world. The Bible has a lot to teach us.

THE DIARY'S SECRETS

Reading diaries not only satisfies our fascination with times of old but can also reveal the causes of particular developments and the typical behavior patterns that characterize us to this day. The same is true of the diary known as the Bible. Read properly, it can tell us a great deal about the interplay of human nature and culture. It can also offer us fundamental insights into the workings of cultural evolution and show us the origins of many of the difficulties we all continue to struggle with in modern life.

Our primary finding is as follows: We must no longer think of religion as an entity that has always existed. Instead we must differentiate between two components. The first is intuitive-individual religion. It has no doctrines of its own, for its rites and practices are age-old components of our second nature. It rests on our religiosity—the intrinsic, genetically anchored part of the human condition that governs the way we perceive and interpret the world. Just how religiously musical (to use Max Weber's metaphor) we are in this regard depends on the interaction between our individual dispositions and our socialization, and therefore the culture into which we are born.

We have christened the other component—the integral part of our culture intended to help us come to terms with life in a new world—intellectual-institutional religion. This type of religion cannot get by without its own retinue of experts. It is the domain of the cultural-protection hypothesis that we developed based on our anthropological reading of the Bible. In the end it describes nothing more than humanity's attempt to protect itself against the new status quo by means of culture.

At the very beginning we find a primordial cultural soup in which functional differentiations appear in a rudimentary form at best. A key issue in the understanding of intellectual religion is that science and religion were once one and the same—if one could even say they existed at all. Today, we describe this type of culture as religious in nature simply because its practitioners believed that supernatural agents caused everything that happened around them. This worldview was the gift of their first and second natures, and at that time their third nature had not yet developed alternative paradigms to explain the way the world works. Cultural evolution—the bubbling of the primordial soup—would ultimately lead to specialization and differentiation, and so to science. It was a gradual process—indeed, we saw that until well into the nineteenth century, most scientists were convinced that their research would enable them to disclose the beauty of God's creation.

The cultural-protection system took a number of forms. On the one hand we have its protoscientific elements, which are on full display in the pages of the Torah: avoiding bodily fluids, forbidding particular sexual practices, and quarantining the ill. These represented efforts to prevent calamity, seen as God's punishment. Over the course of the following centuries, advances in science and medicine made these measures obsolete. Where they nevertheless persisted, they could only do so because they had become firmly anchored in our second nature or served as a costly signal to help shore up the religious community's cohesion. A particularly telling example of this development is the prohibition of homosexuality, which is nothing more than the product of an outdated, prescientific theory.

We described the second manifestation of the cultural-protection system under the heading apocalypse. Evil powers are responsible for all of the bad things that happen in the world. To protect his culture or, indeed, his life, one has to fight back against these demonic forces and seek out the aid of the forces of good. But this part, too, is outdated. We now have a much better idea of what causes events around us. Only in those areas where science and logic fail does this type of thinking still surface today, such as when people resort to pilgrimages or exorcisms to heal conditions or diseases that conventional medicine cannot cure. It the realm of the worldly, the "apocalyptic matrix" generally takes the form of conspiracy theories. Our tendency to seek out social causalities is deeply rooted in human nature, which is all too ready to believe that some scheming actor is behind everything that happens to us.

The third element of the cultural-protection system is the social aspect that deals with group cohesion, aka *asabiya*. Even a merciful God who calls on us to love our neighbor reflects the attempt to mitigate our experiences of mismatch. The Psalms and New Testament showed us just how pressing the need is for a God prepared to support his followers. We yearn for a God who will stand behind sinners and put mercy before justice. Jesus did exactly that. This is not the merciless judge of the Last Judgment but the hunter-gatherer Jesus at whom we marveled in chapter 17. In the prehistoric groups that shaped our psychologies, there were no such things as abstract laws. Any offense was an outrage, but in the end—and after a great deal of discussion and negotiation—such scandals almost always ended in forgiveness. There was always someone who would leave no stone unturned in an attempt to tilt the scales in the accused's favor. You could always depend on your own people.

And isn't this just what believers continue to wish for to this day—a God who is as merciful or even indulgent with them as Jesus was with the adulteress? "He that is without sin among you, let him first cast a stone at her." God is supposed to love all of us despite our shortcomings. He doesn't give up on anyone, even someone who breaks his laws. The intuitive God is always on the individual's side—unlike the intellectual God, who always takes the side of law, the church, and society as a whole.

We shouldn't forget that our identity is also largely a social identity. The group was always a part of our personality. When the closely bound group disappeared, it was as if a part of our own selves had been amputated, leaving us with phantom pains. This, too, is a reason for the gods' careers, for when the old group solidarity faded away, we began to invest increasingly in heavenly relationships. And we were well advised to do so since the gods were also responsible for the new misfortunes of epidemics and catastrophes. To put it bluntly, God became the surrogate for the old group bonds, because he helped to alleviate our phantom pains.

THE FIRST AND LAST COMMANDMENT

This is why we find the commandment to love God in both the Old and the New Testaments. If we love God, then he will love us. It is the age-old law of reciprocity. And the call to love God goes hand in hand with the commandment to love thy neighbor. Jesus and the Pharisee both agreed that these were the Torah's greatest laws. But as a human universal, the origins of the much older golden rule preceded those of religion by hundreds of thousands of years.

Among the hunter-gatherers, the law of reciprocity was a given. It formed the basis of all social togetherness—it was virtually the first and final commandment of humanity. Primatologists have observed the same mechanism already at work among our relatives. All primate species studied so far place great value on reconciliation, particularly among friends. According to Frans de Waal, we must therefore consider a yearning for fairness as some of the oldest abilities in our evolutionary history: "They derive from the need to preserve harmony in the face of resource competition."[1] When the Bible speaks of loving one's neighbor, it calls for nothing less than for us to behave as the hunter-gatherers of old always had within their own communities.

The fact that the Bible had to demand that people adopt the golden rule—even though it was an absolute given throughout most of human history—proves that it had virtually disappeared from daily life at a time when societies continued to grow ever larger and more anonymous. This is how the appeals to love God and one's neighbor flowed so seamlessly into the cultural-protection system known as the Bible. It sought to restore a long-lost sense of harmony in order to combat the isolation, lack of social control, and resulting social distortions. It provided social glue, but not so much, as often argued, by threatening punishment. In effect, the Bible helped biology reassert itself.

Whereas we today have more appropriate means at our disposal for combating catastrophes and disease, when it comes to altruism and human cooperation, religion can still claim standing. Effective alternatives are few and far between. Even the new state institutions offer only limited aid. This explains why preachers continue to call from the world's pulpits for brotherly love. But we should not forget that religion is only a stand-in for our first nature. Its message is simple: be human again! But this commandment isn't really truly religious.

BACK TO THE ROOTS

Let us remember the exodus from Egypt. Yahweh freed the Israelites from bondage, Moses led them into a glorious future, and all the while the people continued to murmur and murmur. They were never satisfied, and Moses had no sooner set out to climb Mount Sinai than they began once again to dance around the golden calf. For us, this constant murmuring—against which the harshest of God's and his prophet's disciplinary measures were of no avail—symbolizes human nature's attempts to make itself heard. And even after they had made it to the Promised Land, the people continued to demand their old gods; the new one was simply too abstract, too unrealistic, and not feminine enough. In other words, he just wasn't suitable for daily use.

When analyzing the Bible, that diary of 1,000 years of human history, one could argue that the process of cultural evolution it embodies represents a dialectical process. Human hardship, caused especially by recurrent natural and man-made catastrophes, was the pacemaker of cultural evolution. Again and again these disasters laid waste to societies, repeatedly forcing our third nature to come up with new strategies to deliver us from what the prophet Jeremiah called "the sword, the famine, and the pestilence" and allow humanity to live a life of peace. But when these cultural solutions did not sit well with the innate preferences of our first nature, the latter began to "murmur." And it would do so incessantly. Our first nature's distress could only diminish when corrections to the problematic third-nature solution swung the pendulum back in the direction of first nature. This modified solution, more in tune with human needs, would have the potential to become a comfortable part of our second nature, a completely self-evident component of culture.

Every new catastrophe brought the cauldron of cultural evolution to a boil once again. New solutions had to be developed and old ones improved. This dialectical principle operated every time. As a result, cultural evolution is bound by what evolutionary biologists call "path dependence": once a path has been chosen, the process tends to stick to it. Thus changes and corrections cannot deviate too much from this direction. In the case of cultural evolution, our first nature constantly acts as a corrective element, and it expresses its preferences and antipathies in a particularly unflinching way.

In this sense one might say that the history of Western culture—or at least part of it—reflects a movement back to the roots, a keenness to return to the social conventions of the hunter-gatherers, even though the journey may have taken a rather twisted route. We can see this today in Western societies, where greater levels of democracy, more freedoms for women, and more intuitive religion have managed to reassert themselves. All of this is pleasing to our first nature, which is increasingly getting its way.

If this is correct, then it is also possible to describe human history as a dialectical process between biology and culture. It would mean that the human condition is not only to be discovered in "nature" (that is to say, inferred from genes, fossils, and universal behavioral patterns) but can equally be recognized in cultural products. Conversely, we can reconstruct the trajectory of cultural evolution based on the assumption that our genetically anchored psychological dispositions ensure the presence of specific preferences. This *longue durée* perspective represents an extension, rather than an alternative, to the historians' approach. Its broader evolutionary perspective has its advantages: it prevents us from getting bogged down in detail, while at the same time introducing a number of new and productive questions. Without it we would have never been able to write this book.

WHAT WE CAN LEARN FROM THE BIBLE

We leave it to others to judge whether our book has contributed to a better understanding of God. But we do claim that our discussion helps us to better understand ourselves. There is still a widespread reluctance to talk about human nature, and we intend for this book to make clear that human nature can hardly be separated from human culture and is anything but deterministic. Denying modern *Homo sapiens*'s typical, genetically anchored behavioral and perceptual dispositions would not only make each human problem individually unique but also mean that everyone is uniquely responsible for all possible expressions of his or her own unease. We have explained that there really is such a thing as a widely experienced discontent with civilization. This is the mismatch feeling at work, the product of our living in an environment that differs from the one in which natural selection shaped our first nature. Humans turned to cultural solutions to deal with the problems they encountered in the various new environments but never fully eliminated them.

We think that this knowledge can help alleviate some of humanity's worries. Isn't it comforting to know we are not individually to blame for many of the problems we continue to wrestle with today? To know that we aren't personally the reason for their existence and that they are the true "original sin"—the legacy of our species' cultural evolution? They are the price we paid for our cultural successes east of Eden, and together we must develop recipes for dealing with the dilemma posed by a way of life that isn't really appropriate to our species. A return to Paradise is no longer an option: much more than an angel with a flaming sword blocks its entrance.

In our reading of the Bible, we identified a number of our first nature's typical reactions. It always insists on its rights, even though in our post-hunter-gatherer world its desires have for the most part long since become irrational. Psychologist Jonathan Haidt had a point when he compared our rational minds with a rider trying to steer a stubborn and restive elephant.[2] Let us list here a sample of our first nature's idiosyncrasies, all of which we directly encountered in the Bible or can reconstruct from it: We know no peace until we have identified who is responsible for catastrophes, diseases, or conflicts. We feel a great need for credibility-enhancing displays that can prevent us from being taken advantage of, for actions speak louder to us than words. Injustice—in other words, a lack of reciprocity—arouses feelings of righteous indignation in us. We find it easy to demonize outsiders. We try mightily to make sense of everything that happens and explore every corner of the world, even imaginary ones such as heaven and hell. We are more likely to believe explanations involving supernatural actors than those employing abstract processes. We are born animists with a soft spot for magic, spells, and charisma, and even the rationalists among us find it hard to let go of their little superstitious quirks. We want to be in control of things. We hold monogamy to be a good idea, even though we find it rather difficult in practice. We attach major importance to the opinions of others. We are social networkers and highly gifted cooperators—but we are also egoists happy to exploit any available advantage to put down our competitors. We are afraid of dying but are not afraid of death—at least as long as no one tries to rob us of our intuition that it's not the end.

Our anthropological reading of the Bible also tells us what our first nature really enjoys: community, joint performance, egalitarianism, equal rights, and, of course, stories and even better stories. In a nutshell, all of those things give us back a piece of our lost Paradise. And for all of them, we can depend on the Bible, for it congealed thousands of years of human experience.

WHAT THE BIBLE CERTAINLY IS

And now we have come to the end. We did not intend to present a new exegesis of the Bible. We only wished to highlight and explain the rich record of cultural evolution hidden in its pages and so bring to light a hitherto unappreciated dimension. This new reading led to our claim that the Bible really is the most important book in the history of humanity.

In light of the Bible's monumental character, we could only conduct an initial survey—although it certainly sufficed to show just what fantastic material we are dealing with. For anyone interested in the evolution of human culture, the Bible is absolutely worth a closer look. And, of course, we expect that applying the same cultural-evolutionary approach to other religions will be very rewarding as well.

Most of all, we hope that we have awakened readers' curiosity about the Bible. Admittedly, it's not an easy read. We ourselves were relieved to hear Bible scholar Michael Satlow confess that although he has dedicated his life to the Book of Books, he has never read it or even one of its testaments from cover to cover: "The Good Book is very hard to read."[3] We believe that setting out to read the Bible will be much easier if we understand it as humanity's diary—a document written and reworked over the course of nearly 1,000 years, which therefore inevitably contains mistakes and inconsistencies. It offers a testament to humanity's heroic struggle after having been cast out of Paradise.

Such a reading should unburden the Bible of all the weight it has had to bear as the flawless word of a flawless God. This expectation has bewildered many readers. How can the Holy Book be so full of errors and horrors? And why do seemingly insignificant offenses drive God to unleash his wrath at humanity? This expectation is, of course, completely misplaced: in fact, the Bible itself never makes the claim to be perfect. For about a thousand years, it was a work in progress. Try to imagine what would have happened if work on the Bible had never ground to a halt as it did. Perhaps scholars and clergy would now be at work on the Fifth Testament. We suspect the history of the West would have taken a different course.

In the end it is up to each and every one of us to decide for ourselves whether a divine spirit awaits discovery between the lines of the Bible. Regardless, we are confident we have made a convincing case that the Bible most definitely is the Good Book of Human Nature.

ACKNOWLEDGMENTS

IT TOOK US MORE THAN FIVE YEARS TO WRITE THIS BOOK, AND WE MET more than two hundred times, often absorbed in our Bibles. We ended up feeling like Heinrich Schliemann, the mid-nineteenth-century German archaeologist, who after reading Homer's *The Iliad* and *The Odyssey* became convinced that they contained vital cues as to the location of the ancient city of Troy. He struck gold, literally.

Although we could not use shovels, our reading of the Bible to us also felt like digging for treasure. Friends and acquaintances helped us persevere: upon learning about our project, they kept on peppering us with questions. These responses strengthened our conviction that the Bible contains many unanswered questions and even beguiles people who don't believe in God.

Above all, we thank Jared Diamond, who read early drafts of several chapters and provided precious feedback. We also are deeply grateful to our agents John and Max Brockman; their team has made it possible for this book to see the light of day. Right from the start, T. J. Kelleher, vice president and editorial director for the sciences at Basic Books, showed great interest in our research and showered us with clever suggestions. We are very grateful to Mark Willard from Tradukas, as well as our copy editor Jennifer Kelland Fagan and project editor Michelle Welsh-Horst, for their first-rate help in producing and polishing the English-language edition. And the good citizens of Zurich were kind enough to politely ignore us when they saw one of us reading the Bible with rapt attention on the tram or in a restaurant—nowadays a somewhat shady activity in Western Europe.

Above all, we are indebted to our respective families: Maria, Johanna, and Jaap and Katharina, Anna, and Nikolai. They motivated us and gave numerous hints, and they even forgave us that after our five years of digging for treasure, all they got was a heap of words.

NOTES

Introduction

1. Glenday, 2015, 374.
2. Keller, 2015. Finkelstein & Silberman, 2001. Ellens & Rollins, 2004.
3. Janowski, 2005. Wolff, 2010. Frevel, 2010. Staubli & Schroer, 2014.
4. Bellah, 2011, 289.
5. Schnabel, 2008.
6. Norenzayan, 2013. Richerson & Christiansen, 2013.
7. Bulbulia et al., 2013, 398.
8. Pinker, 2011. Douglas, 2003. Harris, 1990. R. Wright, 2009. Teehan, 2010.
9. Dawkins, 2006a, 237.
10. Diamond, 1987. Diamond, 2013.
11. Schmid, 2012a, 158. Frankemölle, 2006, 27–29.
12. Madigan & Levenson, 2008, 235.
13. Zenger & Frevel, 2012, 103f. Halbfas, 2010, 37. Flasch, 2013, 50f.
14. Zenger & Frevel, 2012, 104.
15. Kugel, 2008, 40–42. Schmid, 2012a, 28–30.
16. Schmid, 2012a, 27.
17. Dever, 2012, 373.
18. Keel, 2012, 19.
19. Zenger & Frevel, 2012, 22.

20. Schmid, 2012a, 17–22.
21. Schmid, 2012a, 24.
22. Schmitz, 2011, 157.
23. Schmid, 2012a, 36.
24. Levin, 2005, 28.
25. Qtd. in Halbfas, 2010, 15.
26. Schmid, 2012a, 51.
27. Schmid, 2012a, 173, 89.
28. Staubli, 2010, 9. Keel, 2012, 17.
29. Jacobs, 2007, 201.
30. Richerson & Boyd, 2005. Richerson & Christiansen, 2013.
31. van Schaik et al., 2003. van Schaik, 2004.
32. van Schaik, 2016.
33. Diamond, 2013.
34. Tooby & Cosmides, 1992.
35. Lieberman, 2013.
36. Bourdieu, 1993, 28.
37. Elias, 1997.
38. Boyer, 2001, 250.

Part I: Genesis

1. Lang, 2013, 47–49.

Chapter 1: Adam and Eve

1. Bloch, 1975, 45.
2. Zevit, 2013, 76.
3. Blum, 2004, 12.
4. Ranke-Graves & Patai, 1986, 80.
5. Flasch, 2005, 83.
6. Ego, 2011, 11.
7. Albertz, 2003, 23.
8. Albertz, 2003, 23.
9. Albertz, 2003, 23, 33.
10. Ego, 2011, 22.
11. Kugel, 2008, 49f.
12. Flasch, 2005, 48.
13. Gottschall, 2013.
14. Boyer, 2001, 204.
15. Boyer, 2001, 301.
16. Gottschall, 2013, 103.
17. Dobelli, 2011, 22.

18. Otto, 1996, 189.
19. Flasch, 2013, 189.
20. R. Wright, 2009, 191ff.
21. Otto, 1996, 175.
22. Ego, 2011, 44.
23. Ruppert, 2003, 170.
24. Flasch, 2005, 38ff.
25. Ruppert, 2003, 169.
26. *Katechismus*, 2005, 397.
27. Odil Hannes Steck, qtd. in Albertz, 2003, 33f.
28. Willmes, 2008.
29. Flasch, 2005, 81.
30. Schmid, 2012a, 176.
31. Blum, 2004, 15.
32. Ruppert, 2003, 27.
33. Schmid, 2012a, 175.
34. Otto, 1996, 167ff. Albertz, 2003.
35. Ruppert, 2003, 61.
36. Schmid, 2012a, 174. Albertz, 2003, 1–22.
37. Steymans, 2010, 201–228.
38. Ego, 2011, 16.
39. Pfeiffer, 2001, 7.
40. Korpel & de Moor, 2014.
41. R. Wright, 2009, 125.
42. Keel, 2011, 12.
43. Ruppert, 2003, 55–77.
44. Pfeiffer, 2001, 3. Ruppert, 2003, 128.
45. Weber, 1995, 19.
46. Ruppert, 2003, 145f.
47. Ruppert, 2003, 167.
48. R. Wright, 2009, 104.
49. Keel, 2011, 43.
50. Ruppert, 2003, 167.
51. Ruppert, 2003, 120.
52. Blum, 2004, 13.
53. Keel & Schroer, 2008. Winter, 2002.
54. Ego, 2011, 16.
55. Ruppert, 2003, 119. Pfeiffer, 2000. Pfeiffer, 2001. Pfeiffer, 2006a. Pfeiffer, 2006b.
56. Assmann, 2009, 12ff. Schmidt-Salomon, 2012, 38.
57. Albertz, 2003, 37.

58. Albertz, 2003, 38f.
59. Pfeiffer, 2001, 11.
60. Keel & Schroer, 2008, 138.
61. Pfeiffer, 2001, 16.
62. Diamond, 2012. van Schaik, 2016.
63. Boehm, 2012.
64. Berbesque et al., 2014.
65. Bowles, 2011.
66. Assmann, 2007. Assmann, 2013a.
67. Gatz, 1967, 212. Heinberg, 1995, 161ff.
68. Gottschall, 2013, 67, 58.
69. Marlowe, pers. com.
70. Qtd. in Gazzaniga, 2012, 198.
71. R. Wright, 2009, 58.
72. Diamond, 2012, 524. van Schaik, 2016, 299.
73. Rousseau, 1981, 230.
74. Hill & Hurtado, 1996.
75. Crawford, 2009, 60.
76. Diamond, 2013, 196.
77. Exod 22:18. Lev 18:22–23. Lev 20:13–15.
78. Crawford, 2009, 23. Clark, 2010, 142.
79. Exod 22:16. Deut 22:13–21.
80. Hewlett & Hewlett, 2013.
81. Meyers, 2010, 65.
82. Qtd. in Zingsem, 2009, 30f. Frey-Anthes, 2008. Flasch, 2005, 23f.
83. Wells et al., 2012.
84. Zevit, 2013, 206.
85. Berger & Luckmann, 1998, 94f.

Chapter 2: Cain and Abel

1. Byron, 2011, 2.
2. Kugel, 2008, 60f.
3. *Stuttgarter Erklärungsbibel*, 2007, 15.
4. Brandscheidt, 2010.
5. Kugel, 2008, 63ff.
6. Brandscheidt, 2010.
7. Hensel, 2011, 327, 335.
8. Hendel, 2002, 50.
9. Turchin, 2007, 264.
10. Deut 21:15–17.
11. Byron, 2011, 6. Schmid, 2012a, 173.

12. van Schaik, 2016, 238.
13. Byron, 2011, 138ff.
14. Boehm, 2012.
15. *Stuttgarter Erklärungsbibel*, 2007, 15.

Chapter 3: Sons of Man, Sons of God

1. Hieke, 2010, 150.
2. Hieke, 2010, 65.
3. Diamond, 2012, 50.
4. Hendel, 2002, 49.
5. Atran & Henrich, 2010, 19.
6. Norenzayan, 2013, 116. Slingerland et al., 2013, 344f.
7. Hieke, 2010, 185.
8. Assmann, 2013a, 73–75.
9. Clark, 2010, 167.
10. Cohen & Armelagos, 2013.
11. Schmid, 2012a, 244.
12. Steymans, 2010, 226.
13. Schmid, 2012a, 46.
14. Tilly & Zwickel, 2011, 114.
15. Diamond, 2002. Mummert et al., 2011.
16. *Stuttgarter Erklärungsbibel*, 2007, 17.
17. R. Wright, 2009, 124.
18. Metzger, 2012, 25–34.

Chapter 4: The Flood

1. Pinker, 2011, 10.
2. Wälchli, 2014. Assmann, 2015, 114.
3. Jeremias, 2011, 3. Pinker, 2011, 10.
4. Jeremias, 2011, 5f.
5. Norenzayan, 2013. Johnson, 2015b.
6. Johnson & Kruger, 2004. Johnson & Bering, 2009.
7. Bering, 2012, 201.
8. Bateson et al., 2006.
9. Bering, 2012, 93–99.
10. Sosis & Bressler, 2003.
11. Voland, 2007, 119. Johnson, 2005.
12. Johnson & Bering, 2009, 33.
13. Norenzayan, 2013, 171.
14. Foucault, 1977, 256ff.
15. Diamond, 2012, 368.

16. Norenzayan, 2013, 158. Norenzayan et al., 2014.
17. Diamond, 2012, 336.
18. Bering, 2012, 82.
19. McCauley, 2011.
20. R. Wright, 2009, 12–19.
21. Boyer, 2001, 267–270.
22. R. Wright, 2009, 24f. Bulbulia et al., 2013, 397f.
23. Norenzayan, 2013, 136f.
24. Atran & Henrich, 2010, 25.
25. Boyer, 2001, 7.
26. Durkheim, 2007, 76.
27. Norenzayan, 2013, 7f.
28. Haidt, 2012, 256.
29. Norenzayan, 2013, 8f., 23.
30. Diamond, 2013, 288.
31. Pinker, 2011, 42.
32. Kugel, 2008, 71f.
33. 1 Pet 3:20–21.
34. Herget, 2012, 7.
35. Halbfas, 2010, 67.
36. Baumgart, 2005.
37. Buchner & Buchner, 2011.
38. Albertz, 2003, 49–63. Ruppert, 2003, 309. Kugel, 2008, 74–77. Bosshard-Nepustil, 2005.
39. Nietzsche, 1955, 972f.
40. Hume, 2000, 10. Dennett, 2008, 144.
41. Guthrie, 1993. Boyer, 2001. Barrett, 2004.
42. Boyer, 2001, 145.
43. Nietzsche, 1955, 975.
44. Gopnik, 2000. Bering, 2012, 140ff.
45. R. Wright, 2009, 17. Boyer, 2001, 195ff.
46. Piaget, 1992, 210–225.
47. McCauley, 2011.
48. Bering, 2012, 112.
49. Kirkpatrick, 2005, 247. Lehmann et al., 2005. Steadman et al., 1996. Bloom, 2004.
50. Boyer, 2001, 138.
51. R. Wright, 2009, 21.
52. Boyer, 2001, 214f.
53. Clark, 2010, 174.
54. Boyer, 2001, 227.

55. Diamond, 2013, 221.
56. Crawford, 2009, 59f.
57. Diamond, 2013, 221.
58. Clark, 2010, 6.
59. Boyer, 2001, 330.
60. Brotherton, 2013. Leman & Cinnirella, 2007.
61. Jeremias, 2011, 9. Wälchli, 2014. Assmann, 2015, 393.
62. Burkert, 1996, 102–128. Murdock, 1980, 20. Sonnabend, 1999.
63. Burkert, 1996, 121, 125.
64. Burkert, 1996, 103.
65. Maul, 2013.
66. Weber, 1988, 256.
67. Veyne, 2008, 58.
68. Schulze, 2008, 9.
69. Clark, 2010, 36.
70. Clark, 2010, 11.
71. Clark, 2010, 6.
72. Clark, 2010, 163.
73. Burkert, 1996, 108.
74. Botero et al., 2014.
75. Bentzen, 2013, 1. Bentzen, 2015. Sibley & Bulbulia, 2013.
76. Steinberg, 2006.
77. R. Wright, 2009, 39. Clark, 2010, 166.
78. Taleb, 2013.
79. Nietzsche, 1955, 975.
80. Crawford, 2009, 25.
81. Taleb, 2013, 146–154.

Chapter 5: The Tower of Babel

1. Baumgart, 2006.
2. Parzinger, 2014, 141f. Smith et al., 2014, 1530.
3. Newitz, 2014. Shennan et al., 2014.
4. Kuijt, 2000.
5. Newitz, 2014.
6. Smith et al., 2014, 1530.
7. Halbfas, 2010, 69f.
8. Clark, 2010, 3. Crawford, 2009, 58.
9. Diamond, 2013, 221.
10. Clark, 2010, 3f.
11. Diamond, 2013, 221.
12. Clark, 2010, 3.

Chapter 6: Patriarchs and Matriarchs

1. Mühling, 2009. Gertz, 2010b, 269ff. I. Fischer et al., 2010, 240f. I. Fischer, 2013. Moore & Kelle, 2011, 43.
2. *Stuttgarter Erklärungsbibel*, 2007, 23.
3. Michel, 2007.
4. Schmitz, 2011, 125ff.
5. Finkelstein & Silberman, 2001, 36–38.
6. Schmitz, 2011, 130.
7. Schmitz, 2011, 131. Schmid, 2012a, 138.
8. Meyers, 2013b, 109.
9. Otto, 1994, 49.
10. Deut 21:15.
11. Henrich et al., 2012, 664f.
12. Michel, 2007.
13. Keel & Schroer, 2006, 24.
14. Hieke, 2010, 182.
15. Goethe, 1998, Vol. 9, 137.
16. van Schaik, 2016, 187.
17. Henrich et al., 2012, 660f.
18. Pinker, 2011, 7.
19. 2 Kings 3:26–27.
20. Henrich, 2009.
21. Henrich et al., 2012, 666.
22. Hieke, 2010, 171.
23. Dyma, 2010.
24. Num 36.
25. Alt et al., 2013.
26. Eberhart, 2008.
27. Eberhart, 2008.
28. Lev 20:17.
29. Marlowe, 2003.
30. Henrich et al., 2012, 657.
31. Finkelstein & Silberman, 2001, 47.
32. Jer 29:18.

Part II: Moses and the Exodus

1. Schmid, 2012b, 5.
2. Köckert, 2013, 15.
3. Schmitz, 2011, 132.
4. Assmann, 2015, 19, 389.

Chapter 7: Moses

1. Num 20:7–11.
2. Exod 33:11.
3. Josh 1:7–8.
4. Deut 34:1–11.
5. Moroni & Lippert, 2009. Keller, 2015. Kugel, 2008.
6. Frevel, 2012, 717f.
7. Finkelstein & Silberman, 2001, 61.
8. Freud, 1993, 59.
9. Assmann, 2009, 5. Assmann, 1997.
10. Otto, 2006, 26.
11. Gertz, 2008.
12. Otto, 2006, 30.
13. Num 31:17.
14. Otto, 2006, 31–33.
15. Schmitz, 2011, 134. Jericke, 2012.
16. Otto, 2006, 33.
17. Blum, 2012.
18. Otto, 2006, 35.
19. Schmitz, 2011, 135.
20. Schmid, 2012a, 92f.
21. Crüsemann, 1996, 59.
22. Goethe, 1998, Vol. 2, 208f. Zenger & Frevel, 2012, 100.
23. Crüsemann, 1996, 10.
24. Albertz, 2003, 188.
25. Assmann, 2009, 52. Albertz, 2003, 190.
26. Lev 26:1–40. Deut 28:1–68.
27. Lev 26:14–17.
28. Freuling, 2008.
29. Weber, 1988, 231.
30. Otto, 2007, 118ff. Keel, 2011, 78ff.
31. Berlejung, 2012. Burkert, 1996. Murdock, 1980. Sonnabend, 1999. Walter, 2010.
32. Crüsemann, 1996, 75f.
33. Köckert, 2013, 12.
34. D. Wright, 2009.
35. Albertz, 2003, 187ff. Schmid, 2012a, 108. Weber, 1988, 81.
36. Exod 21:12.
37. Exod 21:18–19.
38. Exod 21:21.

39. Exod 21:7.
40. Crüsemann, 1996, 151–159.
41. Exod 22:26–27.
42. Exod 22:27.
43. Exod 22:21.
44. Crüsemann, 1996, 182–185.
45. Crüsemann, 1996, 253.
46. Exod 21:15.
47. Lev 18:15–16.
48. Gertz, 2010b, 225.
49. Crüsemann, 1996, 73.
50. Exod 21:24–25.
51. Gertz, 2010b, 228f.
52. Dawkins, 2006a, 248.
53. Wesel, 1997, 62.
54. Otto, 1994, 64f.
55. Lev 19:18
56. Moenikes, 2012.
57. Baumard & Boyer, 2013.
58. Deut 17:17–20.
59. Exod 22:3.
60. Crüsemann, 1996, 164.
61. Deut 22:22.
62. Crüsemann, 1996, 256.
63. Clark, 2010, 2.
64. Murdock, 1980, 20, xiii.
65. Berlejung, 2010b, 203, 212.
66. Exod 15:26.
67. Deut 28:27–28.
68. Deut 28:58–61.
69. Berlejung, 2010b, 212.
70. Thornhill & Fincher, 2014, 237–264.
71. Ego, 2007.
72. Parker, 1983.
73. Maier, 1997, 476.
74. Rozin et al., 2008. Schaller, 2014.
75. Ego, 2007.
76. Lev 15:31.
77. Douglas, 2003, 30. James, 2014, 46f.
78. Clark, 2010, 38.
79. Lev 15:2–18.

80. Frey-Anthes, 2007.
81. Lev 15:21–24.
82. Strassmann et al., 2012.
83. Deut 23:9–14.
84. Crüsemann, 1996, 265.
85. Clark, 2010, 116. Crawford, 2009, 74.
86. 2 Kings 16:35.
87. Flavius Josephus, 2011, 461.
88. Schmitz, 2011, 107.
89. Veit, 2013.
90. Clark, 2010, 115.
91. Diamond, 2013, 221.
92. Lev 18:22–23. Lev 20:13–16.
93. Exod 22:18.
94. Crawford, 2009, 23. Zehnder, 2008.
95. Frey-Anthes, 2007.
96. Lev 13:1–8.
97. Olyan, 2008, 9.
98. Lev 21:18–21.
99. Lev 21:13.
100. Num 19:1–13.
101. Schaller, 2011. Schaller et al., 2015.
102. Num 25:1–9.
103. Frey-Anthes, 2007.
104. Frey-Anthes, 2007.
105. R. Wright, 2009, 124.
106. Jacobs, 2007, 43.
107. Deut 22:9–10.
108. Deut 23:2.
109. Weber, 1988, 329, 231, 178.
110. Lev 26:14–39.
111. Lev 19:19.
112. Crüsemann, 1996, 264.
113. Deut 22:5.
114. Geertz, 1987, 9.
115. Staubli, 2010, 92f.
116. Lev 11:2.
117. Gies, 2012. Keel, 2011, 110f.
118. Qtd. in Douglas, 2003, 46.
119. Qtd. in Douglas, 2003, 32.
120. Harris, 1990, 69–76.

121. Finkelstein & Silberman, 2001, 119–120.
122. Gies, 2012.
123. Douglas, 2003, 56f. Ego, 2007.
124. Douglas, 2003, 40–46.
125. Staubli, 2001, 46.
126. Crüsemann, 1996, 263.
127. Staubli, 2001, 46.
128. Staubli, 2001, 47.
129. Lev 17:14.
130. Eberhart, 2006.
131. Schmitz, 2011, 38f.
132. Lev 19:27.
133. Deut 22:12.
134. Berlejung, 2010a, 171.
135. U. Zimmermann, 2012.
136. Lev 12:3. Tilly & Zwickel, 2011, 106.
137. U. Zimmermann, 2012.
138. Berlejung, 2010a, 170f.
139. Num 16:22.
140. Num 5:12–31.
141. Deut 21:4–8.
142. Num 15:22–29.
143. Lev 16:21–34.
144. Crüsemann, 1996, 314.
145. Tylor, 1871, Vol. 2., 375.
146. Mauss, 1996. Mauss, 2012.
147. Burkert, 1996, 137f.
148. Exod 39:2.
149. Boyer, 2001, 238–240.
150. Plaut, 2000, 173.

Chapter 8: Yahweh

1. Zenger & Frevel, 2012, 98.
2. Kugel, 2008, 216.
3. Colpe, 2007. Bauks, 2011.
4. Hazony, 2015. Barton, 2013, 192. Niehr, 2013, 31.
5. Schmid, 2003, 29. Becker, 2005, 3.
6. Niehr, 2003, 245.
7. Becker, 2005, 5f.
8. Kaiser, 2013, 25.
9. Kottsieper, 2013.

10. Grätz, 2006.
11. Keel, 2011, 43.
12. Dever, 2006. Keel & Uehlinger, 2010. Keel & Schroer, 2006. Keel, 2008.
13. Weber, 1988, 281, 177, 262.
14. Keel, 2011, 51.
15. Pakkala, 2006, 240.
16. Weber, 1988, 267, 301.
17. Isa 10:5–6.
18. Becker, 2005, 6.
19. Aurelius, 2003, 150f.
20. Becker, 2005, 7.
21. Schmid, 2003. Bauks, 2011.
22. Kessler, 2008, 77.
23. R. Wright, 2009, 86.
24. Schmitz, 2011, 75.
25. Frahm, 2011, 267, 283.
26. Keel, 2011, 79f.
27. Schmid, 2012a, 90.
28. Crüsemann, 1996, 131.
29. Lehnhart, 2009.
30. Num 25:3–4.
31. Keel, 2011, 78.
32. Deut 13:6–11.
33. Frahm, 2011, 283f.
34. Deut 6:4–5.
35. Deut 6:4–5. Mk 12:28–31. Avalos, 2015, 39–49.
36. Otto, 1999, 362f.
37. Otto, 2007, 129.
38. Finkelstein, 2014, 9.
39. Crüsemann, 1996, 197.
40. Otto, 1999, 364–366.
41. Schmitz, 2011, 31.
42. Finkelstein & Silberman, 2001, 10.
43. Finkelstein & Silberman, 2001, 10. Otto, 2006, 43.
44. Schmitz, 2011, 34–37.
45. Levin, 2005, 26.
46. Schmitz, 2011, 18.
47. Niehr, 2003, 228f.
48. Niehr, 2003, 230. Köckert, 2009.
49. Pakkala, 2006, 245.
50. Niehr, 2003, 236.

51. Frevel, 2003, 75.
52. R. Wright, 2009, 165f.
53. Keel, 2011, 83.

Chapter 9: The Murmuring People

1. Aurelius, 2003, 148f.
2. Assmann, 2015, 319f. Achenbach, 2007.
3. Schmitz, 2011. Berlejung, 2010a.
4. Dever, 2012, 287.
5. Stavrakopoulou & Barton, 2013, 1.
6. Stavrakopoulou, 2013, 38.
7. Berlejung, 2010a, 71.
8. Tilly & Zwickel, 2011, 78.
9. Dever, 2012, 250.
10. van der Toorn, 1996. Albertz & Schmitt, 2012.
11. Meyers, 2013b, 103.
12. Dever, 2012, 293.
13. Dever, 2006, 3.
14. Gudme, 2010, 79–81.
15. Meyers, 2013b, 170.
16. Bauks, 2011.
17. Niehr, 2003, 234f.
18. Schroer, 2006, 24f.
19. Bellah, 2011, 289.
20. Dever, 2012, 291, 279f.
21. R. Wright, 2009, 150.
22. Boyer, 2001, 245f.
23. Gudme, 2010, 85.
24. Meyers, 2013a, 130.
25. Niehr, 2013, 29. Schmitt, 2006.
26. Meyers, 2013b, 3.
27. Meyers, 2010, 97. Albertz, 2013, 138.
28. Dever, 2012, 287.
29. R. Wright, 2009, 80f.
30. Barrett, 2011, 135–138.
31. Barrett, 2011, 137.
32. Slone, 2005, 122.
33. Boyer, 2001, 285.
34. Assmann, 2009, 1.
35. Assmann, 2009, 114.
36. Boyer, 2001, 142.

37. Norenzayan, 2013, 18.
38. Meyers, 2013b, 118f.
39. James, 2014, 63.
40. Weber, 1980, 257.
41. Beit-Hallahmi, 2015, 234.
42. Henrich, 2009, 244. Atran & Henrich, 2010.
43. Henrich, 2009, 247.
44. Norenzayan, 2013, 98.
45. R. Wright, 2009, 36f.
46. Ernest Thomas Lawson, Robert McCauley, qtd. in Boyer, 2001, 284.
47. Exod 14:2–4.
48. Exod 14:15–18.
49. Num 16:32.
50. Popper, 2014.
51. Deut 34:10–11.
52. Deut 13:1–3.
53. Exod 20:18–19.
54. Trivers, 2011.
55. Whitehouse, 2004.
56. Berlejung, 2010a, 79.
57. Schmitz, 2011, 132. Meyers, 2012, 155f.
58. Deut 6:4–9.
59. Berger & Luckmann, 1998, 65.
60. Jer 31:33.

Chapter 10: The Torah's Legacy

1. Assmann, 2009, 10.
2. Lev 23:19. Lev 24:26. Deut 14:21.
3. Gies, 2012.
4. Keel, 2011, 110.
5. Clark, 2010, 52.
6. Weber, 1988, 231.
7. Durkheim, 2007, 613.
8. Norenzayan, 2013, 172f.
9. Otto, 1994, 11.
10. Schieder, 2014.
11. Assmann, 2009, 2. Assmann, 2015, 106.
12. Assmann, 2009, 23.
13. Assmann, 2013b, 23.
14. Assmann, 2009, 4.
15. Assmann, 2009, 21f.

16. Burkert, 1996, 8.
17. Burkert, 1998, 47.
18. Burkert, 1996, 171. Burkert, 1998, 47.
19. Jenkins, 2010, 3.
20. Deut 34:7.
21. Num 20:1–12.

Part III: Kings and Prophets

1. Gladwell, 2013.
2. Gintis et al., 2015.
3. Dawkins, 2006a, 31.

Chapter 11: Judges and Kings

1. Assmann, 2015, 11f., 103ff.
2. Zenger & Frevel, 2012, 271.
3. Gertz, 2010b, 292.
4. Finkelstein & Silberman, 2006, 6f.
5. Keel, 2011, 48.
6. 2 Sam 21:19.
7. Schmid, 2012a, 68f.
8. Hentschel, 2012b, 294.
9. Finkelstein & Silberman, 2001, 134f.
10. Finkelstein, 2014, 16f.
11. Finkelstein & Silberman, 2006, 264–265.
12. Finkelstein & Silberman, 2006, 21–22.
13. Hentschel, 2012c, 310f.
14. Biale, 2002, 1150.
15. Anderson, 2006, 6.
16. Finkelstein & Silberman, 2001, 81.
17. Hentschel, 2012a, 264.
18. Weber, 1988, 331.
19. Bering, 2012, 189.
20. Henrich, 2009.
21. Burkert, 1996, 171. Burkert, 1998, 47.
22. Teehan, 2010, 42.
23. Norenzayan, 2013, 7f.
24. Johnson, 2015b.
25. Giese, 2011, 53.
26. Qtd. in Enz, 2012, 43.
27. Qtd. in Enz, 2012, 44.
28. Qtd. in Giese, 2011, 54.

29. Enz, 2012, 9.
30. Turchin, 2007, 92.
31. Finkelstein & Silberman, 2006, 31–59. Keel, 2011, 50.
32. Ibn Khaldun, 2011, 179.
33. 2 Sam 23:8–39.
34. Turchin, 2007, 92.
35. Turchin, 2007, 93.
36. Enz, 2012, 54f.
37. Rilke, 1955, 486.
38. Schäfer, 2004, 300.
39. Finkelstein, 2014, 184f.
40. Finkelstein, 2014, 178–180.
41. Sand, 2014, 193f.
42. Deut 17:16–20.
43. Noort, 2012, 21–50.
44. Finkelstein & Silberman, 2006, 5.

Chapter 12: The Prophets

1. Nissinen, 2010, 16.
2. Kratz, 2003, 7.
3. Zenger & Frevel, 2012, 509f.
4. Assmann, 2010, 15.
5. Kelle, 2014, 286.
6. Grabbe, 2010, 130.
7. Schmid, 2010a, 314.
8. Kratz, 2003, 12–14.
9. Schmid, 2010a, 313.
10. Kratz, 2003, 16f.
11. Kratz, 2003, 7.
12. Weber, 1988, 300f.
13. Schart, 2014. Weber, 1988, 301.
14. Poser, 2012. Morrow, 2004. Joyce, 2010.
15. Kelle, 2014, 290. Wolff, 1987, 19.
16. Zenger & Frevel, 2012, 513.
17. Kelle, 2014, 275.
18. Schmid, 2010a, 315.
19. Schmid, 2010a, 323. Kratz, 2003, 26.
20. Zenger & Frevel, 2012, 510.
21. Deut 18:21–22.
22. Taleb, 2013, 146.
23. Kratz, 2003, 41, 46–49. Schmid, 2012a, 195f.

24. Kratz, 2003, 94.
25. Schmid, 2012a, 224f. Kratz, 2003, 98.
26. Kratz, 2003, 49f.
27. Burkert, 1996, 169f.
28. Boyer, 2001, 198.
29. van Schaik, 2016, 424.
30. Isa 51:16.
31. Schmitt, 2004, 383.
32. Weber, 1988, 327.
33. Wolff, 1987, 27.
34. Wolff, 1987, 22. Weber, 1980, 269.
35. Wolff, 1987, 19.
36. Albertz, 2006, 7.
37. Weber, 1980, 141, 268ff.
38. Nissinen, 2010, 18. Kelle, 2014, 304.
39. Kratz, 2003, 17f.
40. Zenger & Frevel, 2012, 518.
41. Assmann, 2009, 48–56.

Chapter 13: How Can a Good God Be So Bad?

1. Assmann, 2000, 53.
2. Nissinen, 2010, 17.
3. Pinker, 2011, 10.
4. Dawkins, 2006a, 31.
5. Janowski, 2013, VII.
6. Janowski, 2013, 7.
7. Janowski, 2013, 344.
8. Dawkins, 2006a, 235–278.
9. Johnson, 2015a, 292f.
10. Boyer, 2001, 51–92.
11. Boyer, 2001, 51–92.
12. Timo Veijola, qtd. in Janowski, 2013, 144.
13. Timo Veijola, qtd. in Janowski, 2013, 144.
14. Pinker, 2011, 11.
15. *Katechismus*, 2005, 42.
16. Dawkins, 2006a, 226.
17. de Waal, 2013. van Schaik, 2016.
18. Assmann, 2009, 54, 50.
19. Norenzayan, 2013. Norenzayan et al., 2014.
20. Norenzayan, 2014, 64.
21. Stark, 2001, 620.

22. Baumard & Boyer, 2013.
23. 2 Sam 24. 1 Chr 21.
24. Rosenzweig, 2007.
25. Hazony, 2015.
26. Berlejung, 2012, 22.
27. Bering, 2012, 144.
28. Murdock, 1980.
29. Schmitt, 2011, 215.
30. Sonnabend, 1999, 119ff.
31. Alain Cabantous, qtd. in Walter, 2010, 29.
32. Gray & Wegner, 2010, 11.
33. Schaudig, 2012, 425f.
34. Maul, 2013, 316–323.
35. Schlögl, 2006, 43–52.
36. Assmann, 2001, 114f.
37. Linke, 2014, 22–27.
38. Burkert, 1996, 152.
39. Veyne, 2008, 20.
40. Dahlheim, 2014, 326.
41. Sonnabend, 1999, 126, 161.
42. Veyne, 2008, 87–89.
43. Veyne, 2008, 51.
44. Gladwell, 2013.

Part IV: Psalms and Co.

1. Bloch, 1975, 53.
2. Witte, 2010, 414.
3. Staubli, 2010, 296.
4. Witte, 2010, 414.

Chapter 14: The Psalms

1. Zenger & Frevel, 2012, 450.
2. Zenger & Frevel, 2012, 452.
3. Witte, 2010, 415. Staubli, 2010, 307. Zenger & Frevel, 2012, 431.
4. Zenger & Frevel, 2012, 435.
5. Zenger & Frevel, 2012, 447–459.
6. Albertz, 2005, 16, 95.
7. Albertz, 2013, 135.
8. Feldmeier & Spiekermann, 2011, 51f.
9. Albertz, 2005, 94.
10. Kirkpatrick, 2005, 52f. Kaufman, 1981, 67.

11. Bowlby, 1988.
12. Qtd. in Lang, 2012, 24.
13. James, 2014, 61f.
14. Lang, 2012, 23.
15. Robarchek, 1990, 66.
16. Kirkpatrick, 2005, 247. Lehmann et al., 2005. Steadman et al., 1996.
17. Barrett, 2004, 59.
18. Kirkpatrick, 2005, 248.
19. Meyers, 2013b, 170.
20. Rosenthal, 1985.
21. Steadman et al., 1996, 73.
22. Barrett, 2013, 247.
23. Meyers, 2013b, 151.
24. Otto, 2005, 222.
25. Avalos, 2010, 45.
26. Avalos, 1995, 419.
27. *Stuttgarter Erklärungsbibel*, 2007, 677.
28. Madigan & Levenson, 2008, 46f.
29. Clark, 2010, 192.
30. Albertz, 2005, 92f. Albertz, 2013, 137.
31. Albertz, 2005, 49.
32. Steadman et al., 1996, 74. Kirkpatrick, 2005, 248.
33. Eberhardt, 2007, 396.
34. Sand, 2014, 257. Stern, 2005, 347.

Chapter 15: Job

1. Schwienhorst-Schönberger, 2012, 426f.
2. Schmid, 2010b, 19.
3. Witte, 2010, 440–445.
4. Schmid, 2010b, 8.
5. von Stosch, 2013, 10.
6. von Stosch, 2013, 7f.
7. Hoerster, 2005, 87–113.
8. Hazony, 2015, 2f.
9. von Stosch, 2013, 7.
10. Loichinger & Kreiner, 2010, 7.
11. Weber, 1988, 231.
12. Isa 45:6–7.
13. Schmid, 2010b, 24f.
14. Schmid, 2010b, 21.

15. Witte, 2010, 441f. Schwienhorst-Schönberger, 2012, 419f. Schmid, 2010b, 56–62.
16. 2 Kings 23:25–29.
17. Jer 31:29–30. Hes 18:1–9.

Chapter 16: Daniel

1. Boyer, 2001, 203–207. Bering, 2012, 113f.
2. Barrett, 2004, 56. Boyer, 2001, 225.
3. Bering, 2012, 121–124.
4. Bloom, 2004, 191, 207.
5. Steadman et al., 1996.
6. Dever, 2012, 291.
7. A. Fischer, 2011b.
8. A. Fischer, 2011b.
9. Janowski, 2009, 466.
10. A. Fischer, 2011b.
11. Witte, 2010, 505f.
12. Niehr, 2012, 611.
13. Janowski, 2009, 471. Liess, 2005.
14. Madigan & Levenson, 2008, 59.
15. Janowski, 2009, 451.
16. Kühn, 2011.
17. Boyer, 2001, 207, 227.
18. Bering, 2012, 129, 88.
19. Bellah, 2011, 102. Lang, 2009, 19.
20. Assmann, 2001, 526.
21. Veyne, 2008, 138.
22. Assmann, 2006, 269. Boyer, 2001, 267.
23. van der Toorn, 1996, 218–230. Schmitt, 2006.
24. Janowski, 2009, 447f.
25. Janowski, 2009, 458–461. Riede, 2014. Eberhardt, 2007, 396.
26. van der Toorn, 1996, 225. Kühn, 2011.
27. A. Fischer, 2011b. Lang, 2009, 10ff.
28. van der Toorn, 1996, 225.
29. Madigan & Levenson, 2008, 69f.
30. Madigan & Levenson, 2008, 71–80.
31. Eberhardt, 2007, 398.
32. Tilly & Zwickel, 2011, 114.
33. Kratz, 2003, 106–110. Berlejung, 2010a, 184.
34. Oppenheimer, 2008, 34–36.

35. Madigan & Levenson, 2008, 3.
36. Lang, 2009, 22. A. Fischer, 2011a.
37. Antonacci, 2000, 107.
38. Dahlheim, 2014, 35–52.
39. Bieberstein, 2009, 443.
40. Lang, 2009, 21.
41. Lang, 2009, 26–28.
42. *Katechismus*, 2005, 291–300.
43. Sagan, 2013, 206.
44. A. Fischer, 2011a.
45. Lang, 2009, 39–42.
46. Lang, 2009, 54–61.
47. Lang, 2009, 61.
48. Dawkins, 2006a, 319f.
49. *Johannes B. Kerner*, *ZDF* [German TV talk show], November 15, 2007.
50. Madigan & Levenson, 2008, 257.
51. Bronner, 2011, 182f.
52. Flasch, 2013, 252.

Part V: The New Testament

1. Frankemölle, 2006, 23–29.
2. Veyne, 2011, 44.
3. Frankemölle, 2006, 42.
4. Stern, 2005, 353–355. Sand, 2014, 199–288.
5. Sand, 2014, 257.
6. Aslan, 2013, 202–212.
7. Veyne, 2011, 13.
8. Dawkins, 2006b.

Chapter 17: Jesus of Nazareth

1. Lauster, 2014, 31.
2. R. Wright, 2009, 247.
3. Roloff, 2007, 12. Lauster, 2014, 20f. Aslan, 2013, 152.
4. Mk 1:10–13.
5. Lauster, 2014, 23f. Roloff, 2007. Theissen & Merz, 2011.
6. Merz, 2009, 43.
7. Lauster, 2014, 34f.
8. Ratzinger, 2007, 10f.
9. Roloff, 2007, 120–127.
10. Linder, 2009, 12.
11. Ratzinger, 2007, 18.

12. Skiena & Ward, 2014, 5.
13. Aslan, 2013, 103. Theissen & Merz, 2011, 458.
14. Kratz, 2003, 36.
15. Burkert, 1996, 171.
16. Jenkins, 2010, 3.
17. Jn 20:30.
18. Aslan, 2013, 164f.
19. Mt 28:13–15. Aslan, 2013, 176f.
20. Barrett, 2004, 21–30. Boyer, 2001.
21. Madigan & Levenson, 2008, 7, 32.
22. Flasch, 2013, 109.
23. *Stuttgarter Erklärungsbibel*, 2007, 1403.
24. Roloff, 2007, 58.
25. Mt 1:18–23.
26. Flasch, 2013, 112. Pagels, 1998, 120.
27. Mk 6:3.
28. Mt 13:55.
29. Lk 4:22.
30. Aslan, 2013, 36f. Roloff, 2007, 59. Posener, 2007, 11–15. Pagels, 1998, 70.
31. Posener, 2007, 18.
32. Pagels, 1998, 121.
33. Posener, 2007, 91. Schreiner, 2003, 108.
34. Schreiner, 2003, 8.
35. Flasch, 2013, 113.
36. Aslan, 2013, 36.
37. Lk 2:4–5.
38. Roloff, 2007, 58.
39. Dahlheim, 2014, 37–47. Roloff, 2007, 36.
40. Lauster, 2014, 94.
41. Dahlheim, 2014, 47–52.
42. Bringmann, 2005, 258.
43. Aslan, 2013, 212.
44. Frankemölle, 2006, 118–125.
45. Pagels, 1998, 67–101. Flasch, 2015, 79.
46. Pagels, 1998, 247.
47. Lauster, 2014, 20.
48. Clark, 2010, 180.
49. Avalos, 2010. Lauster, 2014, 79.
50. Ehrman, 2014, 99.
51. Stern, 2005, 356.
52. Madigan & Levenson, 2008, 7f. Frankemölle, 2006, 96f.

53. Lk 11:20.
54. Rev 1:3.
55. Pagels, 2014, 167.
56. Mk 13:7–8.
57. Mk 13:24–27.
58. Roloff, 2007, 39f.
59. Mk 13:13.
60. Dahlheim, 2014, 74. Theissen & Merz, 2011, 468.
61. Mt 10:34.
62. Aslan, 2013, 120.
63. Frankemölle, 2006, 116–118. Roloff, 2007, 50f.
64. Lauster, 2014, 26. Merz, 2009, 30. Pagels, 1998, 25–66.
65. Aslan, 2013, 31f. Theissen & Merz, 2011, 467.
66. Merz, 2009, 40.
67. Theissen & Merz, 2011, 470–480.
68. Theissen & Merz, 2011, 481f. Aslan, 2013, 136.
69. Theissen & Merz, 2011, 458f.
70. Moeller, 2011, 16.
71. Theissen & Merz, 2011, 468. Merz, 2009, 41.
72. Merz, 2009, 36, 42. Moeller, 2011, 14.
73. Mk 9:1.
74. Mt 6:9. Aslan, 2013, 116.
75. Mt 5:17.
76. Mt 15:11.
77. Mk 2:27.
78. Mt 5:21–44.
79. Lauster, 2014, 28.
80. Theissen & Merz, 2011, 351.
81. Lauster, 2014, 28.
82. Theissen & Merz, 2011, 353.
83. Mt 6:24.
84. van Schaik, 2016. Boehm, 2012. Bernhard et al., 2006.
85. Boehm, 2012, 135.
86. Wilson et al., 2014.
87. Goodall, 1990, 210.
88. de Waal, 2006, 185.
89. Keeley, 1996. Gat, 2006. Pinker, 2011. Diamond, 2013.
90. Diamond, 2013, 158.
91. Wiessner, 2006.
92. Zimbardo, 2007. Smith, 2011.
93. Mt 22:37–39.

94. Baumard & Boyer, 2013.
95. Aslan, 2013, 121.
96. Moenikes, 2012.
97. Aslan, 2013, 122.
98. Mt 10:5–6.
99. Roloff, 2007, 94.
100. Theissen & Merz, 2011, 348. Moenikes, 2012.
101. R. Wright, 2009, 260.
102. Mt 7:3.
103. Jn 8:3–11.
104. Theissen & Merz, 2011, 331.
105. Diamond, 2013, 86–90.
106. Mt 12:30.
107. Mt 13:29–43.
108. Pagels, 1998, 39, 46.
109. Turchin, 2007, 264.
110. Mt 6:19.
111. Mk 10:25.
112. Merz, 2009, 51.
113. Mt 6:24.
114. Merz, 2001, 90.
115. Boehm, 2012, 46.
116. Mt 7:24–30.
117. Mt 5:3.
118. Theissen & Merz, 2011, 232, 247.
119. Lk 6:24–25.
120. Mt 5:4–6.
121. R. Wright, 2009, 262.
122. Linder, 2009, 62.
123. Ebner & Schreiber, 2013, 597.
124. Roloff, 2007, 79f.
125. Crossan, 1991. Mack, 2001. Lang, 2010.
126. Theissen & Merz, 2011, 29. Bilde, 2013, 263f.
127. Roloff, 2007, 91.
128. Roloff, 2007, 96.
129. Mk 10:42–44.
130. Mk 3:31–35. Roloff, 2007, 96.
131. Avalos, 2015, 51.
132. Mt 6:26.
133. Mt 11:19.
134. Lang, 2010, 176.

135. Diamond, 2013, 457.
136. Theissen & Merz, 2011, 203–208.
137. Roloff, 2007, 68f.
138. Dever, 2012, 287.
139. Meyers, 2013b, 4f.
140. Meyers, 2013b, 170.
141. Beit-Hallahmi, 2015, 89.
142. Jud 19.
143. Lk 15:4–7.
144. Veyne, 2011, 27.
145. Boehm, 2012, 47.
146. Ratzinger, 2007, 11.
147. Räisänen, 2012, 382.
148. Veyne, 2011, 31f.
149. Pagels, 1998, 251.
150. Trimondi & Trimondi, 2006, 13.
151. Schmidt-Salomon, 2012, 26–33.

Chapter 18: When Jesus Stayed in Heaven

1. Madigan & Levenson, 2008, 21.
2. Moss, 2013. Dahlheim, 2014, 388. Clauss, 2015, 75.
3. *Katechismus*, 2005, 1022.
4. Weber, 1980, 255.
5. Veyne, 2011, 28f.
6. David Flusser, qtd. in Frankemölle, 2006, 130f.
7. Lauster, 2014, 20.
8. Lauster, 2014, 13.
9. Theissen & Merz, 2011, 482.
10. Ehrman, 2014, 353.
11. Bultmann, 1960, 20.
12. Lauster, 2014, 119. Dahlheim, 2014, 366–372.
13. Böttrich, 2009, 112.
14. Gal 4:4.
15. Posener, 2007. Schreiner, 1994. Schreiner, 2003.
16. Schreiner, 2003, 26.
17. Keel, 2008.
18. Imbach, 2008, 17.
19. Keel, 2008, 62.
20. Weber, 1980, 255.
21. *Katechismus*, 2005, 391.
22. Lang, 2009, 54.

23. Rev 20:10.
24. Flasch, 2015.
25. Moeller, 2011, 68.
26. Lang, 2009, 32.
27. Moeller, 2011, 85.
28. *Ökumenisches Heiligenlexikon.*
29. Knoblauch, 2009, 172.
30. Ehrman, 2014, 212.
31. Flasch, 2015, 26, 399f.
32. Dahlheim, 2014, 357.
33. Veyne, 2008, 87f.
34. Hazony, 2015, 2f.
35. Veyne, 2011, 29.
36. Grabner-Haider, 2007, 323–334.
37. Jn 1:1–14.
38. Beierwaltes, 2014, 205–214.
39. James, 2014, 432–440.
40. C. Zimmermann, 2010.
41. Lauster, 2014, 125.
42. Veyne, 2011, 25.
43. Norenzayan, 2013. Johnson, 2015b.
44. *Katechismus,* 2005, 213.
45. Otto, 2007, 171.
46. Hazony, 2015.
47. Miles, 1995.
48. 1 Cor 15:3.
49. Bultmann, 1960, 20.
50. Bultmann, 1960, 18.
51. *Katechismus,* 2005, 614.
52. Jn 3:16.
53. Jn 1:29.
54. *Katechismus,* 2005, 614.
55. Metzger, 2012, 67f.
56. Flasch, 2013, 212f.
57. Lauster, 2014, 46, 49.
58. Dahlheim, 2014, 355.
59. Lauster, 2014, 62.
60. Veyne, 2011, 42.
61. 1 Cor 11:17–34.
62. Berger & Luckmann, 1998, 58.
63. Veyne, 2011, 46.

64. Dahlheim, 2014, 381f.
65. Flasch, 2005, 34–42.
66. Lauster, 2014, 66f.
67. Weber, 1995, 18f.
68. Berger & Luckmann, 1998, 120.
69. Lauster, 2014, 97.

Chapter 19: The Book of Nature

1. Harrison, 2015, 22.
2. Henry, 2008, 985.
3. Luckmann, 1991. Berger, 1997. Graf, 2004. Joas & Wiegandt, 2007. Knoblauch, 2009.
4. Harrison, 2015.
5. Brooke, 1991, 42.
6. Weber, 1995, 19.
7. Lindberg, 2010.
8. Nobis, 1971, 957–959. Vanderjagt & van Berkel, 2005.
9. Beierwaltes, 2014, 209.
10. Henry, 2008, 86f.
11. E. P. Fischer, 2007, 306.
12. Henry, 2008, 87.
13. Nobis, 1971, 957–959. Henry, 2008, 86. E. P. Fischer, 2007, 304.
14. K. Fischer, 2015, 13. Harrison, 2015, 172.
15. Galilei, 2012, 115.
16. Galilei, 2012, 67.
17. Dorn, 2000, 22–25, 103–106.
18. Galilei, 2012, 58.
19. Galilei, 2012, 67.
20. K. Fischer, 2015, 13f.
21. Henry, 2008, 85.
22. Singh, 2008, 86.
23. Harrison, 2010, 23.
24. Harrison, 2010, 23, 32.
25. Qtd. in E. P. Fischer, 2007, 300.
26. Numbers, 2009, 1f.
27. Harrison, 2010, 28.
28. Darwin, 1989, 119.
29. Dray, 2005.
30. Masci, 2007.
31. Weber, 1995, 41.
32. Joas, 2007, 14. Graf, 2004. Knoblauch, 2009.

33. Graf, 2004, 64f.
34. Casanova, 2007.
35. Luckmann, 1991. Knoblauch, 2009, 265.
36. Casanova, 2007, 340. Graf, 2004, 29.
37. Paloutzian & Park, 2013.
38. Veyne, 2011, 34f.
39. Darwin, 1989, 123.
40. Flasch, 2013, 252.
41. Knoblauch, 2009, 267.
42. Lübbe, 2004. Franz, 2009, 42–53.
43. Flasch, 2013, 252.
44. Knoblauch, 2009, 267.
45. de Waal, 2013, 207.
46. Bering, 2012, 114.
47. Dawkins, 2006a, 354–357.
48. de Waal, 2013, 204. Gottschall, 2013.
49. Graf, 2004, 65.
50. Singh, 2008, 87f.

Epilogue

1. de Waal, 2013, 234.
2. Haidt, 2006, 17–22.
3. Satlow, 2014, 1.

BIBLIOGRAPHY

Achenbach, Reinhard (2007). Murren. *Das wissenschaftliche Bibellexikon im Internet* (www.wibilex.de) (11/11/2015).

Albertz, Rainer (2003). *Geschichte und Theologie. Studien zur Exegese des Alten Testaments und zur Religionsgeschichte Israels.* Berlin: Walter de Gruyter.

Albertz, Rainer (2005/1978). *Persönliche Frömmigkeit und offizielle Religion. Religionsinterner Pluralismus in Israel und Babylon.* Atlanta, GA: Society of Biblical Literature.

Albertz, Rainer (2006). *Elia. Ein feuriger Kämpfer für Gott.* Leipzig: Evangelische Verlagsanstalt.

Albertz, Rainer (2013/2010). Personal Piety. In: Francesca Stavrakopoulou & John Barton (Eds.). *Religious Diversity in Ancient Israel and Judah.* London: Bloomsbury. 135–146.

Albertz, Rainer & Schmitt, Rüdiger (2012). *Family and Household Religion in Ancient Israel and the Levant.* Winona Lake, IN: Eisenbrauns.

Alkier, Stefan, Bauks, Michaela & Koenen, Klaus (Eds.) (2007ff.). *Das wissenschaftliche Bibellexikon im Internet* (www.wibilex.de) (11/11/2015).

Alt, Kurt W., Benz, Marion, Müller, Wolfgang, Berner, Margit E., Schultz, Michael, Schmidt-Schultz, Tyede H., Knipper, Corina, Gebel, Hans-Georg, Nissen, Hans J. & Vach, Werner (2013). Earliest Evidence for Social Endogamy in the 9,000-Year-Old-Population of Basta, Jordan. *Public Library of Science (PLOS) ONE* 8. 65649.

Anderson, Benedict (2006/1983). *Imagined Communities: Reflections on the Origin and Spread of Nationalism.* 2nd ed. London: Verso.

Antonacci, Mark (2000). *The Resurrection of the Shroud: New Scientific, Medical, and Archeological Evidence.* New York: M. Evans and Company.

Aslan, Reza (2013). *Zealot: The Life and Times of Jesus of Nazareth.* New York: Harper Element.

Assmann, Jan (1997). *Moses the Egyptian: The Memory of Egypt in Western Monotheism.* Cambridge, MA: Harvard University Press.

Assmann, Jan (2000). *Herrschaft und Heil. Politische Theologie in Altägypten, Israel und Europa.* Munich: Hanser.

Assmann, Jan (2001). *Tod und Jenseits im Alten Ägypten.* Munich: C. H. Beck (*Death and Salvation in Ancient Egypt.* Ithaca, NY: Cornell University Press, 2005).

Assmann, Jan (2006). Kulte und Religionen. Merkmale primärer und sekundärer Religion(serfahrung) im Alten Ägypten. In: Andreas Wagner (Ed.). *Primäre und sekundäre Religion als Kategorie der Religionsgeschichte des Alten Testaments.* Berlin: Walter de Gruyter. 269–280.

Assmann, Jan (2007/2000). *Religion und kulturelles Gedächtnis. Politische Theologie in Altägypten, Israel und Europa.* Munich: Hanser (*Religion and Cultural Memory.* Stanford, CA: Stanford University Press, 2005).

Assmann, Jan (2009/2000). *The Price of Monotheism.* Stanford, CA: Stanford University Press.

Assmann, Jan (2010). Zur Einführung: Die biblische Einstellung zu Wahrsagerei und Magie. In: Jan Assmann & Harald Strohm (Eds.). *Magie und Religion.* Munich: Wilhelm Fink Verlag. 11–22.

Assmann, Jan (2013a/1992). *Das kulturelle Gedächtnis. Schrift, Erinnerung und politische Identität in frühen Hochkulturen.* Munich: C. H. Beck (*Cultural Memory and Early Civilization: Writing, Remembrance, and Political Imagination.* Cambridge, MA: Harvard University Press, 2011).

Assmann, Jan (2013b/2006). *Monotheismus und die Sprache der Gewalt.* Vienna: Picus.

Assmann, Jan (2015). *Exodus. Die Revolution der Alten Welt.* Munich: C. H. Beck.

Atran, Scott (2002). *In Gods We Trust: The Evolutionary Landscape of Religion.* Oxford: Oxford University Press.

Atran, Scott & Henrich, Joseph (2010). The Evolution of Religion. How Cognitive By-Products, Adaptive Learning Heuristics, Ritual Displays, and Group Competition Generate Deep Commitments to Prosocial Religions. *Biology Theory* 5. 18–30.

Aurelius, Erik (2003). Die fremden Götter im Deuteronomium. In: Manfred Oeming und Konrad Schmid (Eds.). *Der eine Gott und die Götter. Polytheismus und Monotheismus im antiken Israel.* Zurich: Theologischer Verlag. 145–169.

Avalos, Hector (1995). *Illness and Health Care in the Ancient Near East: The Role of the Temple in Greece, Mesopotamia, and Israel.* Atlanta, GA: Scholars Press.

Avalos, Hector (2010/1999). *Health Care and the Rise of Christianity.* Grand Rapids, MI: Baker.

Avalos, Hector (2015). *The Bad Jesus: The Ethics of New Testament Ethics.* Sheffield: Sheffield Phoenix Press.

Barrett, Justin L. (2004). *Why Would Anyone Believe in God?* Walnut Creek, CA: AltaMira Press.

Barrett, Justin L. (2011). *Cognitive Science, Religion, and Theology: From Human Minds to Divine Minds.* West Conshohocken, PA: Templeton Press.

Barrett, Justin L. (2013). Exploring Religion's Basement: The Cognitive Science of Religion. In: Raymond D. Paloutzian & Chrystal L. Park (Eds.). *Handbook of the Psychology of Religion and Spirituality.* New York: Guilford Press. 234–255.

Barton, John (2013/2010). Reflecting on Religious Diversity. In: Francesca Stavrakopoulou & John Barton (Eds.). *Religious Diversity in Ancient Israel and Judah.* London: Bloomsbury. 191–193.

Bateson, Melissa, Nettle, Daniel & Roberts, Gilbert (2006). Cues of Being Watched Enhance Cooperation in a Real-World Setting. *Biology Letters* 2. 412–414.

Bauks, Michaela (2011). Monotheismus. *Das wissenschaftliche Bibellexikon im Internet* (www.wibilex.de) (11/11/2015).

Baumard, Nicolas & Boyer, Pascal (2013). Explaining Moral Religions. *Trends in Cognitive Science* 17. 272–280.

Baumgart, Norbert Clemens (2005). Sintflut/Sintfluterzählung. *Das wissenschaftliche Bibellexikon im Internet* (www.wibilex.de) (11/11/2015).

Baumgart, Norbert Clemens (2006). Turmbauerzählung. *Das wissenschaftliche Bibellexikon im Internet* (www.wibilex.de) (11/11/2015).

Becker, Uwe (2005). Von der Staatsreligion zum Monotheismus. Ein Kapitel israelitisch-jüdischer Religionsgeschichte. *Zeitschrift für Theologie und Kirche* 102. 1–16.

Beierwaltes, Werner (2014/1998). *Platonismus im Christentum.* 3rd Edit. Frankfurt am Main: Vittorio Klostermann.

Beit-Hallahmi, Benjamin (2015). *Psychological Perspectives on Religion and Religiosity.* London: Routledge.

Bellah, Robert (2011). *Religion in Human Evolution: From the Paleolithic to the Axial Age.* Cambridge, MA: Belknap Press of the Harvard University Press.

Ben-Sasson, Haim-Hillel (Ed.) (2005/1976). *Geschichte des Jüdischen Volkes. Von den Anfängen bis zur Gegenwart.* 5th Edit. Munich: C. H. Beck (*A History of the Jewish People.* Cambridge, MA: Harvard University Press, 1985/1976).

Bentzen, Jeanet Sinding (2013). Origins of Religiousness: The Role of Natural Disasters. University of Copenhagen, Department of Economics. Discussion Paper 13-02 (http://cope.ku.dk/publications/workingpaper_SSRN-id2221859.pdf) (11/11/2015).

Bentzen, Jeanet Sinding (2015). Acts of God? Religiosity and Natural Disasters Across Subnational World Districts. University of Copenhagen, Department of Economics. Discussion Paper 15-06 (www.economics.ku.dk/research/publications/wp/dp_2015/1506.pdf) (11/11/2015).

Berbesque, J. Colette, Marlowe, Frank W., Shaw, Peter & Thompson, Peter (2014). Hunter-Gatherers Have Less Famine Than Agriculturalists. *Biology Letters* 10. 8–53.

Berger, Peter L. (1997/1992). *Sehnsucht nach Sinn. Glauben in einer Zeit der Leichtgläubigkeit.* Frankfurt/New York: Campus (*A Far Glory: The Quest for Faith in an Age of Credulity.* New York: The Free Press, 1992).

Berger, Peter L. & Luckmann, Thomas (1998/1966). *Die gesellschaftliche Konstruktion der Wirklichkeit.* 5th Edit. Frankfurt am Main: Fischer (*The Social Construction of Reality.* New York: Doubleday, 1966).

Bering, Jesse (2012/2011). *The God Instinct: The Psychology of Souls, Destiny, and the Meaning of Life.* New York: Norton.

Berlejung, Angelika (2010a). Geschichte und Religionsgeschichte des antiken Israel. In: Jan Christian Gertz (Ed.). *Grundinformation Altes Testament.* Göttingen: Vandenhoeck & Ruprecht. 59–192.

Berlejung, Angelika (2010b). Auf den Leib geschrieben. Körper und Krankheit in der physiognomischen Tradition des Alten Orients und des Alten Testaments. In: Gregor Etzelmüller & Annette Weissenrieder (Eds.). *Religion und Krankheit.* Darmstadt: Wissenschaftliche Buchgesellschaft. 185–216.

Berlejung, Angelika (Ed.) (2012). *Disaster and Relief Management. Katastrophen und ihre Bewältigung.* Tübingen: Mohr Siebeck.

Berlejung, Angelika & Janowski, Bernd (Eds.) (2009). *Tod und Jenseits im alten Israel und in seiner Umwelt. Theologische, religionsgeschichtliche, archäologische und ikonographische Aspekte.* Tübingen: Mohr Siebeck.

Bernhard, Helen, Fischbacher, Urs & Fehr, Ernst (2006). Parochial Altruism in Humans. *Nature* 442. 912–915.

Biale, David (Ed.) (2002). *Cultures of the Jews: A New History.* New York: Schocken Books.

Bieberstein, Klaus (2009). Jenseits der Todesschwelle. Die Entstehung der Auferweckungshoffnungen in der alttestamentlichen-frühjüdischen Literatur. In: Angelika Berlejung & Bernd Janowski (Eds.). *Tod und Jenseits im alten Israel und in seiner Umwelt. Theologische, religionsgeschichtliche, archäologische und ikonographische Aspekte.* Tübingen: Mohr Siebeck. 423–446.

Bilde, Per (2013). *The Originality of Jesus: A Critical Discussion and a Comparative Attempt.* Göttingen: Vandenhoeck & Ruprecht.

Bloch, Ernst (1975). Sinn der Bibelkritik. In: Hans Jürgen Schultz (Ed.). *Sie werden lachen–die Bibel. Überraschungen mit dem Buch.* Stuttgart: Kreuz. 43–56.

Bloom, Paul (2004). *Descartes' Baby: How the Science of Child Development Explains What Makes Us Human.* New York: Basic Books.

Blum, Erhard (2004). Von Gottesunmittelbarkeit zu Gottähnlichkeit. Überlegungen zur theologischen Anthropologie der Paradieserzählung. In: Gönke Eberhardt & Kathrin Liess (Eds.). *Gottes Nähe im Alten Testament.* Stuttgart: Katholisches Bibelwerk. 9–29.

Blum, Erhard (2012). Der historische Mose und die Frühgeschichte Israels. *Hebrew Bible and Ancient Israel* 1. 37–63.

Boehm, Christopher (2012). *Moral Origins: The Evolution of Virtue, Altruism, and Shame.* New York: Basic Books.

Bosshard-Nepustil, Erich (2005). *Vor uns die Sintflut. Studien zu Text, Kontexten und Rezeption der Fluterzählung Genesis 6–9.* Stuttgart: Kohlhammer.

Botero, Carlos A., Gardner, Beth, Kirby, Kathryn R., Bulbulia, Joseph, Gavin, Michael C. & Gray, Russell D. (2014). The Ecology of Religious Beliefs. *Proceedings of the National Academy of Sciences (PNAS)* 111. 16784–16789.

Böttrich, Christfried (2009). Jesus und Maria im Christentum. In: Christfried Böttrich, Beate Ego & Friedmann Eissler (Eds.). *Jesus und Maria in Judentum, Christentum und Islam.* Göttingen: Vandenhoeck & Ruprecht. 60–119.

Bourdieu, Pierre (1993). *Soziologische Fragen.* Frankfurt am Main: Suhrkamp.

Bowlby, John (1988). *A Secure Base: Parent-Child Attachment and Healthy Human Development.* New York: Basic Books.

Bowles, Samuel (2011). Cultivation of Cereals by the First Farmers Was Not More Productive Than Foraging. *Proceedings of the National Academy of Sciences of the United States of America (PNAS)* 108. 4760–4765.

Boyer, Pascal (2001). *Religion Explained: The Evolutionary Origins of Religious Thought.* New York: Basic Books.

Brandscheidt, Renate (2010). Kain und Abel. *Das wissenschaftliche Bibellexikon im Internet* (www.wibilex.de) (11/11/2015).

Bringmann, Klaus (2005). *Geschichte der Juden im Altertum. Vom babylonischen Exil bis zur arabischen Eroberung.* Stuttgart: Klett-Cotta.

Bronner, Leila Leah (2011). *Journey to Heaven: Exploring Jewish Views of the Afterlife.* Jerusalem: Urim Publications.

Brooke, John Hedley (1991). *Science and Religion: Some Historical Perspectives.* Cambridge, UK: Cambridge University Press.

Brotherton, Robert (2013). *The President Is Dead: Why Conspiracy Theories About the Death of JFK Endure* (http://conspiracypsychology.com/2013/11/21/jfk-conspiracy-theories) (11/11/2015).

Buchner, Norbert & Buchner, Elmar (2011). *Klima und Kulturen. Die Geschichte von Paradies und Sintflut.* Weinstadt: Bernhard Albert Greiner.

Bulbulia, Joseph, Geertz, Armin W., Atkinson, Quentin D., Cohen, Emma, Evans, Nicholas, François, Pieter, Gintis, Herbert, Gray, Russell D., Henrich, Joseph, Jordon, Fiona M., Norenzayan, Ara, Richerson, Peter J., Slingerland, Edward, Turchin, Peter, Whitehouse, Harvey, Widlok, Thomas, & Wilson, David S. (2013). The Cultural Evolution of Religion. In: Peter J. Richerson & Morton H. Christiansen (Eds.). *Cultural Evolution: Society, Technology, Language, and Religion.* Cambridge, MA: MIT Press. 381–404.

Bultmann, Rudolf (1960/1941). Neues Testament und Mythologie. Das Problem der Entmythologisierung der neutestamentlichen Verkündigung. In: Hans-Werner Bartsch (Ed.). *Kerygma und Mythos. Ein theologisches Gespräch.* Vol. 1. 4th Edit. Hamburg: Herbert Reich. Evangelischer Verlag. 15–48.

Burkert, Walter (1996). *Creation of the Sacred: Tracks of Biology in Early Religions.* Cambridge, MA: Harvard University Press.

Burkert, Walter (1998). *Kulte des Altertums. Biologische Grundlagen der Religion.* Munich: C. H. Beck.

Byron, John (2011). *Cain and Abel in Text and Tradition: Jewish and Christian Interpretations of the First Sibling Rivalry.* Leiden: Brill.

Casanova, José (2007). Die religiöse Lage in Europa. In: Hans Joas & Klaus Wiegandt (Eds.). *Säkularisierung und die Weltreligionen.* Frankfurt am Main: Fischer. 322–358.

Clark, David P. (2010). *Germs, Genes, & Civilization: How Epidemics Shaped Who We Are Today.* Upper Saddle River, NJ: FT Press.

Clauss, Manfred (2015). *Ein neuer Gott für die alte Welt. Die Geschichte des frühen Christentums.* Berlin: Rowohlt–Berlin Verlag.

Cohen, Mark N. & Armelagos, George J. (2013/1984). *Paleopathology at the Origins of Agriculture.* Gainesville: University Press of Florida.

Colpe, Carsten (2007). Monotheismus. In: Gesine Palmer (Ed.). *Fragen nach dem einen Gott. Die Monotheismusdebatte im Kontext.* Tübingen: Mohr Siebeck. 23–28.

Crawford, Dorothy H. (2009/2007). *Deadly Companions: How Microbes Shaped Our History.* Oxford: Oxford University Press.

Crossan, John Dominic (1991). *The Historical Jesus: The Life of a Mediterranean Jewish Peasant.* New York: HarperCollins.

Crüsemann, Frank (1996). *The Torah: Theology and Social History of Old Testamental Law.* Edinburgh: T&T Clark.

Dahlheim, Werner (2014). *Die Welt zur Zeit Jesu.* Munich: C. H. Beck.

Darwin, Charles (1989). *The Works of Charles Darwin.* Vol. 29. New York: New York University Press.

Dawkins, Richard (2006a). *The God Delusion.* Boston: Houghton Mifflin.

Dawkins, Richard (2006b). *Atheists for Jesus* (www.rationalresponders.com/atheists_for_jesus_a_richard_dawkins_essay) (11/11/2015).

Day, John (Ed.) (2010). *Prophecy and Prophets in Ancient Israel: Proceedings of the Oxford Old Testament Seminar.* New York: T&T Clark.

de Waal, Frans (2006). *Der Affe in uns. Warum wir sind, wie wir sind.* Munich: Hanser (*Our Inner Ape: A Leading Primatologist Explains Why We Are Who We Are.* New York: Penguin, 2006).

de Waal, Frans (2013). *The Bonobo and the Atheist: In Search of Humanism Among the Primates.* New York: W. W. Norton & Company.

Dekkers, Midas (1994). *Geliebtes Tier. Die Geschichte einer innigen Beziehung.* Munich: Hanser.

Dennett, Daniel C. (2008/2006). *Den Bann brechen. Religion als natürliches Phänomen.* Frankfurt am Main: Verlag der Weltreligionen (*Breaking the Spell: Religion as a Natural Phenomenon.* New York: Viking, 2006).

Dever, William G. (2006). *Did God Have a Wife? Archaeology and Folk Religion in Ancient Israel.* Grand Rapids, MI: W. B. Eerdmans.

Dever, William G. (2012). *The Lives of Ordinary People in Ancient Israel: When Archaeology and the Bible Intersect.* Grand Rapids, MI: W. B. Eerdmans.

Diamond, Jared (1987). The Worst Mistake in the History of the Human Race. *Discover Magazine,* May. 64–66.

Diamond, Jared (2002) Evolution, Consequences and Future of Plant and Animal Domestication. *Nature* 418. 700–707.

Diamond, Jared (2012). *The World Until Yesterday: What Can We Learn from Traditional Societies?* New York: Penguin.

Diamond, Jared (2013/1997). *Guns, Germs, and Steel: The Fate of Human Societies.* London: Vintage.

Dobelli, Rolf (2011). *Die Kunst des klaren Denkens. 52 Denkfehler, die Sie besser anderen überlassen.* Munich: Hanser (*The Art of Thinking Clearly.* New York: Harper, 2013).

Dorn, Matthias (2000). *Das Problem der Autonomie der Naturwissenschaften bei Galilei.* Stuttgart: Franz Steiner Verlag.

Douglas, Mary (2003/1966). *Purity and Danger: An Analysis of Concepts of Pollution and Taboo.* New York: Routledge.

Dray, Philip (2005). *Stealing God's Thunder: Benjamin Franklin's Lightning Rod and the Invention of America.* New York: Random House.

Durkheim, Émile (2007/1912). *Die elementaren Formen des religiösen Lebens.* Frankfurt am Main: Verlag der Weltreligionen (*The Elementary Forms of the Religious Life.* New York: Free Press, 1995).

Dyma, Oliver (2010). Ehe (AT). *Das wissenschaftliche Bibellexikon im Internet* (www.wibilex.de) (11/11/2015).

Eberhardt, Gönke (2007). *JHWH und die Unterwelt. Spuren einer Kompetenzausweitung JHWHs im Alten Testament.* Tübingen: Mohr Siebeck.

Eberhart, Christian (2006). Schlachtung/Schächtung. *Das wissenschaftliche Bibellexikon im Internet* (www.wibilex.de) (11/11/2015).

Eberhart, Christian (2008). Blutschande. *Das wissenschaftliche Bibellexikon im Internet* (www.wibilex.de) (11/11/2015).

Ebner, Martin & Schreiber, Stefan (Eds.) (2013/2008). *Einleitung in das Neue Testament.* 2nd Edit. Stuttgart: Kohlhammer.

Ego, Beate (2007). Reinheit/Unreinheit/Reinigung (AT). *Das wissenschaftliche Bibellexikon im Internet* (www.wibilex.de) (11/11/2015).

Ego, Beate (2011). Adam und Eva im Judentum. In: Christfried Böttrich, Beate Ego & Friedmann Eissler (Eds.). *Adam und Eva in Judentum, Christentum und Islam.* Göttingen: Vandenhoeck & Ruprecht. 11–78.

Ehrman, Bart (2014). *How Jesus Became God: The Exaltation of a Jewish Preacher from Galilee.* New York: HarperOne.

Elias, Norbert (1997/1939). *Über den Prozess der Zivilisation.* 2 vols. Frankfurt am Main: Suhrkamp (*The Civilizing Process: Sociogenetic and Psychogenetic Investigations.* Malden, MA: Blackwell, 2000).

Ellens, J. Harold & Rollins, Wayne G. (Eds.) (2004). *Psychology and the Bible: A New Way to Read the Scriptures.* 4 vols. Westport, CT: Praeger.

Enz, Peter (2012). *Der Keim der Revolte. Militante Solidarität und religiöse Mission bei Ibn Khaldun.* Freiburg: Alber.

Feldmeier, Reinhard & Spiekermann, Hermann (2011). *Der Gott der Lebendigen. Eine biblische Gotteslehre.* Tübingen: Mohr Siebeck.

Finkelstein, Israel (2014). *Das vergessene Königreich. Israel und die verborgenen Ursprünge der Bibel.* Munich: C. H. Beck (*The Forgotten Kingdom: The Archaeology and History of Northern Israel.* Atlanta, GA: Society of Biblical Literature, 2013).

Finkelstein, Israel & Silberman, Neil A. (2001). *The Bible Unearthed: Archaeology's New Vision of Ancient Israel and the Origin of Its Sacred Texts.* New York: Free Press.

Finkelstein, Israel & Silberman, Neil A. (2006). *David and Solomon: In Search of the Bible's Sacred Kings and the Roots of Western Tradition.* New York: Free Press.

Fischer, Alexander (2011a). Auferweckung. *Das wissenschaftliche Bibellexikon im Internet* (www.wibilex.de) (11/11/2015).

Fischer, Alexander (2011b). Tod. *Das wissenschaftliche Bibellexikon im Internet* (www.wibilex.de) (11/11/2015).

Fischer, Ernst Peter (2007). Die Wissenschaft zittert nicht. Die säkularen Naturwissenschaften und das moderne Lebensgefühl. In: Hans Joas & Klaus Wiegandt (Eds.). *Säkularisierung und die Weltreligionen.* Frankfurt am Main: Fischer. 284–321.

Fischer, Irmtraud (2013/1995). *Gottesstreiterinnen. Biblische Erzählungen über die Anfänge Israels.* Stuttgart: Kohlhammer.

Fischer, Irmtraud, Navarro Puerto, Mercedes & Taschl-Erber, Andrea (Eds.). (2010). *Tora. Die Bibel und die Frauen. Eine exegetisch-kulturgeschichtliche Enzyklopädie.* Vol. 1.1. Stuttgart: Kohlhammer.

Fischer, Klaus (2015). *Galileo Galilei. Biographie seines Denkens.* Stuttgart: Kohlhammer.

Flasch, Kurt (2005/2004). *Eva und Adam. Wandlungen eines Mythos.* Munich: C. H. Beck.

Flasch, Kurt (2013). *Warum ich kein Christ bin. Bericht und Argumentation.* Munich: C. H. Beck.

Flasch, Kurt (2015). *Der Teufel und seine Engel. Die neue Biographie.* Munich: C. H. Beck.

Flavius Josephus (2011/94 CE). *Jüdische Altertümer.* Wiesbaden: Marixverlag.

Foucault, Michel (1977). *Überwachen und Strafen. Die Geburt des Gefängnisses.* Frankfurt am Main: Suhrkamp (*Discipline and Punish: The Birth of the Prison.* London: Vintage, 1995).

Frahm, Eckart (2011). Mensch, Land und Volk: Assur im Alten Testament. In: Johannes Renger (Ed.). *Assur: Gott, Stadt und Land.* Wiesbaden: Harrassowitz. 267–285.

Frankemölle, Hubert (2006). *Frühjudentum und Urchristentum. Vorgeschichte, Verlauf, Auswirkungen.* Stuttgart: Kohlhammer.

Franz, Jürgen H. (2009). *Religion in der Moderne. Die Theorien von Jürgen Habermas und Hermann Lübbe.* Berlin: Frank & Timme.

Freud, Sigmund (1993/1914). *Der Moses des Michelangelo. Schriften über Kunst und Künstler.* Frankfurt am Main: Fischer.

Freud, Sigmund (2010/1930). *Das Unbehagen in der Kultur. Und andere kulturtheoretische Schriften.* Frankfurt am Main: Fischer (*Civilization and Its Discontents.* New York: Penguin, 2002).

Freuling, Georg (2008). Tun-Ergehen-Zusammenhang. *Das wissenschaftliche Bibellexikon im Internet* (www.wibilex.de) (11/11/2015).

Frevel, Christian (2003). YHWH und die Göttin bei den Propheten. In: Manfred Oeming und Konrad Schmid (Eds.). *Der eine Gott und die Götter. Polytheismus und Monotheismus im antiken Israel.* Zurich: Theologischer Verlag. 49–75.

Frevel, Christian (Ed.) (2010). *Biblische Anthropologie. Neue Einsichten aus dem Alten Testament.* Freiburg: Herder.

Frevel, Christian (2012). Grundriss der Geschichte Israels. In: Erich Zenger & Christian Frevel (Eds.). *Einleitung in das Alte Testament.* Stuttgart: Kohlhammer. 701–870.

Frey-Anthes, Henrike (2007). Krankheit und Heilung (AT). *Das wissenschaftliche Bibellexikon im Internet* (www.wibilex.de) (11/11/2015).

Frey-Anthes, Henrike (2008). Lilit. *Das wissenschaftliche Bibellexikon im Internet* (www.wibilex.de) (11/11/2015).

Galilei, Galileo (2012). *Selected Writings.* New York: Oxford University Press.

Gat, Azar (2006). *War in Human Civilization.* New York: Oxford University Press.

Gatz, Bodo (1967). *Weltalter, goldene Zeit und sinnverwandte Vorstellungen.* Hildesheim: Olms.

Gazzaniga, Michael (2012/2011). *Die Ich-Illusion. Wie Bewusstsein und freier Wille entstehen.* Munich: Hanser (*Who's in Charge? Free Will and the Science of the Brain.* New York: Ecco, 2011).

Geertz, Clifford (1987/1973). *Dichte Beschreibung. Beiträge zum Verstehen kultureller Systeme.* Frankfurt am Main: Suhrkamp (*Interpretation of Culture: Selected Essays.* New York: Basic Books, 1973).

Gertz, Jan Christian (2008). Mose. *Das wissenschaftliche Bibellexikon im Internet* (www.wibilex.de) (11/11/2015).

Gertz, Jan Christian (2010a/2006). *Grundinformation Altes Testament.* 4th Edit. Göttingen: Vandenhoeck & Ruprecht.

Gertz, Jan Christian (2010b). Tora und Vordere Propheten. In: Jan Christian Gertz (Ed.). *Grundinformation Altes Testament.* Göttingen: Vandenhoeck & Ruprecht. 193–311.

Gies, Kathrin (2012). Speisegebote (AT). *Das wissenschaftliche Bibellexikon im Internet* (www.wibilex.de) (11/11/2015).

Giese, Alma (2011). Ibn Khaldun—Leben und Werk. In: Ibn Khaldun. *Die Muqaddima. Betrachtungen zur Weltgeschichte.* Munich: C. H. Beck. 13–62.

Gintis, Herbert, van Schaik, Carel & Boehm, Christopher (2015). Zoon Politikon: The Evolutionary Origins of Human Political Systems. *Current Anthropology* 56. 327–353.

Gladwell, Malcolm (2013). *David and Goliath: Underdogs, Misfits and the Art of Battling Giants.* New York: Little, Brown and Company.

Glenday, Craig (2015). *Guinness World Records.* New York: Bantam Books.

Goethe, Johann Wolfgang von (1998/1981). *Werke. Hamburger Ausgabe.* Munich: Deutscher Taschenbuch Verlag.

Goodall, Jane (1990). *Through a Window: My Thirty Years with the Chimpanzees of Gombe.* Boston/New York: Houghton Mifflin.

Gopnik, Alison (2000). Explanation as Orgasm and the Drive for Causal Understanding: The Function, Evolution, and Phenomenology of the

Theory Formation System. In: Frank C. Keil & Robert A. Wilson (Eds.). *Cognition and Explanation.* Cambridge, MA: MIT Press. 299–323.

Gottschall, Jonathan (2013). *The Storytelling Animal: How Stories Make Us Human.* New York: Mariner.

Grabbe, Lester M. (2010). Shaman, Preacher, or Spirit Medium? The Israelite Prophet in the Light of Anthropological Models. In: John Day (Ed.). *Prophecy and Prophets in Ancient Israel: Proceedings of the Oxford Old Testament Seminar.* New York: T&T Clark. 117–132.

Grabner-Haider, Anton (Ed.) (2007). *Kulturgeschichte der Bibel.* Göttingen: Vandenhoeck & Ruprecht.

Graf, Friedrich Wilhelm (2004). *Die Wiederkehr der Götter. Religion in der modernen Kultur.* Munich: C. H. Beck.

Grätz, Sebastian (2006). Baal. *Das wissenschaftliche Bibellexikon im Internet* (www.wibilex.de) (11/11/2015).

Gray, Kurt & Wegner, Daniel M. (2010). Blaming God for Our Pain: Human Suffering and the Divine Mind. *Personality and Social Psychology Review* 14. 7–16.

Gudme, Anne Katrine de Hemmer (2010). Modes of Religion: An Alternative to "Popular/Official" Religion. In: Emanuel Pfoh (Ed.). *Anthropology and the Bible: Critical Perspectives.* Piscataway, NJ: Gorgias Press. 77–90.

Guthrie, Stewart (1993). *Faces in the Clouds: A New Theory of Religion.* New York: Oxford University Press.

Haidt, Jonathan (2006). *The Happiness Hypothesis: Finding Modern Truth in Ancient Wisdom.* New York: Basic Books.

Haidt, Jonathan (2012). *The Righteous Mind: Why People Are Divided by Politics and Religion.* London: Allen Lane.

Halbfas, Hubertus (2010/2001). *Die Bibel.* Ostfildern: Patmos.

Harris, Marvin (1990/1985). *Wohlgeschmack und Widerwillen. Die Rätsel der Nahrungsmitteltabus.* Stuttgart: Klett-Cotta (*Good to Eat: Riddles of Food and Culture.* New York: Simon & Schuster, 1985).

Harrison, Peter (2010). "Science" and "Religion": Constructing the Boundaries. In: Thomas Dixon, Geoffrey Cantor & Stephen Pumfrey (Eds.). *Science and Religion: New Historical Perspectives.* New York: Cambridge University Press. 23–49.

Harrison, Peter (2015). *The Territories of Science and Religion.* Chicago: University of Chicago Press.

Hazony, Yoram (2015). *The Question of God's Perfection* (http://bibleand philosophy.org/wp-content/uploads/2015/01/hazony-question-of-gods -perfection.pdf) (11/11/2015).

Heinberg, Richard (1995). *Memories and Visions of Paradise: Exploring the Universal Myth of a Lost Golden Age.* Wheaton: Quest Books.

Hendel, Ronald S. (2002). Israel Among the Nations: Biblical Culture in the Ancient Near East. In: David Biale (Ed.). *Cultures of the Jews: A New History.* New York: Schocken Books. 43–76.

Hendel, Ronald S. (2010). *Reading Genesis: Ten Methods.* New York: Cambridge University Press.

Henrich, Joseph (2009). The Evolution of Costly Displays, Cooperation and Religion: Credibility Enhancing Displays and Their Implications for Cultural Evolution. *Evolution and Human Behavior* 30. 244–260.

Henrich, Joseph, Boyd, Robert & Richerson, Peter J. (2012). The Puzzle of Monogamous Marriage. *Philosophical Transactions of the Royal Society B* 367. 657–669.

Henry, John (2008/1997). *The Scientific Revolution and the Origins of Modern Science.* 3rd ed. New York: Palgrave Macmillan.

Hensel, Benedikt (2011). *Die Vertauschung des Erstgeburtssegens in der Genesis.* Berlin: Walter de Gruyter.

Hentschel, Georg (2012a). Das Buch Josua. In: Erich Zenger & Christian Frevel (Eds.). *Einleitung in das Alte Testament.* Stuttgart: Kohlhammer. 257–268.

Hentschel, Georg (2012b). Die Samuelbücher. In: Erich Zenger & Christian Frevel (Eds.). *Einleitung in das Alte Testament.* Stuttgart: Kohlhammer. 290–300.

Hentschel, Georg (2012c). Die Königsbücher. In: Erich Zenger & Christian Frevel (Eds.). *Einleitung in das Alte Testament.* Stuttgart: Kohlhammer. 301–312.

Herget, Jürgen (2012). *Am Anfang war die Sintflut. Hochwasserkatastrophen in der Geschichte.* Primus: Darmstadt.

Hewlett, Bonnie L. & Hewlett, Barry S. (2013). Hunter-Gatherer Adolescence. In: Bonnie L. Hewlett (Ed.). *Adolescent Identity: Evolutionary, Developmental and Cultural Perspectives.* London: Routledge. 73–104.

Hieke, Thomas (2010). Genealogie als Mittel der Geschichtsdarstellung in der Tora und die Rolle der Frauen im genealogischen System. In: Irmtraud Fischer, Mercedes Navarro Puerto & Andrea Taschl-Erber (Eds.). *Tora. Die Bibel und die Frauen. Eine exegetisch-kulturgeschichtliche Enzyklopädie.* Vol. 1.1. Stuttgart: Kohlhammer. 149–185.

Hill, Kim & Hurtado, A. Magdalena (1996). *Aché Life History: The Ecology and Demography of a Foraging People.* New York: Aldine de Gruyter.

Hoerster, Norbert (2005). *Die Frage nach Gott.* Munich: C. H. Beck.

Hume, David (2000/1757). *Die Naturgeschichte der Religion.* Hamburg: Felix Meiner Verlag (*Dialogues and Natural History of Religion.* New York: Oxford University Press, 1993).

Ibn Khaldun (2011/1377). *Die Muqaddima. Betrachtungen zur Weltgeschichte.* Munich: C. H. Beck.

Imbach, Josef (2008). *Marienverehrung zwischen Glaube und Aberglaube.* Düsseldorf: Patmos.

Jacobs, A. J. (2007). *The Year of Living Biblically.* New York: Simon & Schuster.

James, William (2014/1902). *Die Vielfalt religiöser Erfahrung. Eine Studie über die menschliche Natur.* Frankfurt am Main: Verlag der Weltreligionen (*The Varieties of Religious Experience: A Study in Human Nature.* New York: Penguin, 1982).

Janowski, Bernd (2005). Der Mensch im alten Israel. Grundfragen alttestamentlicher Anthropologie. *Zeitschrift für Theologie und Kirche* 102. 143–175.

Janowski, Bernd (2009). JHWH und die Toten. Zur Geschichte des Todes im Alten Israel. In: Angelika Berlejung & Bernd Janowski (Eds.). *Tod und Jenseits im alten Israel und in seiner Umwelt. Theologische, religionsgeschichtliche, archäologische und ikonographische Aspekte.* Tübingen: Mohr Siebeck. 447–477.

Janowski, Bernd (2013). *Ein Gott, der straft und tötet? Zwölf Fragen zum Gottesbild des Alten Testaments.* Neukirchen-Vluyn: Neukirchener Theologie.

Jenkins, Philip (2010). *Jesus Wars: How Four Patriarchs, Three Queens and Two Emperors Decided What Christians Would Believe for the Next 1500 Years.* New York: HarperOne.

Jeremias, Jörg (2011/2009). *Der Zorn Gottes im Alten Testament. Das biblische Israel zwischen Verwerfung und Erwählung.* Neukirchen-Vluyn: Neukirchener Theologie.

Jericke, Delef (2012). Hebräer/Hapiru. *Das wissenschaftliche Bibellexikon im Internet* (www.wibilex.de) (11/11/2015).

Joas, Hans (2007). Die religiöse Lage in den USA. In: Hans Joas & Klaus Wiegandt (Eds.). *Säkularisierung und die Weltreligionen.* Frankfurt am Main: Fischer. 358–375.

Joas, Hans & Wiegandt, Klaus (Eds.) (2007). *Säkularisierung und die Weltreligionen.* Frankfurt am Main: Fischer.

Johnson, Dominic (2005). God's Punishment and Public Goods: A Test of the Supernatural Punishment Hypothesis in 186 World Cultures. *Human Nature* 16. 410–446.

Johnson, Dominic (2015a). Big Gods, Small Wonder: Supernatural Punishment Strikes Back. *Religion, Brain & Behavior* 5. 290–298.

Johnson, Dominic (2015b). *God Is Watching You: How the Fear of God Makes Us Human.* New York: Oxford University Press.

Johnson, Dominic & Bering, Jesse (2009). Hand of God, Mind of Man: Punishment and Cognition in the Evolution of Cooperation. In: Jeffrey Schloss & Michael Murray (Eds.). *The Believing Primate: Scientific, Philosophical, and Theological Reflections on the Origin of Religion.* New York: Oxford University Press. 26–43.

Johnson, Dominic & Kruger, Oliver (2004). The Good of Wrath: Supernatural Punishment. *Political Theology* 5. 159–176.

Joyce, Paul (2010). The Prophets and Psychological Interpretation. In: John Day (Ed.). *Prophecy and Prophets in Ancient Israel: Proceedings of the Oxford Old Testament Seminar.* New York: T&T Clark. 133–148.

Kaiser, Otto (2013). *Der eine Gott Israels und die Mächte der Welt. Der Weg Gottes im Alten Testament vom Herrn seines Volkes zum Herrn der ganzen Welt.* Göttingen: Vandenhoeck & Ruprecht.

Katechismus der Katholischen Kirche (2005). Munich: Oldenbourg.

Kaufman, Gordon D. (1981). *The Theological Imagination: Constructing the Concept of God.* Philadelphia: Westminster.

Keel, Othmar (2008). *Gott weiblich. Eine verborgene Seite des biblischen Gottes.* Gütersloh: Gütersloher Verlagshaus.

Keel, Othmar (2011). *Jerusalem und der eine Gott. Eine Religionsgeschichte.* Göttingen: Vandenhoeck & Ruprecht.

Keel, Othmar (2012). Die Heilige Schrift ist eine Bibliothek und nicht ein Buch. In: Steymans, Hans Ulrich & Staubli, Thomas (Eds.). *Von den Schriften zur (Heiligen) Schrift. Keilschrift. Hieroglyphen, Alphabete und Tora.* Freiburg: Bibel+Orient Museum. 14–18.

Keel, Othmar & Schroer, Silvia (2006/2004). *Eva–Mutter alles Lebendigen. Frauen- und Göttinnenidole aus dem Alten Orient.* 2nd Edit. Freiburg: Academic Press.

Keel, Othmar & Schroer, Silvia (2008). *Schöpfung. Biblische Theologien im Kontext altorientalischer Religionen.* Göttingen: Vandenhoeck & Ruprecht.

Keel, Othmar & Uehlinger, Christoph (2010/2001). *Göttinnen, Götter und Gottessymbole. Neue Erkenntnisse zur Religionsgeschichte Kanaans und Israels aufgrund bislang unerschlossener ikonographischer Quellen.* 6th Edit. Freiburg: Academic Press.

Keeley, Lawrence H. (1996). *War Before Civilization.* New York: Oxford University Press.

Kelle, Brad E. (2014). The Phenomenon of Israelite Prophecy in Contemporary Scholarship. *Currents in Biblical Research* 12. 275–320.

Keller, Werner (2015/1955). *The Bible as History.* 2nd ed. New York: William Morrow.

Kessler, Rainer (2008). *Sozialgeschichte des alten Israel. Eine Einführung.* Darmstadt: Wissenschaftliche Buchgesellschaft.

Kirkpatrick, Lee A. (2005). *Attachment, Evolution, and the Psychology of Religion.* New York: Guilford Press.

Knoblauch, Hubert (2009). *Populäre Religion. Auf dem Weg in eine spirituelle Gesellschaft.* Frankfurt am Main: Campus.

Köckert, Matthias (2009). Vom Kultbild Jahwes zum Bilderverbot. Oder: Vom Nutzen der Religionsgeschichte für die Theologie. *Zeitschrift für Theologie und Kirche* 106. 371–406.

Köckert, Matthias (2013/2007). *Die Zehn Gebote.* Munich: C. H. Beck.

Korpel, Marjo & de Moor, Johannes (2014). *Adam, Eve, and the Devil: A New Beginning.* Sheffield: Sheffield Phoenix Press.

Kottsieper, Ingo (2013). El. *Das wissenschaftliche Bibellexikon im Internet* (www.wibilex.de) (11/11/2015).

Kratz, Reinhard Gregor (2003). *Die Propheten Israels.* Munich: C. H. Beck.

Kugel, James L. (2008/2007). *How to Read the Bible: A Guide to Scripture: Then and Now.* New York: Free Press.

Kühn, Dagmar (2011). Totenkult (Israel). *Das wissenschaftliche Bibellexikon im Internet* (www.wibilex.de) (11/11/2015).

Kuijt, Ian (2000). People and Space in Early Agricultural Villages: Exploring Daily Lives, Community Size, and Architecture in the Late Pre-pottery Neolithic. *Journal of Anthropological Archaeology* 19. 75–102.

Lang, Bernhard (2002). *JAHWE der biblische Gott. Ein Porträt.* Munich: C. H. Beck.

Lang, Bernhard (2009/2003). *Himmel und Hölle. Jenseitsglaube von der Antike bis heute.* 2nd Edit. Munich: C. H. Beck.

Lang, Bernhard (2010). *Jesus der Hund. Leben und Lehre eines jüdischen Kynikers.* Munich: C. H. Beck.

Lang, Bernhard (2012). Persönliche Frömmigkeit. Vier Zugänge zu einer elementaren Form des religiösen Lebens. In: Wiebke Friese, Anika Greve, Kathrin Kleibl, Kristina Lahn & Inge Nielsen (Eds.). *Persönliche Frömmigkeit. Funktion und Bedeutung individueller Gotteskontakte im interdisziplinären Dialog.* Münster: LIT Verlag. 19–36.

Lang, Bernhard (2013). *Die Bibel. Die 101 wichtigsten Fragen.* Munich: C. H. Beck.

Lauster, Jörg (2014). *Die Verzauberung der Welt. Eine Kulturgeschichte des Christentums.* Munich: C. H. Beck.

Lehmann, Arthur C., Myers, James E. & Moro, Pamela A. (2005/1985). Ghosts, Souls, and Ancestors: Power of the Dead. In: Arthur C. Lehmann, James E. Myers & Pamela A. Moro (Eds.). *Magic, Witchcraft, and Religion: An Anthropological Study of the Supernatural.* 6th ed. New York: McGraw-Hill. 301–304.

Lehnhart, Bernhart (2009). Leiche/Leichenschändung. *Das wissenschaftliche Bibellexikon im Internet* (www.wibilex.de) (11/11/2015).

Leman, Patrick J. & Cinnirella, Marco (2007). A Major Event Has a Major Cause: Evidence for the Role of Heuristics in Reasoning About Conspiracy Theories. *Social Psychological Review* 9. 18–28.

Levin, Christoph (2005). *The Old Testament: A Brief Introduction.* Princeton, NJ: Princeton University Press.

Lieberman, Daniel E. (2013). *The Story of the Human Body: Evolution, Health, and Disease.* New York: Pantheon Books.

Liess, Kathrin (2005). Auferstehung (AT). *Das wissenschaftliche Bibellexikon im Internet* (www.wibilex.de) (11/11/2015).

Lindberg, David, C. (2010). The Fate of Science in Patristic and Medieval Christendom. In: Peter Harrison (Ed.). *The Cambridge Companion to Science and Religion.* New York: Cambridge University Press. 21–38.

Linder, Leo G. (2009). *Das Unternehmen Jesus. Wahrheit und Wirklichkeit des frühen Christentums.* Köln: Fackelträger.

Linke, Bernhard (2014). *Antike Religion.* Munich: Oldenbourg.

Loichinger, Alexander & Kreiner, Armin (2010). *Theodizee in den Weltreligionen.* Paderborn: Ferdinand Schöningh.

Lübbe, Hermann (2004). *Religion nach der Aufklärung.* Munich: Wilhelm Fink Verlag.

Luckmann, Thomas (1991/1967). *Die unsichtbare Religion.* Frankfurt am Main: Suhrkamp (*The Invisible Religion: The Problem of Religion in Modern Society.* New York: MacMillan Company, 1967).

Mack, Burton, L. (2001). *The Christian Myth: Origins, Logic, and Legacy.* New York: Continuum.

Madigan, Kevin J. & Levenson, Jon D. (2008). *Resurrection: The Power of God for Christians and Jews.* New Haven, CT: Yale University Press.

Maier, Bernhard (1997). Reinheit I. Religionsgeschichtlich. In: Gerhard Müller, Horst Balz & Gerhard Krause (Eds.). *Theologische Realenzyklopädie.* Vol. 28. Berlin: Walter de Gruyter. 473–477.

Marlowe, Frank W. (2003). The Mating System of Foragers in the Standard Cross-Cultural Sample. *Cross-Cultural Research* 37. 282–306.

Masci, David (2007). How the Public Resolves Conflicts Between Faith and Science. *Pew Research Center Publications* (www.pewforum.org/2007/08/27/how-the-public-resolves-conflicts-between-faith-and-science) (11/11/2015).

Maul, Stefan (2013). *Die Wahrsagekunst im Alten Orient. Zeichen des Himmels und der Erde.* Munich: C. H. Beck.

Mauss, Marcel (1996/1950). *Die Gabe. Form und Funktion des Austauschs in archaischen Gesellschaften.* Frankfurt am Main: Suhrkamp (*The Gift: The Form and Reason for Exchange in Archaic Societies.* London: Routledge, 2010).

Mauss, Marcel (2012/1968). *Schriften zur Religionssoziologie.* Frankfurt am Main: Suhrkamp.

McCauley, Robert N. (2011). *Why Religion Is Natural and Science Is Not.* Oxford: Oxford University Press.

Merz, Annette (2001). Mammon als schärfster Konkurrent Gottes—Jesu Vision vom Reich Gottes und das Geld. In: Severin J. Lederhilger (Ed.). *Gott oder Mammon. Christliche Ethik und die Religion des Geldes.* Frankfurt am Main: Peter Lang. 34–90.

Merz, Annette (2009). Der historische Jesus—faszinierend und unverzichtbar. In: Friedrich Wilhelm Graf & Klaus Wiegandt (Eds.). *Die Anfänge des Christentums.* Frankfurt am Main: Fischer. 26–56.

Metzger, Paul (2012). *Der Teufel.* Wiesbaden: Marixverlag.

Meyers, Carol (2010). Archäologie als Fenster zum Leben von Frauen in Alt-Israel. In: Irmtraud Fischer, Mercedes Navarro Puerto & Andrea Taschl-Erber (Eds.). *Tora. Die Bibel und die Frauen. Eine exegetisch-kulturgeschichtliche Enzyklopädie.* Vol. 1.1. Stuttgart: Kohlhammer. 63–109.

Meyers, Carol (2012). The Function of Feasts: An Anthropological Perspective on Israelite Religious Festivals. In: Saul M. Olyan (Ed.). *Social Theory and the Study of Israelite Religion: Essays in Retrospect and Prospect.* Atlanta, GA: Society of Biblical Literature. 141–168.

Meyers, Carol (2013a). Household Religion. In: Francesca Stavrakopoulou & John Barton (Eds.). *Religious Diversity in Ancient Israel and Judah.* New York: Bloomsbury. 118–135.

Meyers, Carol (2013b). *Rediscovering Eve: Ancient Israelite Women in Context.* Oxford: Oxford University Press.

Michel, Andreas (2007). Isaak. *Das wissenschaftliche Bibellexikon im Internet* (www.wibilex.de) (11/11/2015).

Miles, Jack (1995). *God: A Biography.* New York: Alfred A. Knopf.

Moeller, Bernd (2011). *Geschichte des Christentums in Grundzügen.* Göttingen: Vandenhoeck & Ruprecht.

Moenikes, Ansgar (2012). Liebe/Liebesgebot. *Das wissenschaftliche Bibellexikon im Internet* (www.wibilex.de) (11/11/2015).

Moore, Megan Bishop & Kelle, Brad (2011). *Biblical History and Israel's Past: The Changing Study of the Bible and History.* Grand Rapids, MI: Eerdmans.

Moroni, Claudia & Lippert, Helga (2009). *Die Biblischen Plagen. Zorn Gottes oder Rache der Natur.* Munich: Piper.

Morrow, William (2004). Post-traumatic Stress Disorder and Vicarious Atonement in the Second Isaiah. In: J. Harold Ellens & Wayne G. Rollins (Eds.). *Psychology and the Bible: A New Way to Read the Scriptures.* Vol. 1. Westport, CT: Praeger. 167–183.

Moss, Candida (2013). *The Myth of Persecution: How Early Christians Invented a Story of Martydom.* New York: HarperOne.
Mühling, Anke (2009). Erzeltern. *Das wissenschaftliche Bibellexikon im Internet* (www.wibilex.de) (11/11/2015).
Mummert, Amanda, Esche, Emely, Robinson, Joshua & Armelagos, George J. (2011). Stature and Robusticity During the Agricultural Transition: Evidence from the Bioarchaeological Record. *Economics & Human Biology* 8. 284–301.
Murdock, George Peter (1980). *Theories of Illness: A World Survey.* Pittsburgh, PA: University of Pittsburgh Press.
Newitz, Annalee (2014). How Farming Almost Destroyed Ancient Human Civilization (http://io9.com/how-farming-almost-destroyed-human-civilization-1659734601) (11/11/2015).
Niehr, Herbert (2003). Götterbilder und Bilderverbot. In: Manfred Oeming und Konrad Schmid (Eds.). *Der eine Gott und die Götter. Polytheismus und Monotheismus im antiken Israel.* Zurich: Theologischer Verlag. 227–247.
Niehr, Herbert (2012). Das Buch Daniel. In: Erich Zenger & Christian Frevel (Eds.). *Einleitung in das Alte Testament.* Stuttgart: Kohlhammer. 610–621.
Niehr, Herbert (2013/2010). "Israelite" Religion and "Canaanite" Religion. In: Francesca Stavrakopoulou & John Barton (Eds.). *Religious Diversity in Ancient Israel and Judah.* New York: Bloomsbury. 23–36.
Nietzsche, Friedrich (1955). Götzen-Dämmerung. Oder: Wie man mit dem Hammer philosophiert. 1889. In: Friedrich Nietzsche. *Werke.* 2, Munich: Hanser (*The Twilight of the Idols and the Anti-Christ: Or How to Philosophize with a Hammer.* New York: Penguin, 1990).
Nissinen, Martti (2010). Comparing Prophetic Sources: Principles and a Test Case. In: John Day (Ed.). *Prophecy and Prophets in Ancient Israel: Proceedings of the Oxford Old Testament Seminar.* New York: T&T Clark. 3–24.
Nobis, Heribert Maria (1971). *Buch der Natur.* In: Joachim Ritter (Ed.). *Historisches Wörterbuch der Philosophie.* Vol. 1. Basel: Schwabe. 957–959.
Noort, Ed (2012). Josua im Wandel der Zeiten: Zu Stand und Perspektiven der Forschung am Buch Josua. In: Ed Noort (Ed.). *The Book of Joshua.* Leuven: Peeters. 21–50.
Norenzayan, Ara (2013). *Big Gods: How Religion Transformed Cooperation and Conflict.* Princeton, NJ: Princeton University Press.
Norenzayan, Ara (2014). Does Religion Make People Moral? *Behaviour* 151. 365–384.
Norenzayan, Ara, Shariff, Azim. F., Gervais, Will M., Willard, Ayana K., McNamara, Rita A., Slingerland, Edward & Henrich, Joseph (2014). The Cultural Evolution of Prosocial Religions. *Behavioral and Brain Sciences* (forthcoming). 1–86.

Numbers, Ronald L. (Ed.) (2009). *Galileo Goes to Jail and Other Myths About Science and Religion.* Cambridge, MA: Cambridge University Press.

Oeming, Manfred & Schmid, Konrad (Eds.) (2003). *Der eine Gott und die Götter. Polytheismus und Monotheismus im antiken Israel.* Zurich: Theologischer Verlag.

Ökumenisches Heiligenlexikon (www.heiligenlexikon.de) (11/11/2015).

Olyan, Saul M. (2008). *Disability in the Hebrew Bible. Interpreting Mental and Physical Differences.* Cambridge, MA: Cambridge University Press.

Olyan, Saul M. (Ed.) (2012). *Social Theory and the Study of Israelite Religion: Essays in Retrospect and Prospect.* Atlanta, GA: Society of Biblical Literature.

Oppenheimer, Aharon (2008). Heilige Kriege im antiken Judentum. Monotheismus als Anlass zum Krieg? In: Klaus Schreiner (Ed.). *Heilige Kriege. Religiöse Begründungen militärischer Gewaltanwendung: Judentum, Christentum und Islam im Vergleich.* Munichburg: Oldenbourg. 31–42.

Otto, Eckart (1994). *Theologische Ethik des Alten Testaments.* Stuttgart: Kohlhammer.

Otto, Eckart (1996). Die Paradieserzählung Genesis 2–3: Eine nachpriesterschriftliche Lehrerzählung in ihrem religionshistorischen Kontext. In: Anja A. Diesel, Reinhard G. Lehmann, Eckart Otto & Andreas Wagner (Eds.). *Jedes Ding hat seine Zeit . . . Studien zur israelitischen und altorientalischen Weisheit.* Berlin: Walter de Gruyter. 167–192.

Otto, Eckart (1999). *Das Deuteronomium. Politische Theologie und Rechtsreform in Juda und Assyrien.* Berlin: Walter de Gruyter.

Otto, Eckart (2005). Magie—Dämonen—göttliche Kräfte. In: Werner H. Ritter & Bernhard Wolf (Eds.). *Heilung—Energie—Geist: Heilung zwischen Wissenschaft, Religion und Geschäft.* Göttingen: Vandenhoeck & Ruprecht. 208–225.

Otto, Eckart (2006). *Mose. Geschichte und Legende.* Munich: C. H. Beck.

Otto, Eckart (2007). *Das Gesetz des Mose.* Darmstadt: Wissenschaftliche Buchgesellschaft.

Pagels, Elaine (1994/1988). *Adam, Eva und die Schlange. Die Geschichte der Sünde.* Reinbek: Rowohlt (*Adam, Eve, and the Serpent.* New York: Random House, 1988).

Pagels, Elaine (1998/1995). *Satans Ursprung.* Frankfurt am Main: Suhrkamp (*The Origin of Satan.* New York: Random House, 1995).

Pagels, Elaine (2014/2012). *Apokalypse. Das letzte Buch der Bibel wird entschlüsselt.* Munich: Deutscher Taschenbuch Verlag (*Revelations: Visions, Prophecy, and Politics in the Book of Revelation.* New York: Viking, 2012).

Pakkala, Juha (2006). Die Entwicklung der Gotteskonzeptionen in den deuteronomistischen Redaktionen von polytheistischen zu monotheistischen Vorstellungen. In: Jan Christian Gertz, Doris

Prechel, Konrad Schmid & Markus Witte (Eds.). *Die deuteronomistischen Geschichtswerke. Redaktions- und religionsgeschichtliche Perspektiven zur "Deuteronomismus"—Diskussion in Tora und Vorderen Propheten*. Berlin: Walter de Gruyter. 239–248.

Paloutzian, Raymond D. & Park, Crystal L. (Eds.) (2013). *Handbook of the Psychology of Religion and Spirituality*. New York: Guilford Press.

Parker, Robert (1983). *Miasma: Pollution and Purification in Early Greek Religion*. Oxford: Clarendon Press.

Parzinger, Hermann (2014). *Die Kinder des Prometheus. Eine Geschichte der Menschheit vor der Erfindung der Schrift*. Munich: C. H. Beck.

Pfeiffer, Henrik (2000). Der Baum in der Mitte des Gartens. Zum überlieferungsgeschichtlichen Ursprung der Paradieserzählung. Teil I: Analyse. *Zeitschrift für die Alttestamentliche Wissenschaft* 112. 487–500.

Pfeiffer, Henrik (2001). Der Baum in der Mitte des Gartens. Zum überlieferungsgeschichtlichen Ursprung der Paradieserzählung. Teil II: Prägende Traditionen und theologische Akzente. *Zeitschrift für die Alttestamentliche Wissenschaft* 113. 2–16.

Pfeiffer, Henrik (2006a). Eden. *Das wissenschaftliche Bibellexikon im Internet* (www.wibilex.de) (11/11/2015).

Pfeiffer, Henrik (2006b). Paradies/Paradieserzählung. *Das wissenschaftliche Bibellexikon im Internet* (www.wibilex.de) (11/11/2015).

Piaget, Jean (1992/1926). *Das Weltbild des Kindes*. Munich: Deutscher Taschenbuch Verlag (*The Child's Conception of the World*. London: Routledge, 1929).

Pinker, Steven (2011). *The Better Angels of Our Nature: Why Violence Has Declined*. New York: Penguin.

Plaut, W. Gunther (Ed.) (2000/1981). *Die Tora in jüdischer Auslegung*. Vol. 2: *Schemot/Exodus*. Gütersloh: Gütersloher Verlagshaus.

Popper, Karl (2014/1963). *Conjectures and Refutations: The Growth of Scientific Knowledge*. London: Routledge.

Posener, Alan (2007/1999). *Maria*. Reinbek: Rowohlt.

Poser, Ruth (2012). *Das Ezechielbuch als Trauma-Literatur*. Leiden: Brill.

Räisänen, Heikki (2012). Jesus and Hell. In: Tom Holmén (Ed.). *Jesus in Continuum*. Tübingen: Mohr Siebeck. 355–385.

Ranke-Graves, Robert & Patai, Raphael (1986). *Hebräische Mythologie. Über die Schöpfungsgeschichte und andere Mythen aus dem Alten Testament*. Reinbek: Rowohlt.

Ratzinger, Joseph—Benedikt XVI (2007). *Jesus von Nazareth*. Vol. 1. Freiburg im Breisgau: Herder (*Jesus of Nazareth: From the Baptism in the Jordan to the Transfiguration*. New York: Doubleday, 2007).

Richerson, Peter J. & Boyd, Robert (2005). *Not by Genes Alone: How Culture Transformed Human Evolution.* Chicago: University of Chicago Press.

Richerson, Peter J. & Christiansen, Morton H. (2013). *Cultural Evolution: Society, Technology, Language, and Religion.* Cambridge, MA: MIT Press.

Riede, Peter (2014). Jenseitsvorstellungen. *Das wissenschaftliche Bibellexikon im Internet* (www.wibilex.de) (11/11/2015).

Rilke, Rainer Maria (1955). *Sämtliche Werke.* Vol. 1. Wiesbaden: Insel.

Robarchek, Clayton (1990). Motivations and Material Causes: On the Explanation of Conflict and War. In: Jonathan Haas (Ed.). *The Anthropology of War.* Cambridge, MA: Cambridge University Press. 56–76.

Roloff, Jürgen (2007/2000). *Jesus.* Munich: C. H. Beck.

Rose, Jeffrey I. (2010). New Light on Human Prehistory in the Arabo-Persian Gulf Oasis. *Current Anthropology* 51. 849–883.

Rosenthal, Carolyn J. (1985). Kinkeeping in the Familial Division of Labor. *Journal of Marriage and the Family.* 965–974.

Rosenzweig, Phil (2007). *The Halo Effect . . . and the Eight Other Business Delusions That Deceive Managers.* New York: Free Press.

Rousseau, Jean-Jacques (1981). *Schriften.* Vol. 1. Berlin: Ullstein.

Rozin, Paul, Haidt, Jonathan & McCauley, Clark (2008). Disgust. In: Michael Lewis, Jeanette M. Haviland-Jones & Lisa Feldman Barrett (Eds.). *Handbook of Emotions.* 3rd ed. New York: Guilford Press. 757–776.

Ruppert, Lothar (2003). *Genesis. Ein kritischer und theologischer Kommentar. 1. Teilband: Gen 1,1–11,28.* Würzburg: Echter.

Sagan, Carl (2013/1980). *Cosmos.* New York: Random House.

Sand, Shlomo (2014/2008). *Die Erfindung des jüdischen Volkes. Israels Gründungsmythos auf dem Prüfstand.* Berlin: List (*The Invention of the Jewish People.* London: Verso, 2009).

Satlow, Michael L. (2014). *How the Bible Became Holy.* New Haven, CT: Yale University Press.

Schäfer, Daniel (2004). *Alter und Krankheit in der Frühen Neuzeit. Der ärztliche Blick auf die letzte Lebensphase.* Frankfurt: Campus.

Schaller, Mark (2011). The Behavioural Immune System and the Psychology of Human Sociality. *Philosophical Transactions of the Royal Society B.* 366. 3418–3426.

Schaller, Mark (2014). When and How Disgust Is and Is Not Implicated in the Behavioral Immune System. *Evolutionary Behavioral Sciences* 8. 251–256.

Schaller, Mark, Murray, Damian R. & Bangerter, Adrian (2015). Implications of the Behavioural Immune System for Social Behaviour and Human Health in the Modern World. *Philosophical Transactions of the Royal Society B* 370. 2014.0105.

Schart, Aaron (2014). Prophetie AT. *Das wissenschaftliche Bibellexikon im Internet* (www.wibilex.de) (11/11/2015).

Schaudig, Hanspeter (2012). Erklärungsmuster von Katastrophen im Alten Orient. In: Angelika Berlejung (Ed.). *Disaster and Relief Management. Katastrophen und ihre Bewältigung*. Tübingen: Mohr Siebeck.

Schieder, Rolf (Ed.) (2014). *Die Gewalt des einen Gottes: Die Monotheismusdebatte zwischen Jan Assmann, Mischa Brumlik, Rolf Schieder, Peter Sloterdijk und anderen.* Darmstadt: Wissenschaftliche Buchgesellschaft.

Schlögl, Hermann Alexander (2006). *Das alte Ägypten: Geschichte und Kultur von der Frühzeit bis zu Kleopatra.* Munich C. H. Beck.

Schmid, Konrad (2003). Differenzierungen und Konzeptualisierungen der Einheit Gottes in der Religions- und Literaturgeschichte Israels. In: Manfred Oeming & Konrad Schmid (Eds.). *Der eine Gott und die Götter. Polytheismus und Monotheismus im antiken Israel.* Zurich: Theologischer Verlag. 11–38.

Schmid, Konrad (2010a). Hintere Propheten (Nebiim). In: Jan Christian Gertz (Ed.). *Grundinformation Altes Testament.* Göttingen: Vandenhoeck & Ruprecht. 313–412.

Schmid, Konrad (2010b). *Hiob als biblisches und antikes Buch. Historische und intellektuelle Kontexte seiner Theologie.* Stuttgart: Verlag Katholisches Bibelwerk.

Schmid, Konrad (2012a). *The Old Testament: A Literary History.* Minneapolis, MN: Fortress Press.

Schmid, Konrad (2012b). Editorial. *Hebrew Bible and Ancient Israel* 1. 5–6.

Schmidt-Salomon, Michael (2012). *Jenseits von Gut und Böse. Warum wir ohne Moral die besseren Menschen sind.* Munich: Piper.

Schmitt, Rüdiger (2004). *Magie im alten Testament.* Münster: Ugarit-Verlag.

Schmitt, Rüdiger (2006). Hausgott/Terafim. *Das wissenschaftliche Bibellexikon im Internet* (www.wibilex.de) (11/11/2015).

Schmitt, Rüdiger (2011). *Der "Heilige Krieg" im Pentateuch und im deuteronomistischen Geschichtswerk. Studien zur Forschungs-, Rezeptions- und Religionsgeschichte von Krieg und Bann im Alten Testament.* Münster. Ugarit-Verlag.

Schmitz, Barbara (2011). *Geschichte Israels.* Paderborn: Schöningh.

Schnabel, Ulrich (2008). *Die Vermessung des Glaubens. Forscher ergründen, wie der Glaube entsteht und warum er Berge versetzt.* Munich: Blessing.

Schreiner, Klaus (1994). *Maria. Jungfrau, Mutter, Herrscherin.* Munich: Hanser.

Schreiner, Klaus (2003). *Maria. Leben, Legenden, Symbole.* Munich: C. H. Beck.

Schroer, Silvia (2006). Idole—die faszinierende Vielfalt der Frauen- und Göttinnenfigürchen. In: Othmar Keel & Silvia Schroer. *Eva—Mutter alles Lebendigen. Frauen- und Göttinnenidole aus dem Alten Orient.* Freiburg: Academic Press. 8–25.

Schulze, Gerhard (2008). *Die Sünde. Das schöne Leben und seine Feinde.* Frankfurt am Main: Fischer.

Schwienhorst-Schönberger, Ludger (2012). Das Buch Ijob. In: Erich Zenger & Christian Frevel (Eds.). *Einleitung in das Alte Testament.* Stuttgart: Kohlhammer. 414–427.

Shennan, Stephen, Downey, Sean S., Timpson, Adrian, Edinborough, Kevan, Colledge, Sue, Kerig, Tim, Manning, Katie & Thomas, Mark G. (2014). Regional Population Collapse Followed Initial Agriculture Booms in Mid-Holocene Europe. *Nature Communications* 4. 3486.

Sibley, Chris G. & Bulbulia, Joseph (2013). Faith After an Earthquake: A Longitudinal Study of Religion and Perceived Health Before and After the 2011 Christchurch New Zealand Earthquake. *Public Library of Science (PLOS) ONE* 12. 49648.

Singh, Simon (2008/2004). *Big Bang. Der Ursprung des Kosmos und die Erfindung der modernen Naturwissenschaft.* Munich: Deutscher Taschenbuch Verlag (*Big Bang: The Origin of the Universe.* London: Four Estate, 2004).

Skiena, Steven S. & Ward, Charles B. (2014). *Who's Bigger? Where Historical Figures Really Rank.* Cambridge: Cambridge University Press.

Slingerland, Edward, Henrich, Joseph & Norenzayan, Ara (2013). The Evolution of Prosocial Religions. In: Peter J. Richerson & Morton H. Christiansen (Eds.). *Cultural Evolution: Society, Technology, Language, and Religion.* Cambridge, MA: MIT Press. 335–348.

Slone, D. Jason (2005). *Theological Incorrectness: Why Religious People Believe What They Shouldn't.* New York: Oxford University Press.

Smith, David Livingstone (2011). *Less Than Human: Why We Demean, Enslave, and Exterminate Others.* New York: St. Martin's Press.

Smith, Michael E., Ur, Jason & Feinman, Gary M. (2014). Jane Jacobs' "Cities First" Model and Archaeological Reality. *International Journal of Urban and Regional Research* 38. 1525–1535.

Sonnabend, Holger (1999). *Naturkatastrophen in der Antike. Wahrnehmung–Deutung–Management.* Stuttgart: J. B. Metzler.

Sosis, Richard & Bressler, Eric R. (2003). Cooperation and Commune Longevity: A Test of the Costly Signaling Theory of Religion. *Cross-Cultural Research* 37. 211–239.

Stark, Rodney (2001). Gods, Rituals, and the Moral Order. *Journal for the Scientific Study of Religion* 40. 619–636.

Staubli, Thomas (2001). Tiere als Teil menschlicher Nahrung in der Bibel und im Alten Orient. In: Othmar Keel & Thomas Staubli (Eds.). *Im Schatten Deiner Flügel. Tiere in der Bibel und im Alten Orient.* Freiburg: Universitätsverlag Freiburg Schweiz.

Staubli, Thomas (2010/1997). *Begleiter durch das Erste Testament.* Ostfildern: Patmos.

Staubli, Thomas & Schroer, Silvia (2014). *Menschenbilder der Bibel.* Ostfildern: Patmos.

Stavrakopoulou, Francesca (2013/2010). "Popular" Religion and "Official" Religion: Practice, Perception, Portrayal. In: Francesca Stavrakopoulou & John Barton (Eds.). *Religious Diversity in Ancient Israel and Judah.* London: Bloomsbury. 37–58.

Stavrakopoulou, Francesca & Barton, John (Eds.) (2013/2010). *Religious Diversity in Ancient Israel and Judah.* London: Bloomsbury.

Steadman, Lyle B., Palmer, Craig T. & Tilley, Christopher F. (1996). The Universality of Ancestor Worship. *Ethnology* 35. 63–76.

Steinberg, Ted (2006/2005). *Acts of God: The Unnatural History of Natural Disaster in America.* New York: Oxford University Press.

Stern, Menahem (2005/1976). Die Zeit des Zweiten Tempels. In: Haim-Hillel Ben-Sasson (Ed.). *Geschichte des Jüdischen Volkes. Von den Anfängen bis zur Gegenwart.* 5th Edit. Munich: C. H. Beck (*A History of the Jewish People.* Cambridge, MA: Harvard University Press, 1976).

Steymans, Hans Ulrich (2010). Gilgameš und Genesis 1–9. *Biblische Zeitschrift* 54. 201–228.

Steymans, Hans Ulrich & Staubli, Thomas (Eds.) (2012). *Von den Schriften zur (Heiligen) Schrift. Keilschrift, Hieroglyphen, Alphabete und Tora.* Freiburg: Bibel+Orient Museum.

Strassmann, Beverly I., Kurapati, Nikhil T., Hug, Brendan F., Burke, Erin E., Gillespie, Brenda W., Karafet, Tatiana M. & Hammer, Michael F. (2012). Religion as a Means to Assure Paternity. *Proceedings of the National Academy of Sciences of the United States of America (PNAS)* 109. 9781–9785.

Stuttgarter Erklärungsbibel mit Apokryphen. Die Heilige Schrift nach der Übersetzung Martin Luthers (2007). Stuttgart: Deutsche Bibelgesellschaft.

Taleb, Nassim Nicholas (2013/2007). *Der Schwarze Schwan. Die Macht höchst unwahrscheinlicher Ereignisse.* Munich: Deutscher Taschenbuch Verlag (*The Black Swan: The Impact of the Highly Improbable.* New York: Random House, 2007).

Teehan, John (2010). *In the Name of God: The Evolutionary Origins of Religious Ethics and Violence.* Malden, MA: Wiley-Blackwell.

Theissen, Gerd & Merz, Annette (2011/1996). *Der historische Jesus. Ein Lehrbuch.* 4th Edit. Göttingen: Vandenhoeck & Ruprecht (*Historical Jesus: A Comprehensive Guide.* Minneapolis, MN: Fortress Press, 1998).

Thornhill, Randy & Fincher, Corey L. (2014). *The Parasite-Stress Theory of Values and Sociality: Infectious Disease, History and Human Values Worldwide.* Cham: Springer.

Tilly, Michael & Zwickel, Wolfgang (2011). *Religionsgeschichte Israels. Von der Vorzeit bis zu den Anfängen des Christentums.* Darmstadt: Wissenschaftliche Buchgesellschaft.

Tooby, John & Cosmides, Leda (1992). Cognitive Adaptations for Social Exchange. In: Jerome Barkow, Leda Cosmides, and J. Tooby (Eds.). *The Adapted Mind: Evolutionary Psychology and the Generation of Culture.* New York: Oxford University Press. 163–228.

Trimondi, Victor & Trimondi, Victoria (2006). *Krieg der Religionen. Politik, Glaube und Terror im Zeichen der Apokalypse.* Munich: Wilhelm Fink Verlag.

Trivers, Robert (2011). *Deceit and Self-Deception: Fooling Yourself the Better to Fool Others.* New York: Penguin.

Turchin, Peter (2007). *War and Peace and War: The Rise and Fall of Empires.* New York: Plume.

Tylor, Edward B. (1871). *Primitive Culture: Researches into the Early History of Mankind and the Development of Civilization.* 2 vols. London: John Murray.

van der Toorn, Karel (1996). *Family Religion in Babylonia, Syria and Israel: Continuity and Change in the Forms of Religious Life.* Leiden: Brill.

van Schaik, Carel (2004). *Among Orangutans. Red Apes and the Rise of Human Culture.* Cambridge, MA: The Belknap Press of Harvard University Press.

van Schaik, Carel (2016). *The Primate Origins of Human Nature.* Hoboken, NJ: Wiley-Blackwell.

van Schaik, Carel, Ancrenaz, Marc, Borgen, Gwendolyn, Galdikas, Birute, Knott, Cheryl D., Singleton, Ian, Suzuki, Akira, Utami, Sri Suci & Merrill, Michelle (2003). Orangutan Cultures and the Evolution of Material Culture. *Science* 299. 102–105.

Vanderjagt, Arjo & van Berkel, Klaas (Eds.) (2005). *The Book of Nature in Antiquity and the Middle Ages.* Leuven: Peeters.

Veit, Christoph (2013). Krieg und Medizin. *if-Zeitschrift für innere Führung* (www.if-zeitschrift.de) (11/11/2015).

Veyne, Paul (2008). *Die griechisch-römische Religion. Kult, Frömmigkeit und Moral.* Stuttgart: Reclam.

Veyne, Paul (2011/2008). *Als unsere Welt christlich wurde. Aufstieg einer Sekte zur Weltmacht.* Munich: C. H. Beck (*When Our World Became Christian: 312–394.* Malden, MA: Polity, 2010).

Voland, Eckart (2007). *Die Natur des Menschen. Grundkurs Soziobiologie.* Munich: C. H. Beck.

von Stosch, Klaus (2013). *Theodizee.* Paderborn: Schöningh.

Wälchli, Stefan (2014). Zorn AT. *Das wissenschaftliche Bibellexikon im Internet* (www.wibilex.de) (11/11/2015).

Walter, François (2010). *Katastrophen. Eine Kulturgeschichte vom 16. bis ins 21. Jahrhundert.* Stuttgart: Reclam.

Weber, Max (1980/1921). *Wirtschaft und Gesellschaft.* 5th Edit. Tübingen: Mohr Siebeck (*Economy and Society: An Outline of Interpretive Sociology.* Berkeley, CA: University of California Press, 1978).

Weber, Max (1988/1920). *Gesammelte Aufsätze zur Religionssoziologie III.* 8th Edit. Tübingen: Mohr Siebeck.

Weber, Max (1995/1919). *Wissenschaft als Beruf.* Stuttgart: Reclam.

Wells, Jonathan C. K., DeSilva, Jeremy M. & Stock, Jay T. (2012). The Obstetric Dilemma: An Ancient Game of Russian Roulette, or a Variable Dilemma Sensitive to Ecology? *Yearbook of Physical Anthropology* 55. 40–71.

Wesel, Uwe (1997). *Geschichte des Rechts. Von den Frühformen bis zum Vertrag von Maastricht.* Munich: C. H. Beck.

Whitehouse, Harvey (2004). *Modes of Religiosity: A Cognitive Theory of Religious Transmission.* Walnut Creek, CA: AltaMira Press.

Wiessner, Polly (2006). From Spears to M-16s: Testing the Imbalance of Power Hypothesis Among the Enga. *Journal of Anthropological Research* 62. 165–191.

Willmes, Bernd (2008). Sündenfall. *Das wissenschaftliche Bibellexikon im Internet* (www.wibilex.de) (11/11/2015).

Wilson, Michael L., Boesch, Christophe, Fruth, Barbara, Furuichi, Takeshi, Gilby, Ian C., Hashimoto, Chie, Hobaiter, Catherine L., Hohmann, Gottfried, Itoh, Noriko, Koops, Kathelijne, Lloyd, Julia N., Matsuzawa, Tetsuro, Mitani, John C., Mjungu, Deus C., Morgan, David, Muller, Martin N., Mundry, Roger, Nakamura, Michio, Pruetz, Jill, Pusey, Anne E., Riedel, Julia, Sanz, Crickette, Schel, Anne M., Simmons, Nicole, Waller, Michel, Watts, David P., White, Frances, Wittig, Roman M., Zuberbühler, Klaus, & Wrangham, Richard W. (2014). Lethal Aggression in *Pan* Is Better Explained by Adaptive Strategies Than Human Impacts. *Nature* 513. 414–417.

Winter, Urs (2002). Der Lebensbaum im Alten Testament und die Ikonographie des stilisierten Baumes in Kanaan/Israel. In: Ute Neumann-Gorsolke & Peter Riede (Eds.). *Das Kleid der Erde. Pflanzen in der Lebenswelt des Alten Israel.* Stuttgart: Calwer Verlag. 138–162.

Witte, Markus (2010). Schriften (Ketubim). In: Jan Christian Gertz. *Grundinformation Altes Testament.* Göttingen: Vandenhoeck & Ruprecht. 413–534.

Wolff, Hans Walter (1987). *Studien zur Prophetie–Probleme und Erträge.* Munich: Ch. Kaiser Verlag.

Wolff, Hans Walter (2010). *Anthropologie des Alten Testaments.* Gütersloh: Gütersloher Verlagshaus.

Wright, David P. (2009). *Inventing God's Law: How the Covenant Code of the Bible Used and Revised the Laws of Hammurabi.* Oxford: Oxford University Press.

Wright, Robert (2009). *The Evolution of God.* New York: Little, Brown and Company.

Zehnder, Markus (2008). Homosexualität (AT). *Das wissenschaftliche Bibellexikon im Internet* (www.wibilex.de) (11/11/2015).

Zenger, Erich & Frevel, Christian (Eds.). (2012/1995). *Einleitung in das Alte Testament.* 8th Edit. Stuttgart: Kohlhammer.

Zevit, Ziony (2013). *What Really Happened in the Garden of Eden?* New Haven, CT: Yale University Press.

Zimbardo, Philip (2007). *The Lucifer Effect: Understanding How Good People Turn Evil.* New York: Random House.

Zimmermann, Christiane (2010). Vater (NT). *Das wissenschaftliche Bibellexikon im Internet* (www.wibilex.de) (11/11/2015).

Zimmermann, Ulrich (2012). Beschneidung (AT). *Das wissenschaftliche Bibellexikon im Internet* (www.wibilex.de) (11/11/2015).

Zingsem, Vera (2009). *Lilith. Adams erste Frau.* Stuttgart: Reclam.

INDEX

Carel van Schaik is a behavioral and evolutionary biologist. He is professor of biological anthropology at the University of Zurich, where he is also director of the Anthropological Institute and Museum. He was trained in biology at Utrecht University (the Netherlands) and spent over three decades studying rainforest primates, especially orangutans. He was a researcher at Princeton University and, from 1989 until 2004, a professor of biological anthropology at Duke University.

Kai Michel is a historian, literary scholar, and science journalist. Straddling the fields of science, archaeology, history, and religion, Michel has been published by noted German-language magazines. He has held the posts of editor or head of department at *Die Zeit*, *Facts*, and *Die Weltwoche* and reviewed non-fiction books for the *Frankfurter Allgemeine Zeitung*.